A HISTORY OF
METALLOGRAPHY

*Reproduction of direct typographical imprint
from the etched surface of the Elbogen Iron Meteorite.
Reduced to 9/10 original size.
(Schreibers, 1813.) (See page 151.)*

A HISTORY OF METALLOGRAPHY

THE DEVELOPMENT OF IDEAS ON THE STRUCTURE
OF METALS BEFORE 1890

By Cyril Stanley Smith

THE MIT PRESS
Cambridge, Massachusetts   London, England

First MIT Press paperback edition 1988
© 1960, 1988 by Cyril Stanley Smith
Original edition published by The University of
Chicago Press, 1960.

This book was printed and bound in the United
States of America.

Library of Congress Cataloging-in-Publication Data

Smith, Cyril Stanley, 1903–
  A history of metallography.

  Reprint. Originally published: Chicago : University
of Chicago Press, 1960.
  Bibliography: p.
  Includes index.
  1. Metallography.   2. Damascening.   I. Title.
TN690.S57   1988   669'.95   88–12737
ISBN 0–262–69120–5 (paperback)

To those craftsmen

whose intuitive understanding of materials

provided the seed

from which

metallurgical science grew

Table of Contents

List of Illustrations

Preface to the Paperback Edition

This book was first published in 1960. It is being reprinted with few changes, despite the fact that the field of materials science and engineering has burgeoned almost beyond belief in the intervening quarter of a century—very largely as a result of an awakened appreciation of the critical importance of the interaction of the different levels of structure discussed herein. (Though this book deals mainly with metals, the structural principles responsible for their properties are applicable to condensed matter of any kind.)

Advances in physics and chemistry have always depended upon experimental studies of properties associated with particular levels of substructural fit and misfit, but rarely have such techniques been so intimately involved as in the recent development of materials science. On the one hand, the techniques of electron and field microscopy have advanced to the point where the patterns of association between structural components can be studied with continuous variation of magnification down to the very level where individual atoms and structural "imperfections" can be observed. On the other hand, not only has the immense field of organic polymers arisen from the merging of classical chemical molecular theory with more complex structural effects, but the development of new magnetic materials, semiconductors, and superconductors has enriched our understanding of the structural basis of all properties. These developments have both required and enabled a far closer interaction between theory and practice than was previously possible; the industrial techniques for producing material structures to match the requirements of theory have acquired almost unimaginable precision. All this and also the most significant of all structures—those of living organisms—are beyond the scope of the present book; indeed it has little on the broader history of metallurgy and deals mainly with structures based on atoms rather than with electron movements within and XV between them. It is nevertheless a prehistory of much of the modern science of condensed matter.

I wrote the book mainly to satisfy my own curiosity, though with some hope that it would convince professional historians of science to add to their principal concerns an interest in the origins of what is now the most active subdivision of physics, that of the solid state.* The book has received some notice from metallurgists and, unexpectedly, from a number of artist-blacksmiths who found in the discussion of Damascus and Japanese swords some suggestions that enabled them to recreate textured steel for knives and various decorative purposes.

Chapter 8 briefly mentions the work of the seventeenth- and eighteenth-century corpuscular philosophers. This strikes me as being very naive, for I now believe that corpuscular concepts deserve far more attention than they have been given by either scientists or historians. The writings of René Descartes and Emanuel Swedenborg represent the apex of development of medieval views on structure and deal imaginatively with levels of association rather than with atomicity. Notice the many stages of aggregation and directions of dissociation in Figure A, taken from Swedenborg's *Principia Rerum Naturalium* (1734). However, such thoughts were incompatible with Newtonian physics as it triumphantly provided quantitative treatment of simple symmetries and of centers of interaction rather than boundaries. Today we are again coming to suspect that the concave "space" between "atoms" is more important to aggregation than convexity. Fermi surfaces have much in common with medieval corpuscles, with the formation of continuously connected concave nets of pulsating electron interaction in which are embedded convex regions whose insides are inaccessible.

The topological approach to structure is more generally relevant to the structure of materials than is the treatment of lattice symmetries. Indeed, a case could be made that the overemphasis on the idealized quantitative aspects of atom, molecule, and crystals by nineteenth-century physicists and chemists actually delayed the understanding of the true basis of structure, which is far more cellular than it is atomic. (The application of elementary topology to the actual shapes of grains

---

* The studies being commissioned by the International Project on the History of Solid State Physics, initiated in 1981, will provide a major resource when they are published. Two books should be noted: *The Beginnings of Solid State Physics* (London, 1980), the record of a Royal Society conference attended by many physicists who have made major contributions to the subject, and *Solid State Science: Past, Present and Predicted,* edited by D. L. Weaire and C. G. Windsor (Bristol, 1987).

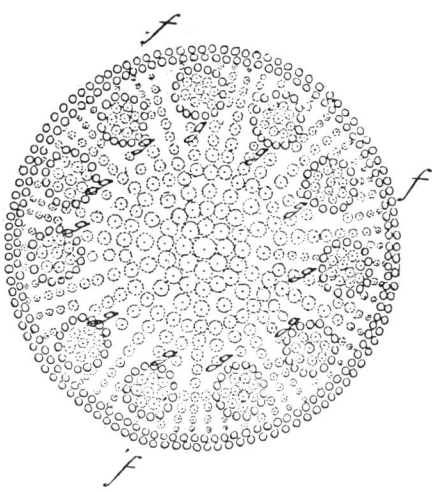

Fig. A.—Diagram illustrating Swedenborg's concept of the structure of matter (1734). The density of packing and the freedom of motion depend on the assembly of elementary units in a hierarchical arrangement of bounded "finites" on different scales.

in polycrystalline materials is discussed in my book *A Search for Structure* [Cambridge, Mass., 1981], pp. 1–32.)

The most interesting properties of useful materials depend upon intricacies of structure that have much in common with the complexities of biological cells and organisms. As with any history, both what did and what did not happen are of interest. Had better observational techniques been available, the early corpuscular views might have developed into something like solid-state physics early in the nineteenth century.

My own interests since 1960 have centered around the interaction between art and science. General speculations on this topic constitute most of *A Search for Structure* and served as the starting point for an exhibition I organized for the Smithsonian Institution and M.I.T., *From Art to Science* (catalogue; Cambridge, Mass., 1980). My conclusions were summarized in the postscript to that catalogue:

Everything that we can talk about involves the relationship of a human mind to other humans, to the things they have made, or to nature. Without internal experience there could be no external communication. The mind is more complex than the brain alone; the response of the whole body influences both experience and memory. Understanding, even scientific understanding as distinct from analytical proof and extension, seems to be in large part emotional, for it involves an interplay between what the senses can feel and what the mind can think about—movement, resistance, and the interlock of parts into patterns. Perception is the discovery of a link between sensable quality and conceivable structure. One cannot know the meaning until one has had the experience.

Both as individuals and as societies we first explore the world sensually (aesthetically at the higher points); then we seek to explain and exploit what we have discovered by finding replicable relationships between as many parts as possible; finally we move beyond this to emotionally satisfying understanding. Both the initial discovery of anything and the final stages of adjustment at maturity seem to involve more of what we call "art" than does the intermediate stage of logical growth. The biological and mental makeup of *Homo sapiens* yields a holistic mechanism that exploits the presence of misfits on one structural level to generate a different balance of order and disorder within larger levels of interconnectedness.

This exhibition shows that art for the sake of art has had a role in the development of useful technology and of logical science. However, neither science nor art needs to find justification in the useful things that sometimes stem from them. It is sufficient that they are simply and wonderfully human. Both are expressions of human potentialities, and if they have any function at all it is to provide chancy concepts with opportunities for new interactions, thus carrying the mechanisms of evolution from the biological realm into those of culture and society. The utility of a thing or an idea depends upon the replication and incorporation at the interactive social level of something initially individual and unique.

Aesthetic curiosity, not perceived purpose, seems to have been the motive for most discoveries that have triggered technological and scientific changes with their profound effect on social organization. The nucleus of this viewpoint can be detected in Section I of the present book.

Perhaps out of all this will grow a kind of philosophy of structure. For the past three centuries scientists have mainly sought regularities in nature and have postulated and then found unitary components of structure, individual particles on ever smaller scales down to the level where time-reversal confers an equal individuality on their absence. Accompanying the success of reductionism came the almost inevitable separation between art and science, while traditional philosophy seemed to have difficulty with both. Perhaps the acceptance of universal heterogeneity and the necessity of change, so central to materials science, will lead to a kind of hierarchical temporal topology of the boundaries of inertia, providing the basis for a renewed and broader form of Natural Philosophy as framework that would aid understanding of complex systems, including social ones, and both inspire and control flights of imagination.

There was a time when science was simple, mainly because it excluded most of the phenomena of nature. In the twentieth century its domain has been vastly extended and enriched, but of necessity at the expense of fragmentation. Nevertheless, details of known repetitive behavior can be dealt with most effectively by computer, and the human mind can be used to seek understanding of the relationships between substructural levels within moving boundaries of attention, which is more like art than analytical science. And it seems probable that scientists will no longer be ashamed to deal with the nuances of the real structures that have interested and guided those whose hands and minds have been applied to the working of materials.

October 1987

This book arose from the desire of a practicing metallurgist to understand the background of his professional activity: It is published in the belief that scientific metallurgy provides a good illustration of the complexity of the development of human knowledge, for it depends on the interaction of theory and empirical knowledge in a way that is not so evident when tracing the growth of the pure sciences. Perhaps the study even has a certain timeliness, for it is abundantly clear that the proper balance between "applied" and "pure" science that is necessary for technological and cultural well-being has yet to be achieved.

It is hoped that this specialized history will provide a few new facts of interest to those concerned with the broader aspects of the history of science. The story touches on the history of art occasionally and on the general history of ideas rather frequently, though its core is simply the growth of concepts of the nature of materials as they developed in the minds of men who used them, men who were concerned little with either aesthetics or philosophy. Ideas on the nature and structure of metals are obviously only part of basic concepts of the nature of matter in general, yet it has not often been the case that "pure" scientists were aware of practical metallurgical phenomena or that metallurgists have directly based their work on rigorous science. An empirical observation of some fact and its utilization by an alert artisan or engineer has generally preceded scientific interest in it, and, conversely, practical metal workers have ever been slow to see the use of theory.

Even with the limited scope of this book it has been necessary to pass over much that is pertinent to it. It should have been preceded by a good history of the constitution of matter, discussing atomic and molecular theory as well as crystallography and biological structure. It would have been desirable, too, to study the lives of some of the men who made important advances in order to find the personal motivation and the detailed genesis of the ideas behind the bare published papers.

The story of the atom has been often told. This work is a study of man's realization of another scale of structure—of groupings of atoms, of the special qualities that result from geometric order and the conflicts of orientation. Although man's eyes early led him to the concept of a significant substructure and his mind to the necessity of particles in space, he was slow to realize that the commonest form of granularity depended only on orientation, and that the essence of crystallinity lay in internal order rather than in external form.

The glitter of gems and the geometric cleavage of minerals seemed to have nothing in common with a shapable, malleable metal. Those aspects of structure that do not depend on microcrystallinity were appreciated relatively early. The seventeenth-century philosophers understood quite clearly that the hardening and softening of metals depended upon some variations in the arrangement of particles and their interactions with their neighbors. That prince of applied scientists, Réaumur, postulated in steel the interdiffusing and interacting of particles in solid solution and their segregation in a manner that would be completely modern except for geometric ordering of the units. The concept of carbon in solid solution in steel and as precipitated carbide was well accepted early in the nineteenth century, and everyone knew that metals could occasionally be crystalline. It was, however, the observation of metals under the microscope—a quarter of a millenium after the discovery of the microscope—that showed the universality of crystallinity in metal. The idea did not lead to the discovery: It was quite the reverse. The background of this seemingly simple observation leads into art, industry, war, philosophy, and nearly every aspect of human activity.

The structure of metal is revealed to some extent by the shrinkage of a casting, by the texture of the broken or deformed surface, and by local variations in the manner of chemical attack. Metals are simple aggregates of innumerable small crystals, packed more or less randomly against each other to fill space. This simple fact was not clearly understood until nearly the end of the nineteenth century, yet long before this the existence of a variable granularity was known to the philosopher and used by the artisan. Differential chemical attack was used in Europe to give a decorative mark of quality to swords as early as the second century A.D. but the technique passed out of use, and only in the Orient was surface texture continually used. The superiority of the famed swords of Damascus and Japan stemmed directly from the fact that the smith related a certain visible texture to the serviceability of the weapon and adjusted his manufacturing process to produce it. He did not, however, speculate about the meaning of these patterns in terms of crystallization and segregation. Conversely, in seventeenth-century Europe, where the early microscopists would have delighted in such structures had they seen them, there was nothing available for study but distorted fractures or brutally burnished or abraded surfaces, and as a consequence metallography was two centuries later in arising than was biological microscopy. Attempts to duplicate Oriental materials did, however, later on inspire a great deal of practical work by European metallurgists and it did lead to important scientific findings. Bergman's work on carbon in steel and cast iron, intimately involved in the demise of phlogiston, was a direct outcome of the establishment of a Swedish factory for making "Damascus" gun barrels, and Tschernoff's famed work on steel transformation was directly inspired by the earlier French and Russian duplication of true Damascus blades.

The structure of steel as shown by its fracture and interpreted in the light of Cartesian corpuscular philosophy was an essential guide to Réaumur's important work on steel and malleable cast iron, though he did not realize the crystallinity of the "parts" he discussed, and in any case others were slow to follow his lead. The growth of mathematical crystallography had almost no effect on the understanding of polycrystalline matter, for the formal crystallographers were concerned mostly with the mathematical symmetry of shape or property vectors, not with internal order. Another scientific triumph, the atomic theory of Dalton, had the effect of temporarily postponing rather than advancing metallographic understanding, for it presented a theory of the molecule which closed men's minds to non-stoichiometric compounds, while providing an adjustable molecular parameter to account for some effects that actually stem from microcrystallinity. For a long time chemistry could not encompass solid solutions and all chemical compounds had to be of fixed, simple composition. While it was easy for the earlier corpuscular aggregates to have a wide range of composition and to allow diffusion in a solid, the early nineteenth-century chemist could imagine such behavior only in non-crystalline material. It was much the same in physics: The development of mathematical elasticity was based on the assumption of a homogeneous isotropic material or, in its most elegant form as crystal elasticity, on uniform directional behavior. The beautiful ideality of the models and the elegance of the mathematics made physicists insensitive to the considerable and growing evidence for the real state of variable microcrystallinity.

The rapid growth of the railroads following the 1830's was accompanied by a proportionate increase in the number of "crystalline" fractures in wrought-iron axles. Although some outstanding engineers and chemists realized immediately that it was the manner of fracture rather than the structure that changed, one important engineering society inconclusively debated the question of vibration-induced crystallization as late as 1893, and even to this day the fallacy that vibration makes metals crystallize constitutes practically all of the common man's "knowledge" of our subject.

The first considerable change from age-old metallurgical practices was occasioned more by pressure for increased production than by the application of scientific knowledge. The second phase was a chemical one, in which improved methods of analysis (themselves largely an outgrowth of the ancient assayers' techniques) gave a degree of control over raw materials that was well exploited. With chemical analysis also came the discovery that many curious effects such as hot and cold shortness in iron were related to the presence of impurities, and it brought with it an overemphasis on chemical specification as a means of control. Finally, with microscopy, structure became important and produced a far deeper understanding of the role of all the factors than had been consciously or unconsciously used in the past.

The extensive but purely qualitative knowledge of the structure of matter that existed in the seventeenth century was partly forgotten in the eighteenth under the influence of a growing preoccupation of physicists with mathematics. The great strength of the Newtonian approach in mechanics and elasticity produced insensitivity to those aspects of the behavior of matter that could not be so treated. Great advances always are obtained at a price, and a successful theory for a time throws out of focus those aspects of the behavior of matter that it does not fit. Conversely, qualitative observations, which usually precede quantitative ones, may overemphasize rather trivial effects. Engineers and metallurgists have always been forced to try things empirically with nothing but a general background of theory to guide them. They have used, and they will continue to use, matter under conditions that cannot be analyzed by precise physical theory, and in so doing they sometimes observe and roughly explain new types of behavior. Indeed, unless physics develops better means of handling systems of intermediate complexity, experiment will always prove better than theory: The best computer for the behavior of an alloy will forever be the alloy itself.

The crystallinity of many rocks is obvious to the naked eye, and much of the work of mineralogists is of direct concern to the metallurgist. Structure in stone was used decoratively ages before structure in metals was seen, and rock textures were catalogued in premicroscopic days. The plutonic theory of the earth at the beginning of the nineteenth century combined with the new atomic theory to inspire attempts to grow in the laboratory chemical compounds that had previously been known only as natural minerals and to understand the manner of their growth. Then Henry Sorby, the brilliant amateur living in the steel-making center of Sheffield, developed thin-section polarized light techniques for studying rocks, and, in 1863, was led, through an interest in the crystalline structure of meteorites (which had been discovered in 1804, but ignored by metallurgists) to study the structure of artificial steels and iron. Though in petrography Sorby found a disciple, Ferdinand Zirkel, to develop the science promptly, his work on metals was not followed up for twenty years, and then only because of parallel developments based on studies of fracture. Yet when metallography once became fashionable, its advance was rapid and it outpaced petrography as a science, perhaps simply because reactions in alloys occur so much more rapidly than in systems of mineralogical interest and are, therefore, more amenable to laboratory study.

As one surveys the whole story of the development of metallography, the analogy with a phase transformation is unavoidable. From an amorphous body of practical knowledge, there had appeared a number of transitional systems of order that disappeared as a result of later transformations (note that it is the arrangement that changes, not the components, for facts once established have the permanence of matter). Each transformation was preceded by a kind of supersaturation of observed facts, and by subcritical suggestions of order, but eventually the facts precipitated upon a new idea which formed the nucleus of order for subsequent rapid growth. Eventually growth slows or stops for lack of new information; however, any given body of knowledge may itself be usefully incorporated into a larger one (like an atom in a molecule, in a crystal, in the whole structure) or it may be entirely transformed in orientation or in inner structure by growth from a new nucleus which presents a more profound order.

Much space has been devoted in this book to the period before there was any recognizable metallurgical science, and almost every available type of record, whether written or wrought, has been looked at for its significance. I have chosen to end at about 1890 because metallography had by then begun to crystallize into a definite discipline. Recent events call for a different historical approach, for perspective is inevitably lacking. Except for the truly epoch-making introduction of X-ray diffraction, it is difficult to see where, in the developments since 1900, lie the germs of the broader materials science of the future, and the sheer mass of literature discourages a comprehensive study. In some respects the period following 1890, though more exciting as science, is less interesting as history than the earlier period. The forces promoting order became overwhelming and the science took over the men. In the earlier periods it was the intellectual interests and accidental contacts of unique individuals that served to produce the bridging between disciplines which forms exciting nuclei. Though it is almost certain that by the end of the nineteenth century some petrologist would have turned a microscope on a properly prepared metal surface, it was Sorby's particular environment that led him to do this in 1863 instead of twenty years later. After 1880 industrial pressures made many men look at the structure of steel, but again it was unique that Osmond happened to be associated with Troost and Le Chatelier and so combined his structural observations with thermal ones.

The story of metallography in no way conforms to the popular picture of an applied scientist using the concepts of a separately developed basic science to achieve practical results. Perhaps one might have expected that when both mathematical crystallography and the atomic theory existed, the polycrystalline state would have been obvious, and would have been extensively studied by physicists in the simple systems that they handle so well. Yet this did not happen. Except for the hidden logic that determines a man's interest, it was illogical. The exciting period of development almost all rotated around the understanding of the hardening of steel, which is one of the most complex things, even today not fully explained. It was the pressure for the understanding of the age-old mystery of this important material that precipitated almost all of the scientific work in metallurgy since the end of the eighteenth century, and not until this was well on the way to a reasonable solution was attention turned to simple materials like copper or to the fascinating problem of the interfaces between crystals.

Even after metallic crystallinity was fully appreciated, it was for a long time assumed that crystals could only form by unrestricted growth into a liquid or into an amorphous material. The growth of one crystal into another (either because its neighbor was less strained or just because chance had given it more than fourteen neighbors) was difficult to grasp, and the twentieth century was well along before it was realized that change occurs always at interfaces—the two-dimensional region of disorder separating different domains of three-dimensional order—or by the movement of one- or zero-dimensional imperfections in the crystal bodies. Theory has oscillated from disorder to order and back again, finally to a realization that the world must have both. When crystallinity was finally accepted, it was embraced so wholeheartedly that men found it difficult to account for phenomena like diffusion and deformation, and it took fifty years to reintroduce some of the earlier variability in the form of lattice imperfections, defects that are insignificant volumetrically but which confer on a crystalline matrix all of the non-rigid properties that had delayed for so long the realization that metals are normally crystalline.

The history of metallography has much in common with its object, for, as theory of structure tends to overemphasize order and to disregard the complexities of real matter, so also the history of theory tends to ignore the lack of logic in any human developments, even intellectual ones. Imperfections and misfits are as inevitable in scientific theory as they are in the structure of a piece of metal and they are just as necessary to full understanding.

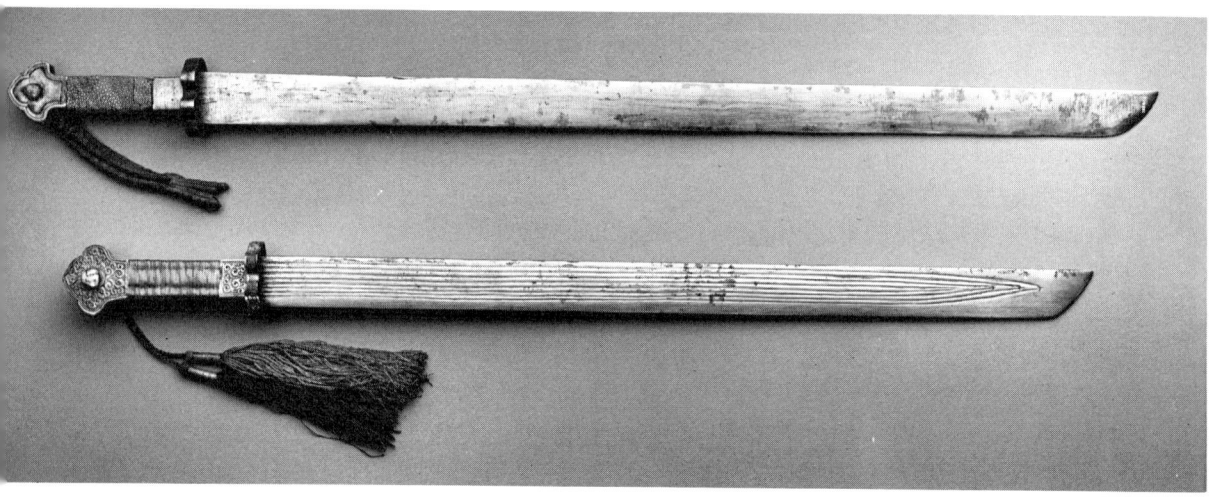

Tibetan swords (cf. p. 9) and Japanese sword guards (cf. p. 57).

THE BACKGROUND
OF METALLOGRAPHY
IN WORKS OF ART

I T IS well known that Egyptian, Minoan, Sumerian, Etruscan, and other early workers of metal made empirical use of most of the features of alloy constitution diagrams, and of hardening by solid solution, transformation, and cold-working. It is less well known that the techniques of the metallographer in developing the structure of metals also have a long background in the arts, and it is the purpose of this section to outline some of the cases where swordsmiths, armorers, and jewelers have made decorative use of surface phenomena depending on structure long before these were used by scientists as an indication of internal order.*

In its Hellenistic origins alchemy was much concerned with color changes in metals, but there is no trace in alchemical writings of any observation of differential corrosion, tarnishing, or staining, depending upon fluctuations of either crystalline orientation or composition. Even the thousands of trials made of the solution of metals in acid for assay or chemical experiment seem to have given rise to no speculations as to the reason for the patently heterogeneous nature of the attack.

Surface patination by natural weathering or chemical reactants was extensively employed by the metal workers of classical antiquity.[2] The contrast of color or texture between inlaid and overlaid metals is the basis of many distinctive styles of goldsmith's and silversmith's work, but in virtually all European examples the metals themselves are intended to be uniform except for superficial shaping or engraving. The artisan's hand produced pleasing shapes and textures, often combining several differently colored metals in intricate inlay or overlay, but the materials might just as well have been structureless. The decorative use of structure is largely an Oriental development, though its origin may have been European.

* A summary of the material in this section was published in *Endeavour*, 1957, **16**, 199–208. It has been remarked that the works of "art" that I discuss are mostly military. Metallurgical ingenuity has always been devoted to weapons on one hand and to items of adornment on the other. It is natural, therefore, that the combination of the two in ceremonial arms and armor should evoke the highest skill and widest range of technique. It is thanks to the interest of collectors and connoisseurs, not to that of practical metallurgists, that such objects have been preserved for examination today, and they undoubtedly give a fragmentary picture of old technologies, for the most numerous common objects, crudely fashioned from the cheapest materials, have a very low probability of survival. The first two volumes of the magnificent *History of Technology* edited by Charles Singer[1] are largely illustrated with drawings or photographs from objects in art museums!

1. The Merovingian Pattern-Welded Blade

The "water" or "damask" pattern of Islamic sword blades is a well known effect of metal structure, but it was probably preceded by a European technique of much interest, that of the Merovingian "pattern-welded" blade.* Although any sword made by hammer-welding and forging together pieces of iron and steel, or even by simply forging a sponge of irregular composition, will have a pleasing pattern of wavy lines on its surface after polishing and etching, there is no mention of such effects in literature of Roman times or earlier, and any iron objects that have come down to us are so badly corroded that the original surface cannot be seen or even inferred. Swords of great interest because of their metallurgical structure have been found in archeological sites, the earliest being from a Roman site in Britain dating from the end of the second century,[1] although they are principally associated with the Merovingian Franks and the Vikings. A Viking sword of this type was described by the Danish archeologist Engelhardt in 1866 and large numbers of them exist, but they created little interest until Salin and France-Lanord[2] reported detailed studies of them. Figure 1 shows the general appearance of a sixth-century blade of this type, in the condition as excavated except for cleaning, while Figure 2† shows the point of a similar one that has been repolished and etched.

The excellent papers of A. France-Lanord[3] discuss both the archeology and the metallurgy of the swords. The principal seat of manufacture

---

* There has been no careful comparative study of the origin of these techniques. It is not impossible that the Oriental crystalline damask was first made by Indian smiths and that the European blades resulted from an attempt to duplicate its texture, using the same erroneous schemes that suggested themselves to French metallurgists of the eighteenth century (see chap. 4).

† Figures 2, 6, 7, 9, 10, 13, 14, 15, 17, 18, 19, 46, 47, and 49 are reproduced from *Endeavour*, Vol. **16,** with permission of the publishers. Figures 7–10, 14, and 17 are reproduced by permission of the trustees of the Wallace Collection.

seems to have been on the Rhine, though they were carried far afield by trade and war. Liestøl[4] believes that the Viking blades were made in Scandinavia, for the technique would not have been beyond the skill of their smiths. However, the sudden disappearance of the blades in the tenth century suggests localized manufacture, for the blades were highly valued and it is unlikely that catastrophe would have suppressed a widely scattered skill. It is possible, of course, that they were displaced simply by the development of cheaper and equally efficient methods of making and heat treating uniform blades, or blades with welded-on edges of steel.

The decorated center portion of these swords was made, according to the reconstructions of France-Lanord as modified by Liestøl, by taking strips of iron and steel (or iron strips carburized on one side) and welding these together to give square rods of laminar structure. The rods were then folded[5] or, much more probably, twisted[6] to produce the desired pattern, and welded together in groups of three or four to form a strip for the middle of the finished sword. The cutting edges were uniform steel of somewhat higher carbon content, often sandwiched between plates of low carbon material, which provided most of the visible surface. On final grinding to shape, the areas of high- and low-carbon metal in the band formed patterns like that of Figures 1 and 2, or variants of this depending on the position of the section in relation to the twist-axis and the incorporation of lengths of untwisted laminate.

A cross-section of a similar sword (a fragment presented to the writer by M. A. France-Lanord) is shown in Figure 3. The contorted medium- and low-carbon areas in the center are clearly seen, as well as the inserted part of material containing carbon to provide the cutting edge.*

* Spectroscopic analysis showed only traces (less than 0.01 per cent) of copper, nickel, cobalt, chromium, and manganese. Chemical determination of phosphorus, oxygen, and nitrogen and microscopic estimation of carbon gave the following results on samples representative of the edge metal and the light-etching body:

	Edge	Body
Carbon	0.15	0.00
Phosphorus	.23	.33
Nitrogen	.0009	.034
Oxygen	0.132	0.291

The local variations of microhardness are shown in Figure 4. A large number of other measurements show that there are many places where carbon-free material is actually harder than that containing carbon (175–200 *vs.* 120–150 VHN). The extreme edge of the blade has a hardness of over 300, and the microstructure is entirely martensite with no free ferrite, showing that the cutting edge had been heated to a moderately high temperature (about 900° C.) and cooled by quenching. The high nitrogen content was undoubtedly responsible

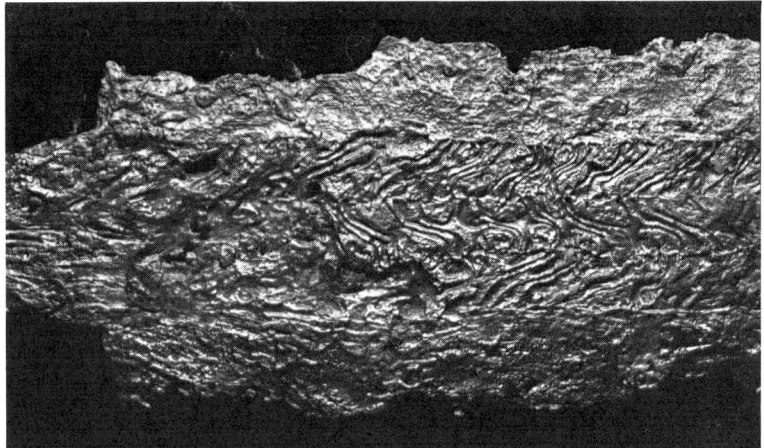

Fig. 1.—Merovingian pattern-welded sword blade. Lorraine, sixth century. Corrosion products removed but otherwise untreated, ×1.

Fig. 2.—Point of pattern-welded sword similar to that in **Figure 1**. Repolished and etched, ×1.5.

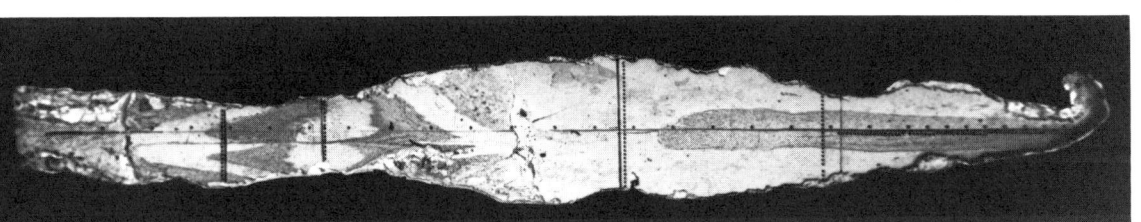

Fig. 3.—Section of sword similar to that in Figure 2. Etched, ×5.5.

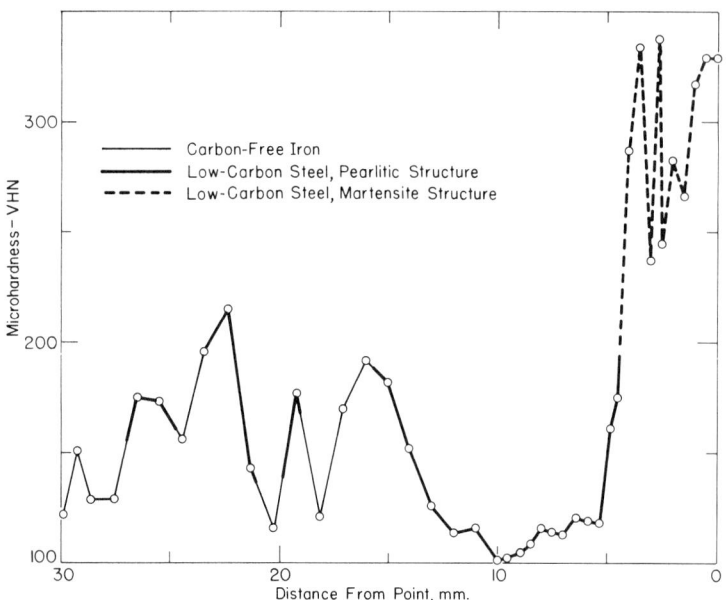

Fig. 4.—*Variation of microhardness along the center line of the section shown in Figure 3.*

Cassiodorus,[7] the sixth-century chronicler, wrote a letter as secretary to Theodoric the Ostrogoth to Thrasamond, King of the Warnorii (Vandals), expressing gratitude for a handsome gift, thus:

Through your brotherly affection swords that will cut even through armor have been forwarded to us, together with pitch-black drums and foreign pageboys of noble birth and fair complexion. These swords are richer for their iron than for value of the gold [which embellishes them]; for there flashes out from them such a polished brilliance that they reflect with the utmost fidelity the faces of those who look at them. Their sides approach the edges with such uniformity that you would think they were not fashioned by files, but cast from fiery furnaces; their centers, hollowed out with beautiful grooves, seem to undulate with worm-like markings; for shadows of such variety play there that you would think the metal was interwoven

for the relative hardness of the low-carbon areas in other parts. Annealing at 275° C. brought out more microscopically-visible nitride particles. The gradient in hardness near the cutting edge was not associated with a carbon gradient, for brief annealing at 900° C. followed by air-cooling showed the carbon content to be reasonably uniform in the whole inserted piece of dark-etching metal. The edge alone had, therefore, been quenched while the body, though heated, had been allowed to cool more slowly—a technique later to be brought to perfection by Japanese swordsmiths (see chap. 6). The sword as a whole was not carburized, but the carbon-content variation is purely a result of forging and welding together masses of more or less uniform material differing in carbon content. The cutting edge was made by piling flat strips, not by folding the iron around the steel. The heat treatment of a very low-carbon steel of this kind is simpler than that of the high-carbon material that later became common, for an interrupted quench or tempering after quenching would be unnecessary.

rather than shining [superficially] with different colors. This metal your whetstone so carefully shapes, this your splendid dust (granted to your country by the bounty of nature) so thoroughly polishes, that it makes the gleam of the iron a very mirror of men. A peculiar opinion has arisen regarding them: that they are swords which from their beauty might be taken for swords made by Vulcan—he who apparently perfected the art of the smith with such elegance that whatever was fashioned by his hands was thought to be, not the work of mortals, but divine. Accordingly, in returning to you our kindest regards through so-and-so your ambassador, we declare that we have received your weapons with pleasure, which they have conveyed to us as earnests of a good peace. In consideration of your munificence we are sending you a gift in return; may it be as acceptable to you as yours was pleasing to us. May these auspicious gifts vouchsafe us friendship, so that in making these heart-felt interchanges between us we may unite our nations and in reciprocal concern bind ourselves together for our mutual advantage.*

Some of the Norse sagas also mention the patterns, and indeed Liestøl believes that the names of many famed old swords refer directly to their patterns. Quite apart from this literary evidence, there can be little doubt that the structure was a visible one, for it is difficult to justify the particular pattern used in the center of the swords on any but aesthetic grounds. The small patches of pattern-welded metal in the lance points illustrated by Salin (his Fig. 6, Part III) are even more obviously unfunctionally decorative than the swords. Poor welding would undoubtedly weaken a contorted structure less than a simple lamellar one, but if avoidance of the ill effects of carelessness had been the aim, more of the weapon, particularly near the cutting edge, would have been patterned. France-Lanord's tests on the blades are not a convincing proof of their superiority, for a slight difference in thickness would more than account for the differences that he observed in elastic bending.

Blades of both earlier[8] and later[9] dates than the pattern-welded ones were obviously made by welding together several plates of slightly varying composition. Some composite fagotting is a natural and almost unavoidable technique of the smith, and, by virtue of the greater deformation, produces a product superior to that resulting from simply flattening out a sponge of iron or steel—unless, of course, the welds are very poorly made and introduce more harmful slag inclusions than those already present in the crude metal from the gangue in the ore. Such a blade would develop a simple wood-grain damask if it were

* Translated from the Latin by Professor John Hawthorne of The University of Chicago.

etched, but the improved mechanical properties resulting from extensive working alone would justify its manufacture. It may be that etching began some time after this method of forging had been established and the association of an etchable figure with superior service led to the production of the intentionally defined pattern of the Merovingian blade as a kind of trade-mark. The type of pattern shown in Figures 2, 5, and 6 is clearly more influenced by art than engineering and the existence of pattern-welded blades is therefore strong presumptive evidence of contemporary knowledge of some means of revealing the structure of metals by a treatment that visibly differentiated portions differing in composition. This is the very essence of the metallographer's technique.

The ancient method would most probably have involved etching (perhaps in a vegetable acid such as fruit juice or sour beer or vinegar, or with a solution of a common corrosive mineral such as copperas or feather alum) which would quickly and strikingly develop contrast between adjacent areas of iron and steel that had been laboriously fashioned to give the final effect. The references that Cassiodorus makes to the polishing "dust" and to the reflection of images makes it certain that abrasives were used to produce nearly flat surfaces on the blade. Conversely, his description of the play of colors in the interwoven pattern strongly suggests an etched texture in the grooved center of the blade, and it may be that the blade had been finished with both etched and polished areas in fine contrast. However, some differentiation can be produced by polishing alone. J. W. Anstee reported[10] that he could see a pattern, supposedly of slag, on the unetched surface of a blade that he had himself forged out of wrought iron. Moreover, Tibetan swords of the nineteenth century were made by welding together two kinds of steel in a far simpler pattern than the Merovingian blades— side-by-side welding of a nested series of hairpin rods—and the structure was developed by scraping alone* (Figs. 5, 6). At its best, however, differential polishing would give a structure that would appeal only to the eyes of a trained connoisseur, and etching would generally be far more effective. Both the reagents themselves and the opportunity for observing their effect would have been available, and the writer believes that this central technique of the metallographer was actually employed on these interesting objects in Merovingian times. It should

---

* It is possible that the present surface of these blades is not the original one but came in part from cleaning with an occidental abrasive, for collectors (and even museum curators) sometimes prefer a bright finish to the "tarnish" on Oriental objects.

Fig. 5.—*Sword from eastern Tibet, nineteenth century or earlier. Pattern of welded hairpins, developed by deep differential scraping. (Chicago Natural History Museum No. 122250.) Reduced to about seven-tenths natural size.*

Fig. 6.—*Sword from Derge, Tibet, nineteenth century or earlier. Pattern developed by abrasion alone. (Chicago Natural History Museum No. 122255.) Natural size.*

perhaps be pointed out that the metallographic implications are only a small part of the whole interest of these complex textured swords. They are intricately associated with the origin and spread of improved siderurgical techniques. Leroi-Gourhan[11] and T. A. Wertime[12] have discussed the possible association between steel with cast and with forged textures, and the spread and diversification of techniques from some location in central or south-central Asia. Just as at the end of the nineteenth century micrography played a central role in the birth of modern metallurgy, so perhaps macrography may have been intimately associated with the first good techniques for making steel. But this is sheer conjecture and further research by archeologists and metallographers is badly needed.

Noted added 1965: Some additional important publications on the pattern-welded blade are listed in the Bibliographic Notes, page 261.

2. The Etching of Armor

Etching for general decorative purposes does not seem to have grown out of the practice with the textured blades described in chapter 1 but to be a separate, later invention. The pickling of metals to remove scale formed on annealing, to blanch silver-bearing coins, and as a preparation for soldering and gilding is very old, but the reagents were too mild to dissolve unoxidized metals at appreciable rates. The first mention in European literature of any etching reagent for metals and the earliest examples of decorative etching (on armor) that have been preserved[1] both date from the fifteenth century. Had etching been used at the time of Theophilus (*ca.* A.D. 1123), it is unlikely that so interesting a process would have escaped his curious and practical mind.

In a manuscript, *Experimenta de coloribus*,[2] which bears a note that it had been transcribed from an earlier one in 1409, occurs the following recipe:

To make a water which corrodes iron.—Take one ounce of sal ammoniac, one ounce of roche alum [*aluminis roche*], one ounce of sublimed silver [*argento sublimato (sic)*, probably a confused reference to mercury sublimate], and one ounce of Roman vitriol. Pound them well, take a glazed earthen vase, pour into it equal parts of vinegar and water, then throw in the above-mentioned articles. Boil the whole until reduced to half a cup or a cup; apply it to such parts of the iron as you may wish to hollow or corrode and the water will corrode them.

The etching reagents in the fifteenth century English sources quoted by Williams[3] consist of salt and vinegar mixed into a paste with finely ground charcoal or of a solution of sal ammoniac, vitriol, and vinegar. The parts to be etched were covered by either a linseed-oil paint or wax, and the design was cut through exactly as is done by a modern etcher. Etching was slow, the mixture being left on for two or three days. Six different etching media are given in the little pamphlet *Stahel und Eysen*,[4] published in 1532 as part of one of the earliest *Kunstbüchlein*, The reagents, all complicated, involved various combinations of verdigris, vitriol, vinegar, salt, sal ammoniac, alum, tartar, and mercury sublimate. All but one were used as a paste, the exception being a liquid solution which was ladled hot over the surface to be etched. Biringuccio (1540)[5] and the pseudonymous *Secrets of Alexis* (1555)[6] refer to similar methods. Thereafter references to the etching of iron and steel in media of this kind become commonplace.

The technique of etched surface decoration was closely related to the preparation of etched plates for printing. According to Mann[7] the first etchings were printed from iron plates,* the making of which seems to have been a part-time activity of armorers, for the sale of prints could scarcely have provided them with a livelihood. In the seventeenth century, however, etching began to be used far more for the preparation of printing plates than for direct decoration on armor. The earliest extant prints are probably those made by Daniel Hopfer of Augsburg in 1503, although the first identifiable date is 1513. Albrecht Dürer in 1515 and shortly thereafter made a number of fine prints in which chemical attack and the burin were both used. Not until copper was commonly used in place of the earlier iron did etching become a popular pictorial medium. Cellini (1568) described the preparation of an etchant for use on this metal. In his classic work, Abraham Bosse in 1645 mentions[8] two different kinds of etchant, to both of which he applies the term *eau forte*, which later came to be limited to the mineral acid. The first of these consists of three pints of vinegar (distilled is better, for it does not harm the varnish), six ounces each of sal ammoniac and common salt and four ounces of verdigris. These were boiled together in a glazed pot and allowed to settle for use. Bosse also mentions the use of parting acid, *eau de depart,* which was made by distilling together vitriol, saltpeter, and alum, an operation that he does not describe, as the acid can be easily obtained from gold refiners.† The acid was diluted with about one-third of water for use. We cannot follow subsequent development of this kind of etching, save to remark that ferric chloride later came into use, and particularly the Dutch mordant (consisting of hydrochloric acid with potassium chlorate) introduced shortly after 1850.

* It is interesting to note in passing that intaglio printing of any kind was an outgrowth of another metallurgically interesting technique, that of decoration with niello. In this, lines engraved with a burin were filled with an approximately eutectic mixture of the sulphides of copper, silver, and lead: it was the use of ink to test the engraving before fusion that suggested its transfer to paper. Engraving considerably antedated the use of chemical etching to bite the lines. Steel again became the most common material for engraving in the nineteenth century, when it displaced copper, partly because of the finer detail that could be cut into it, principally because of its superior wearing qualities.

† Lazarus Ercker is the best early author on the preparation of nitric acid.

Fig. 7.—Detail of the etched breastplate of armor of the Duke of Brunswick, ca. 1540. (Wallace Collection No. 273 [Crown copyright].)

Fig. 8.—Details of etched hunting sword by Ambrosius Gemlich, Munich, 1540. Note the just perceptible striated area in the sky, which results from local variations in composition of the forged metal.

As used by European artisans in decorating armor, the etching re-agent was intended to remove metal uniformly from selected areas. A fine early example of etched armor is shown in Figure 7. The pattern came entirely from the skill of the artist in partially removing a previously applied, uniform protective layer, or, less often, by locally applying the stopoff to form the design in negative. In general, European armorers made no use of the possible decorative effects achievable by differential attack on parts of the metal that differ in composition or structure. At best the etching left a slightly rough surface, which was used as a basis for black paint or gilding to contrast with the surrounding polished or temper-colored (blued) surfaces unaffected by the etching. The action of the etching reagents was not highly sensitive to structure or composition, and, indeed, the designs were usually composed of lines too narrow to show such detail. It is rare to find a large area of etched metal in which structure could be seen if it existed, still rarer to see the structure itself. Search in a number of collections of European arms has disclosed only one such case, the sword shown in Figure 8. This was made by Ambrosius Gemlich of Munich around 1540 and is now in the Wallace Collection in London. The striations in the sky definitely arise from the elongated forged fibrous texture of the metal and add considerably to the pictorial effect, though it seems unlikely that they were introduced, or developed, intentionally. The only other examples of this kind of decoration are Oriental in origin or inspiration.

3. The Damascus Blade

In comparison with the relative neglect of structure by the European metallurgist, the enjoyment and utilization of it in the Orient is impressive. In the Orient, etching to display patterns depending on composition differences was in use contemporaneously with the European pattern-welded blade, and was thereafter continually developed to a high artistic level. The best-known achievement in this direction is the so-called Damascus blade, which takes its name from the locality at which Europeans first encountered it, not from its place of origin.* The best were made in Persia from Indian steel. They were in use even before the Islamic period, for the kingly poet Imru'ulquais (d. *ca.* A.D. 540) refers to a blade having wavy marks like the tracks of ants. A younger contemporary of his, Aus b. Hajar, describes blades in these terms: "It has a water whose wavy streaks are glistening. . . . It is like a pond over whose surface the wind is gliding. . . . The smith has worked out in it a grain as if it were the trail of small black ants that had trekked over it while it was still soft."†

The geographical distribution of these swords seems to have been practically coextensive with the Islamic faith and they continued to be made well into the nineteenth century. Their origin certainly antedates the references given above, possibly by many centuries. It could even be that references to patterned steel in Europe inspired the

---

* A good but incomplete bibliography on the metallurgy of "Damascus" and Persian sword blades is given by K. A. C. Creswell in his *Bibliography of Arms and Armour in Islam*.[1] The most important papers are those by Lenz[2] and Belaiew.[3] Belaiew has discussed[4] the importance of the blades in connection with the development of metallography. Zschokke gives[5] a fine series of photomicrographs of blades, including cross-sections at various magnifications and after a variety of heat treatments. He gives the results of analyses and mechanical tests on several blades, and shows, incidentally, that a properly heat treated modern blade of homogeneous steel has higher hardness, strength, and ductility than the duplex blades of the older smiths. It does not necessarily follow, of course, that modern blades are superior as weapons, as Podoski has claimed.

† The writer is indebted to Professor G. E. von Grunebaum for these early references and their translation.

Oriental metallurgists, or vice versa, although the manner of making the two and the details of their patterns are both so different that independent origin seems more likely. There are some early references in Mohammedan literature to the forging of swords from meteorites, and it has been suggested by J. von Hammer[6] that the visible duplex texture of such swords may have inspired experiments to duplicate it in terrestrial steel. The Austrian mineralogist Widmanstätten actually did forge a pretty knife for his emperor from a meteorite. Meteoric iron would undoubtedly give a blade superior in properties to a low-carbon or unhardened sword blade, but it would hardly compete with a normal Damascus blade of high carbon content. It seems rather that the knowledge of Damascus blades and the etching of them was of more aid in understanding meteorites than vice versa. In any case, the pattern on blades forged from meteorites would be so coarse as to be more closely allied to the pattern-welded sword than to the crystallization damask of the true sword of Islam.

Watered steel with a texture similar to that of the swords was also used in shields and armor of Indian, Turkish, and Persian manufacture. There are fine seventeenth-century examples of it in the Metropolitan Museum of Art, New York, which have been mentioned by Granscay.[7] Inspection of these objects shows that, because of the difficulty of forging the high-carbon steel and the small size of the cakes in which it was made, it was usually employed in the form of plates of a simple shape inserted in larger structures. The cuirass of a north Indian set of armor (ca. A.D. 1700) in the Metropolitan Museum has a rectangular piece of watered steel about 12 x 15 inches, riveted (almost invisibly) into a plain steel surround. A seventeenth-century Turkish shield about 26 inches in diameter has *welded* into it a circular piece about 17 inches in diameter of rather inferior watered steel. There is, however, in the Metropolitan Museum a beautifully wrought sixteenth-century Persian cuirass in three pieces, each apparently a single piece of watered steel skilfully formed into a complex, fluted octagonal shape. These pieces have some gold inlay and extensive deeply chiseled designs which are cut into the watered steel and the plain surround without distinction. It is uncertain whether the steel has been intentionally etched to bring out the structure, which is only faintly visible as a color change. It is interesting that, although the Persian steelworkers were the most extensive etchers of steel to develop its "water," they seem not to have used etching to produce artificially incised designs with a locally applied or removed stopoff until long after its use in Europe.

Damascus blades are made of a very high-carbon steel (about 1.5–2.0 per cent) and owe their beauty and their cutting qualities alike to the inherent structure of the cakes of steel from which they were forged. Different kinds of surface pattern are shown, natural size, in Figures 9, 10, 11, and 15. The light portion contains numerous particles of iron carbide (cementite), for its composition is nearly that of eutectic cast iron, while the dark areas are steel of normal carbon content (approximately eutectoid). The structure, of course, is clearly visible only after etching, which was done with a solution of some mineral sulphate. Blades from different localities tend to have somewhat different colors and occasionally blades are found with adjacent areas showing different colors. This effect has been attributed to the joining together of different metals, although in blades that the author has seen the differences are unmistakably only superficial, resulting from local applications of different etching reagents to continuous structures.

There can be little doubt that the basis of an Islamic damask is the difference in etchability between high-carbon areas, approximating eutectic cast iron in composition, and areas of approximately eutectoid carbon content. There is, however, greater diversity in pattern than can be explained purely on the basis of the structure of a slowly solidified casting. Some structures may originate simply in the distribution of fragments from a proeutectoid cementite network. In some blades of lower than usual carbon content the differential etching probably results from segregation of phosphorus, as Oberhoffer has suggested.[8] In any case, the blades must have been forged at very low temperature in order to avoid solution and redistribution of the carbide, which would both remove the pattern and render the steel more difficult to forge* and give a brittle sword.´

Most of the structures to be seen on choice blades are the result of

---

* It should be pointed out that the forging has to be done in a manner that will elongate the structure nearly equally in all directions, for the structure of the ingot must not be distorted beyond recognition. A piece simply drawn out into a rod would have a structure of invisible fibers and the metal would be as uninteresting as any modern wrought piece of steel. A cast ingot extended in ideal uniaxial compression would have the area of each and every part of the structure increased uniformly proportionately to the decrease in thickness, thus showing magnification without distortion. Cross-rolling produces an effect which is almost too perfect, and the irregularities resulting from local hammer blows lend a pleasing variety. The Oriental smiths did not forge their metal very much. The anisotropy of the microstructures of transverse sections indicates that the ingots were between three and eight times the thickness of the finished swords. The cakes may have been centrally pierced and opened up as De Luynes and Lenz suggest, for this would considerably decrease the necessary drawing out of the metal.

Fig. 9.—Indian dagger, early seventeenth century. Unusual damask showing almost undistorted cast dendrites. (Wallace Collection No. 1384 [Crown copyright].)

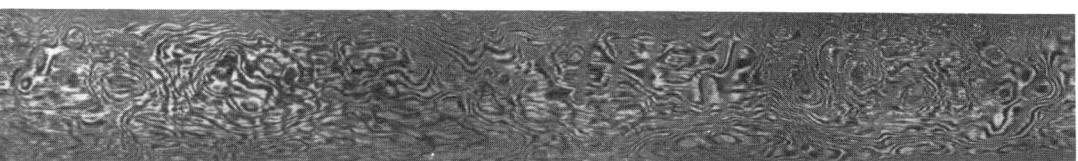

Fig. 10.—Persian straight sword. Ispahan, eighteenth century. Damask shows distorted dendritic structure, ×1.2. (Wallace Collection No. 1434 [Crown copyright].)

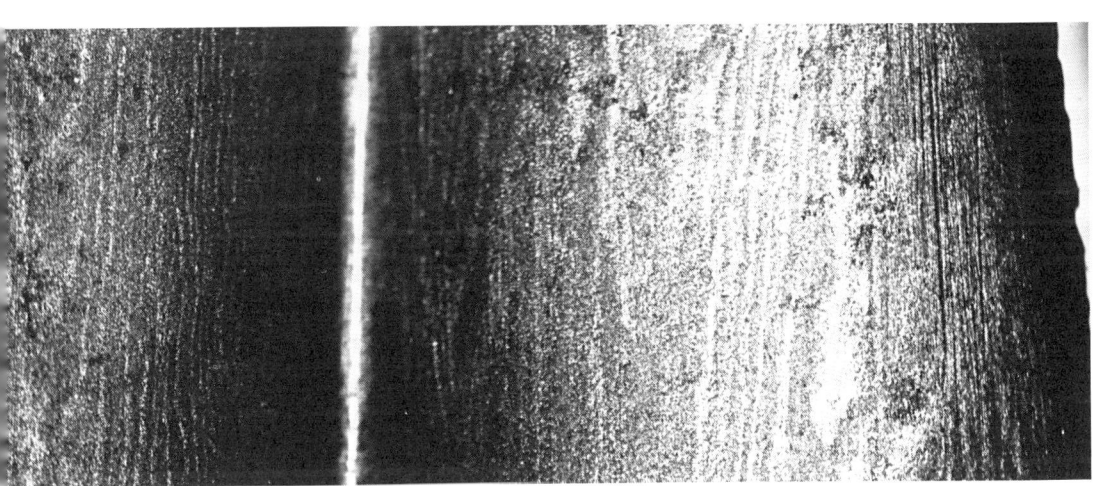

Fig. 11.—Persian dagger, nineteenth century. Damask with no deducible relation to the crystalline structure of a casting, ×1.6. (British Museum [Photograph by Mr. R. M. Organ of the British Museum Laboratory].)

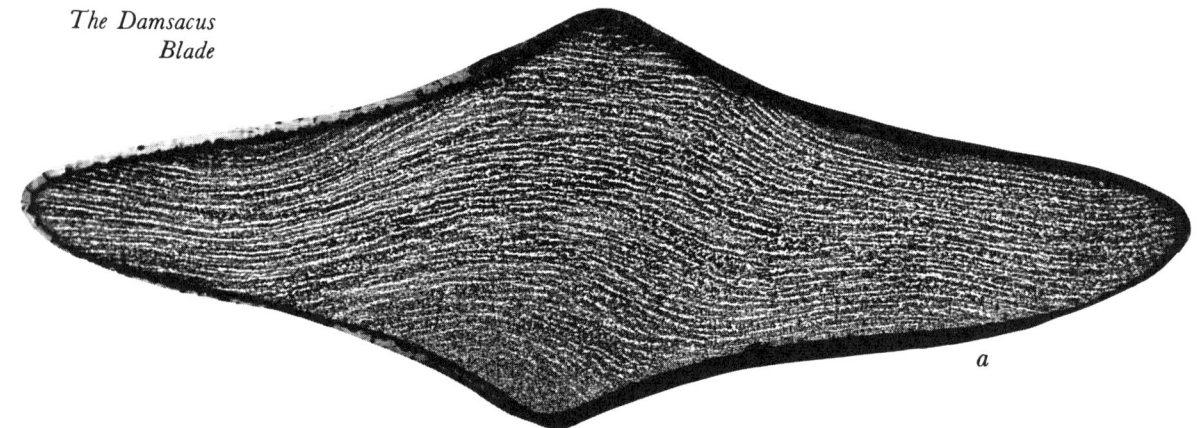

a

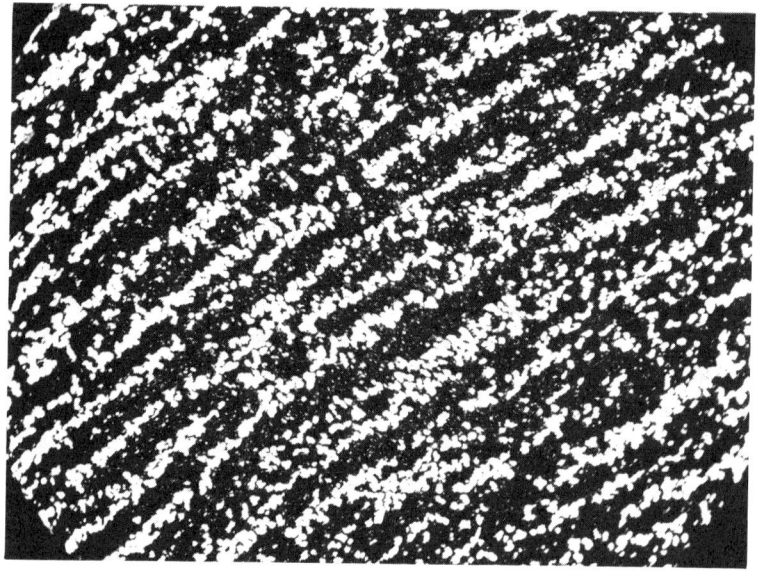

*Fig. 12.—Transverse section
near point of dagger shown
in Figure 11. Etched with
nital, (a) × 5, (b) ×400.
(Photograph by
Mr. R. M. Organ.)*

b

dendritic segregation during solidification, simply distorted by forging, and all recent authorities have taken for granted that the structure is exclusively a crystallization pattern distorted by forging. However, patterns are by no means uncommon which could not possibly be obtained from forging a cast structure. The connectivity of the parts (in a topological sense) cannot be changed by distortion. Figure 9 is a photograph of a blade in the Wallace Collection which is unmistakably a dendritic pattern, indeed, it is so little distorted that the museum cataloguer, not recognizing the pattern and being unfamiliar with the structure of castings, classified it as an artificially watered blade. Figure 10, of another blade in the Wallace Collection, is unmistakably a

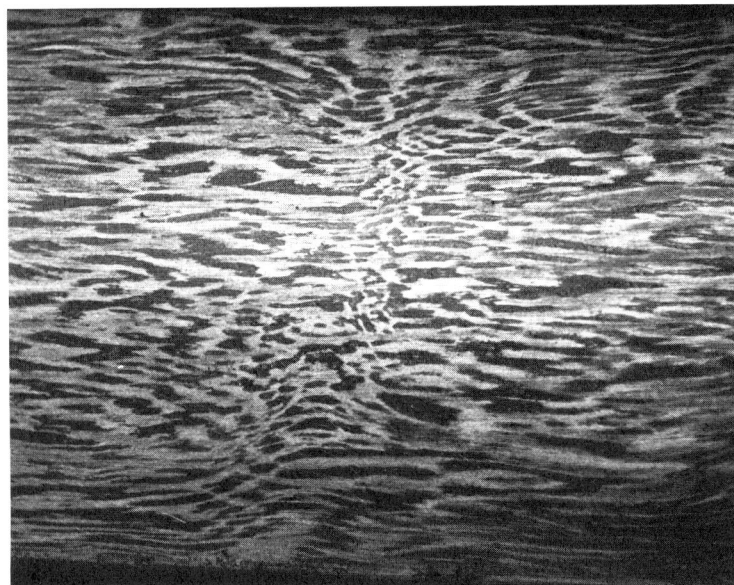

Fig. 13.—Indo-Persian scimitar, first quarter nineteenth century or earlier, ×3. (Author's collection.)

Fig. 14.—Persian sword, late seventeenth century. Natural size. Shows granular damask. (Wallace Collection No. 1404 [Crown copyright].)

result of sectioning a deformed mass of normal branching dendrites. Dendrites, however, no matter how much distorted, cannot give rise to the relatively common patterns of concentric pools of dark and light areas, alternating but not intersecting, such as that shown in Figure 11. This represents a Persian dagger of the nineteenth century photographed by R. M. Organ in the laboratory of the British Museum. The structure seems to be the result of sectioning a stack of alternating lamellac, slightly rumpled but continuous, as can be seen clearly in the transverse section (Fig. 12). The structure is not unlike that of the welded Malayan kris and the Japanese mokumé blades to be discussed later, but its origin must be different, for the high-carbon (white cast iron) layers could hardly have been forged from a separate plate of appreciable thickness, even when sandwiched between more malleable metal. More likely it seems that a stack of iron plates was joined and carburized by allowing molten cast iron to creep in by capillary action.

Figure 14 shows a blade which seems to be a distorted aggregate of large steel granules surrounded by a light-etching network. It is unlikely to have resulted from distorting and sectioning a dendritic cast structure. Zschokke's Plate VI may represent the structure that would result from overheating a blade momentarily, for cooling thereafter would cause pro-eutectoid carbide to precipitate on austenite grain boundaries stretching between the initial carbide-rich patches. Harnecker has shown[9] indeed that a cementite network can be produced even in blister steel, but the resulting damask is not a rich one.

Many blades have local distortions of the general pattern which seem to result either from fortuitous variation in the depth of local hammer imprints or from intentional local manipulation so that the final surface of the blade is not everywhere parallel to the initial surface of the casting. Damascus blades which display the regular cross-marks known as "Mohammed's Ladder" or the "Forty Steps" are much admired (Fig. 15; see also Zschokke's Plate III). Although Lenz[10] and most subsequent writers have accepted the suggestion made by Tschernoff that the steps represent radial dendrites in a steel cake which was subsequently pierced and opened out, this cannot be so. The contrived double-rung ladder pattern illustrated by Granscay[11] is conclusive disproof of a dendritic origin. More likely, the transverse markings result from cutting shallow surface grooves in a nearly finished blade (to change locally the angle of sectioning the various layers of the structure) and then forging the surface flat again. This extra process had been correctly described by Massalski in a report[12] on steelmaking in Persia, and by De Luynes,[13] but it has been ignored by all later writers. It is a well-known technique giving detail to welded damask patterns, and is displayed particularly well in the Indonesian kris, which will be discussed later.

No good Damascus blade has a structure that indicates very extensive working, and they are all remarkably free from slag. Welding cannot have been much used, although some travelers have reported the use of several cakes of steel in making one blade and skilful charlatans have sometimes welded a thin layer of good watered steel onto the surface of an inferior core.

Despite the variation of detail in the structures of Damascus blades,

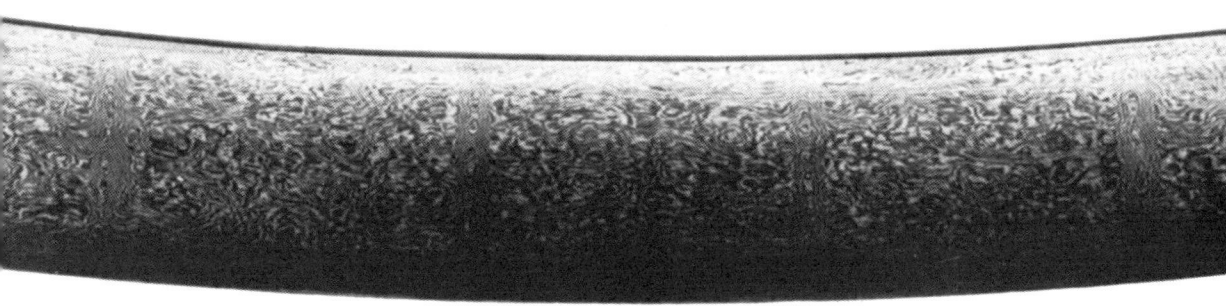

Fig. 15.—Persian sword of Indian form, showing cross-markings known as "Mohammed's Ladder." (British Museum [Photograph courtesy of Mr. Herbert Maryon].)

there can be little doubt that the water is primarily a result of the structure of the cake of wootz or similar steel from which the blades were forged. All the patterns become explicable in their diversity if it be assumed that the cakes were not always entirely melted, but sometimes consisted of an aggregate of unmelted iron granules or little plates cemented together by once-molten cast iron. Such a structure would result from the standard Indian process of steelmaking if the temperature were insufficient to melt the alloy completely. The crucible charge was an iron sponge,[14] sometimes roughly forged into plates,[15] together with wood for carbonization. Cast iron was used in nineteenth-century Persia[16] in place of wood. Sometimes the steel was made directly from ore and wood or charcoal in a single melting. Considering the primitive type of furnace that was used,[17] it is more surprising that complete fusion was ever obtained, particularly since fifteen or more flat-bottomed crucibles were heated together. It is perhaps pertinent that the first-century accounts of Chinese steelmaking disclosed by the recent research of Needham[18] and the sixteenth-century European authors Biringuccio[19] and Agricola[20] all mention the making of steel by soaking iron sponge or plates in molten cast iron at a temperature *below* that of complete fusion.* It would be a simpler technique than prolonged carburization in the solid state (particularly for relatively large pieces), and it does not need the very high temperatures required by true cast steel.

Further support for the viewpoint that the cakes were not necessarily completely melted is provided by the report of the detailed inspection of wootz cakes made by the metallurgist Mushet in 1804.[21] In every

* The name "co-fusion" which Needham gives to this process is misleading, for it implies that the components are melted together. Would not "perfusion" be preferable?

cake that he examined* he found fins and grains of malleable material mixed up with the steely parts and in a footnote he avers that the wavy appearance of Damascus blades probably results from "the various qualities of soft iron and steel which were frequently mixed together in the cakes."

At about the same time both Buchanan[22] and Henry Wilkinson[23] in their informative reports on Indian steel making, observed that cakes of Indian steel were often not fully melted, though they thought this to be a defect. An examination of etched cakes convinced Wilkinson that the pattern on the blade had its origin in the cake and he remarks that

> it would be as impossible to forge a sword blade out of some of these materials, when properly selected, without obtaining the true Damascus figure, as it would be to imitate the pattern by any contortions of iron and steel artificially. These cakes of steel are evidently crystallized, and the future pattern of the sword-blades depends on the size and arrangement of the crystals....

He believed that crystallization was modified by rate of cooling, by minute amounts of earths and alloys as well as carbon, and suspected that electrical action was in some way involved.

The first European reference[24] to the etching of these blades suggests the use of lemon juice, but a seventeenth-century writer—the alert and learned traveler Monconys[25]—recommends the use of an earth from Damascus, for neither the copperas of northern and occidental countries nor the good Cyprian variety (copper sulphate) will work; neither, he says, will any steel but that from India or Persia show water. *Zag* was analyzed in 1816 by Jacquin[26] whose results indicated it to be a naturally occurring impure basic ferric sulphate, like that known

---

* The flat "cakes" of wootz mentioned in all the older descriptions were about 3 to 4 inches in diameter and ¾ inch thick and are quite different from more recent Indian steel samples which have been melted in conical crucibles. Some ingots of this type are included in the complete set of samples illustrating all stages in the production of Indian wootz which was presented by Sir Thomas Holland at the end of the nineteenth century to the Royal School of Mines and are still preserved. The crucibles are round-bottomed, about six inches high, 2½ inches maximum internal diameter, and give a conical ingot about two inches high. A cross-section of one of the ingots made in 1956 shows unmistakably that the steel had been completely melted, for it has a well-defined large dendritic structure (Fig. 16) agreeing with the analyzed carbon content of 1.34 per cent. Although the accompanying knives are devoid of visible structure, the ingot would probably have retained a damask had it been carefully worked. The writer is grateful to the staff of the Royal School of Mines who made this examination at his request and with the encouragement of Professor C. W. Dannatt.

Fig. 16.—*Cross-section of ingot of wootz, late nineteenth century. Etched, ×2. (Royal School of Mines, London.)*

in Europe as *bergbutter* (iron alum, also known as feather alum). The analyses by De Luynes[27] and Zschokke[28] confirm that ferric sulphate is the principal active ingredient, but they suggest that miscellaneous earthy matter rather than aluminum sulphate is the adulterant.

In 1816 the British consul, John Barker,[29] reported the operations used by a cutler in Aleppo to renew the water on a worn blade. This operation inexplicably began by a new heat treatment—heating before sunrise in a charcoal fire to "the color of a soldier's dirty coat" and quenching in a mixture of oils, wax, and fats. The sword was reheated to drive off the fat, then polished and cleaned with lime and tobacco and finally etched with zag dissolved in water, the etch being washed off every two to three minutes and the operation repeated eight to ten times until there was no further change on repetition. Barker remarked that the art of casting the metal had been lost at the time of his visit, but that lumps were still occasionally met with, although the art of forging them had also disappeared.

The blades had a profound effect on European metallurgy, for the secret of both their structure and their superiority was long in being discovered. It was known that they were forged from the Indian steel known as *wootz* but this was regarded as almost unworkable by European smiths. Porta says[30] that it can be worked after it has been annealed in calcined *gypsum* [i.e., lime] and very slowly cooled. Réaumur deplored[31] the skill of Parisian artisans, none of whom succeeded in forging a tool out of a cake of Indian steel. Moxon in 1677 refers[32] to the fact that it very rarely came unwrought into England.

It is the most difficult of any Steel to work at the Forge, for you shall scarce be able to strike upon a blood heat but it will *Redsear;* in so much that these Symeters are by many Workmen thought to be cast Steel;* but when it is wrought it takes the finest and keeps the strongest edge of any other Steel. Workmen set almost an inestimable value on it, to make Punches, Cold Punches, &c of.

* This casual comparison with cast steel, taken with an entry in Robert Hooke's diary for November 21, 1675 (". . . bringing [cemented steel] soe as to melt made the best steel after it had been wrought over againe"), leaves no doubt that cast steel was commonly known in England long before the time of Robert Huntsman. Huntsman's achievement in 1740 was that of establishing a successful business—very important, but not the invention with which he is almost universally credited. It might perhaps be noted that the wide use of high-carbon steels, relatively easy to melt, had to await the introduction of the double process of quenching and tempering. The older direct, or slack, quenching worked best on a relatively low-carbon material. Damascus swords were quenched slowly in oil or by a current of air.

4. European Attempts To Duplicate Damascus Steel

Attempts to duplicate wootz and watered steels generally led to James Stodart's work in association with Faraday* on alloy steels, and to very extensive work in France early in the nineteenth century, most of which was based on the erroneous belief that the texture arose from forging. Many workers described the production of fancy surface patterns by welding, twisting, and forging together variously shaped pieces of iron and steel. The results were both decoratively and mechanically inferior to the Oriental product and, in fact, were merely a return to sword-making of a technique descendent from that of the Merovingian smiths which had meanwhile been extensively used in the Near East and India for making so-called "Damascus" gun barrels, which had nothing but name in common with the sword. (See chap. 5.)

Perret, who was the first European to describe the manufacture of welded-texture blades in 1771, in 1777 seems to have realized[2] that the natural Oriental damask was different. He remarks that it seems to have been made in the crucible and that the grains of steel are welded with those of iron: both the hard steel and the soft iron retain their consistency and the blade was a composite of visible globules which formed a cutting edge with irregular hardness. He did not, however, report any attempts to produce such a structure himself.

* Stodart, an expert cutler, was interested in wootz as early as 1795, long before he commenced his studies in collaboration with the famous scientist. Writing in 1814 he says in a long letter to Benjamin Heyne[1] that wootz is highly variable but that when it has been remelted and cast into bars like English steel it is of superior quality—"decidedly the best I have yet met with"—and he hoped that the East India Company might arrange to teach the Indians to melt the metal better and to use the tilt hammer, when it would become a source of considerable revenue to them. Faraday's association with textured steel may have had something to do with his lifelong concern with structure (see L. P. Williams, "Faraday and the Alloys of Steel," in *The Sorby Centennial Symposium on the History of Metallurgy*, ed. C. S. Smith [New York, 1965], pp. 145–62) .

The first successful duplication of blades of a true Oriental type was achieved in 1821 by Bréant,[3] an inspector of assays at the Paris mint,* whose achievement was an outgrowth of an all-important suggestion made by an associate named Mérimée[7] who was examining wootz cakes brought from England under the stimulus of Stodart and Faraday's 1819 paper. Against the general opinion of the time, Mérimée had advanced the idea that Oriental damask was the result of chemical combination and not a mechanical mixture of metals.

In 1819, Faraday[8] analyzed wootz (with wrong results) and in 1820 in collaboration with Stodart[9] made alloys with a wootz-like damask, retained after forging, by melting carbon-saturated iron with pure alumine (alumina) and remelting with steel. They were sure that the Oriental damask was due to crystallization, but seem not to have suspected the composition differences produced by it. In this and a later paper (1822) they describe alloys of steel with silver, nickel, platinum, chromium, and other metals, which are important in connection with the history of alloy steel. Many of these alloys showed a good damask, but Faraday and Stodart failed to realize the essential simplicity of the Oriental high-carbon cast steel.

Bréant's paper, published in 1823, is one of the minor classics of metallurgy. In it he applies to steel Berzelius' principles of chemical combination, and suggests that carbon added to iron, at the proper temperature, first forms increasing amounts of steel, as a separate "compound," and that as more carbon is added beyond the ideal composition of steel, iron carbide commences to form. On slow cooling a steel of very high carbon content, the "steel" and the carbide will crystallize separately, the crystals being larger the slower the cooling. The details of the damask depend on the manner of forging, being in the form of longitudinal streaks if forged in only one direction, crystalline if forged equally in all.† Bréant developed the structure by immersing in "acidulated water."

---

* A cutler named Damemme of Caen[4] claims to have been the first to make a true Damascus steel, and that he sent a description of it to the *Société d'Encouragement pour l'Industrie Nationale* in 1811. Nothing was published by the society, and his claim cannot be verified. *Note added 1965:* The story of Bréant's work and its background is given in full in a recent paper.[5] The Société d'Encouragement gave a research grant to Bréant, who studied hundreds of alloys before he came to realize that alloying was unnecessary and that a high carbon content and slow cooling were both necessary for a good texture. A comprehensive report comparing textured steels of all kinds was presented by De Thury[6] in April, 1822, and includes illustrations of several blades, including one made from Bréant's steel.

† Compression, if strictly uniaxial, would produce a uniform enlargement of structural features, the increase in area being directly proportional to the reduction in thickness.

The Russian general Paul Anossoff[10] achieved some fame for establishing the manufacture of the blades following Bréant's procedures, though he gives no credit to the French pioneer and the photograph of one of his blades given by Lenz in his excellent general article on damask hardly substantiates[11] the claim that they were equal in beauty to Oriental ones. Wilkinson, however, spoke approvingly of Russian watered blades that reached England in 1839.

Belaiew has called attention[12] to the metallographic interest of Anossoff's paper which was published in 1841 in the *Russian Mining Journal,* and reprinted in French in the *Annuaire du Journal des Mines de Russie* in 1843. Except for its importance in inspiring the work of Tschernoff (see chap. 3), the paper seems to the writer to be less significant than Belaiew implies, for it reflects considerably less understanding of the nature of Damascus steel than had Bréant's earlier one. After nearly two hundred trials in the manner of Faraday of various compositions including alloys of iron with platinum, manganese, chromium, titanium, silver, and gold, and melting à la Clouet with various carbonaceous materials and fluxes, Anossoff settled upon a method of manufacture that involved melting iron with graphite and a flux, slowly cooling, and finally forging and etching. Though he realized that the best damask was related to a high carbon content, he thought that only certain forms of carbon would work and insisted on using only certain kinds of graphite. He observed a kind of inferior damask in metals of quite low carbon content treated with hammer scale, and claimed to be able to develop a texture even in soft steel by a very long annealing treatment, which may possibly have given an extremely large-grained steel containing hypoeutectoid constituents. He did not anneal his preferred product of graphite steel. The structures of finished blades were best developed by etching with a hot solution of Persian iron sulphate (believed to contain aluminum sulphate) or almost equally well with lemon juice or beer vinegar, which are slower in action and easier to use. It was necessary, he remarked, for an etchant to dissolve the iron without dissolving the carbon and a good golden reflectivity can be obtained only when there is actually a combination of iron and carbon. Ordinary carbon (i.e., carbon in the uncombined state) does not give a clear reflection. Anossoff clearly recognized that the final damask was related to structures visible on the surface of the ingot, or even better in the slag that had solidified in contact with the ingot, and he describes the various patterns of parallel or intersecting lines without recognizing that they were crystalline in origin. His paper is followed by a summary of his notebook, consisting of 185 separate sheets with details of the materials charged, the duration of the melt and remarks on forgeability, the appearance of the damask and some summaries of series of tests. He occasionally includes comments such as "the damask [*moiré*] is hardly visible with the microscope" or "in places it shows a fine damask visible under the microscope," but there is nothing to indicate systematic use of the microscope or any understanding of the nature of the structures involved. The most interesting part of his paper is a paragraph in which, after remarking that overheating will destroy the damask, he points out how the knowledge of it can be used to control forging. To quote:

Europeans have not had the experience of the Orientals in recognizing the modification which steel undergoes during forging. Their inferiority comes from the fact that there are no visible signs of these modifications before their eyes. But when they will have to treat Damascus steel, they will soon come to understand the disadvantage of their inexperience, and each of them will know that the disappearance of *moiré* at the forge represents an alteration of the metal attributable to the lack of skill on the part of the smith.

Of perhaps equal importance to Anossoff's paper is the nearly contemporaneous one by De Luynes.* De Luynes[13] analyzed some actual swords as well as several cakes of Persian and Turkish steel. Though his chemical results are impossible to believe—the carbon content ranges between 8 and 13 per cent (perhaps because he extracted and reported carbide as carbon) and many of the samples showed large amounts of nickel, tungsten, and manganese—he noted that the cast steel often had unmelted "nails" imbedded in it, and concluded that Orientals first made an easily melted hard iron and remelted it with wrought-iron nails. (The "nails" were probably dendrites, but they may have been residual rods from a melt-carburization process.) He believed that an alloying element was needed to give a good damask and preferred manganese. He successfully made swords by melting iron nails imbedded in a mixture of manganese oxide and sawdust, breaking up the resulting ingot into nut-size pieces and remelting with an equal weight of iron nails, *"pointes de Paris."* The resulting metal, which seems to have been allowed to solidify as a lump in the crucible, weighed two to three kilograms. It was forged at not higher than a clear red heat into a cake of two-thirds of its original thickness, then pierced into a ring which was cut, opened, and forged out to a bar which

* The writer is grateful to Mr. S. V. Granscay for calling his attention to De Luynes' paper.

was then filed, cleaned, and etched to test the damask. If satisfactory, it was forged to the shape of the blade, but a little thicker, and the surface was then cut with a chisel or file to give the general distribution of the damask that is wanted—wormlike markings, zigzag streaks, transverse markings, etc.—whereafter the blade was forged flat and smooth, polished, hardened, repolished, and finally etched in Oriental *zag* or in dilute nitric acid. De Luynes' emphasis on cutting the markings on the surface of the nearly finished blade suggests that his material may have been rather uninteresting without this. It is difficult to tell whether De Luynes obtained complete fusion in his second melting, and the damask resulted from a distribution of unmelted steel particles, or whether his success came from the ease of obtaining the proper carbon content by the double melting operation.

At about the same time, the Swiss metallurgist J. C. Fischer was experimenting with Damascus-patterned steel. De Luynes includes his name in a list of men who have achieved beautiful results, though he gives no reference to any publication. Nevertheless, Fischer displayed at the London exhibition of 1851 a regulus of Meteor steel "cut across, to shew the Damask of the interior part" and some "bars of the same steel damaskined on one end."[14] The drawing of the exhibit, though lacking in detail, illustrates a rather convincing large-scale damask on one of the bars, but Fischer's notes indicate that Meteor steel was a high-nickel, low-carbon steel that could hardly display a pattern visible to the naked eye.

Interest in the duplication of the blade declined as European steelmakers developed their own techniques and the introduction of the Bessemer and Siemens processes gave homogeneous steel more adaptable to large-scale production. Anossoff's work inspired Tschernoff to his study of structure, but metals capable of showing a damask have been little appreciated by metallurgists for over a century. The mechanical properties of such duplex materials are now being studied in the form of compound fiber-reinforced materials of many kinds as well as mixed granular "cermets" for service at high temperatures. It would seem, moreover, that the aesthetic qualities of mixed metal patterns could well be used in modern jewelry, perhaps not of Damascus steel, but of pewter and of copper or precious metal alloys selected for good crystallographic segregation patterns and worked by cross-rolling and hammering.

5. Welded "Damascus" Gun Barrels and Swords

There is another beautiful Oriental technique, quite different from the damask of the swords, though often confused with it. This is the welding technique used to give strength and texture to gun barrels, and is in many respects a reversion to that of the pattern-welded blade.

A fine deeply etched barrel of seventeenth-century Indian workmanship is shown in Figure 17 and an eighteenth-century Turkish one, with a more regular but less deeply etched pattern, in Figure 18. The technique seems to have originated in the Near East in the sixteenth century, and such guns were a famous product of the Kashmir smiths in the early nineteenth century.[1] Massalski describes[2] in detail the manufacture of these barrels in Constantinople by a celebrated Persian smith, Hajji Mustafa, and gives drawings of the twisting mill and coiled strip.

The earliest detailed description of the manufacture of barrels of this type in Europe is by Wäsström[3] in 1773. Perret had in 1771 used a rather similar technique for making swords, which will be referred to later. Wäsström appreciated all the essential steps involved in the Oriental procedures, though his patterns must have been relatively uncomplex, for he used a faggot of only seven strips (four iron strips sandwiching three of steel), forged and coiled but not twisted. As etchant he used a solution of salt, alum, and sal ammoniac, sometimes followed with aqua fortis. In the discussion following Wäsström's paper, the great metallurgist Sven Rinman made the pertinent comment that the advantage of damask lay only in that it insured proper forging of the steel, for equally good properties were obtainable with homogeneous metal if it were adequately worked. Rinman's interest was deeply aroused by this paper and he proceeded to an extensive study of etching which he published in 1774 in a paper entitled "Experiments of Etching upon Iron and Steel,"[4] a paper of great importance to metallographists.

Rinman studied the action of a wide variety of reagents on several kinds of iron and steel, both separately and forged into damasked bars, and commented on the carbonaceous residue obtained from gray cast iron. It was this work that inspired Bergman's more famous studies on carbon in iron and steel, and provides one more example of the remarkable impact of Oriental metallurgy on European science. This will be discussed in detail in chapter 12, along with other early uses of etching for scientific purposes.

The striking appearance of the barrels became very popular and all the best European sporting guns were of this type from the mid–eighteenth century until near the end of the nineteenth. The etchable differences in composition in the products of the Kashmir and Constantinople forges apparently resulted from oxidation of the strip before coiling; in Europe it resulted from a composite mixture of iron and

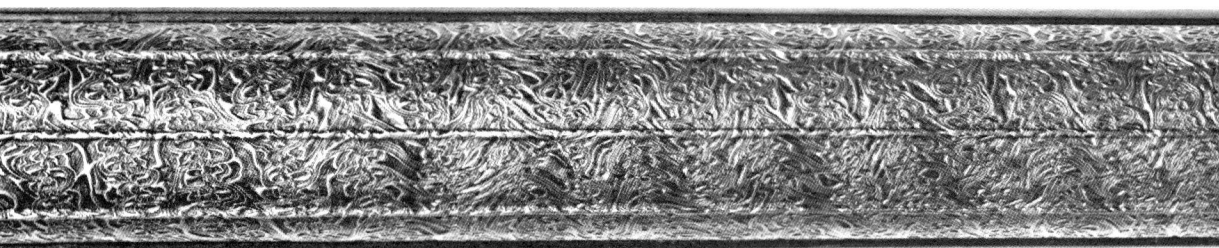

Fig. 17.—Detail of barrel of flintlock gun, Indian, ca. 1650. (Wallace Collection No. 2055 [Crown copyright].)

steel.[5] In England, according to Greener,[6] "stub-twist" barrels with only a slightly mottled surface were made from old horseshoe nails and chopped-up coach springs—a third to a half of the latter—which were barreled to clean them, then heated and welded together and rolled to strip. A better copy of the Oriental texture resulted from welding about 25 bars of iron and steel piled alternatively, drawing out to a 3/8-inch square rod, which was twisted to half its original length (12–14 turns per inch), then two or three such rods were welded together, side by side, and rolled flat to give a suitable strip of skelp. English smiths then coiled the strip into a helix on a mandrel and, after heating, welded the coils together by "jumping"; Oriental smiths used the hammer. After finishing, they etched the barrels with a paste of iron sulphate, the first two etches being for 4 hours, and afterward for 24 hours, repeated for 20–30 days.[7] It was noted that after 24 hours' use, a batch of etchant would cause general rusting rather than selective corrosion. After this prolonged attack, mild abrasion made the promi-

Fig. 18.—Patterns on barrel of a Turkish carbine, eighteenth century. (Victoria and Albert Museum, No. 971–1884.)

nences bright in contrast to the dark background of the more corroded parts. The English practice was to use a milder attack, known as browning, produced either by damp soot or by a mixture of ingredients containing ferric chloride and copper sulphate with some nitrates.

William Greener's description[8] of the manufacture of these barrels, published in 1835, is of added interest to the metallurgist because of the testing procedure that he advocated. Greener advises the gunsmith to determine the quality of the welded skelp by subjecting it both to macroetching test and to a tensile test. He has engravings showing finished etched barrels of various types, and illustrates the appearance of the etched skelp before and after twisting and rewelding. Two of his woodcuts are shown in Figure 19.

Similar techniques were used in the manufacture of textured swords, and many European smiths seem to have been under the impression that they were thereby duplicating a true Damascus steel sword. Despite their unimportance as weapons, an extensive literature grew up around them. The earliest seems to be that of the cutler Perret, who writes in 1771[9] as though he thought himself to be the European inventor of the technique and the discoverer of the secret of the true damask. He welded together six iron and five steel strips 1 inch wide and 1/12 inch thick, twisted the resulting bar, and forged it into a blade. Veins of

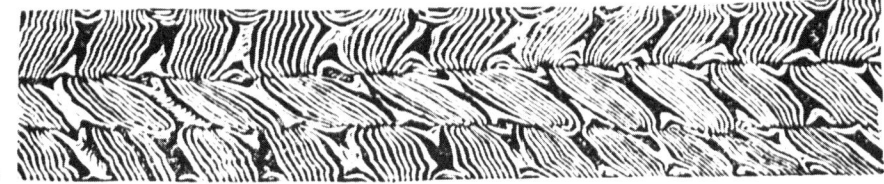

Fig. 19.—(a) Appearance of welded composite strip for gun-barrel making. (b) Three such strips twisted, welded together, and rolled to skelp. Both etched in sulphuric acid. (Greener, 1835.)

steel and iron can, he says, be seen by a connoisseur on a polished blade, but the best effect is developed by spreading on nitric acid with a feather and leaving it for six or seven minutes before washing off. In a later discussion he reports[10] (probably under the influence of Rinman) that the action of dilute nitric acid or lemon juice causes the materials each to take its proper nuance: "The steel becomes a bluish black, the soft iron shows itself a dark grey, and the iron a harsh, whitish grey."

In 1798, in an interesting anticipation of modern powder metallurgy, William Nicholson made a Damascus-textured metal by compressing mixed filings of steel and wrought iron in a die, restriking the compact at a welding heat and forging it into a plate. He reports the texture as being not fibrous, but wavy (*Nicholson's Journal*, 1798, **1**, 468–71). Also in 1798, a paper was presented before the Academy of Sciences in St. Petersburg by Hermann,[11] a German who was living in Russia and had established a steel plant in Siberia. He was aided by Habesci, an Arab with Western leanings, who had seen swordsmiths at work in Damascus. Habesci had clearly not seen the making of true Damascus steel, for Hermann experimented only with welded patterns. He concluded that to make a good patterned sword there should be not over one-third of steel in the welded strip and that twisting and at least six times folding and welding were necessary to give good "flowers" and mechanical qualities. He preferred to etch in dilute nitric acid, or with a mixture of vitriol or copper sulphate and chalk.

The main interest in Hermann's paper lies in the illustrations, two

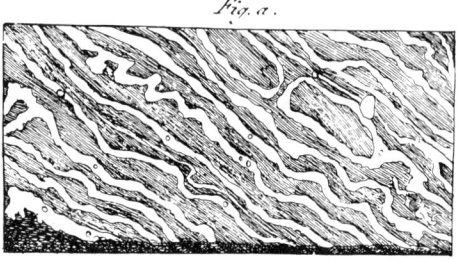

Fig. 20.—Artificial damask made by
welding and forging. (Hermann, 1798.)
These are believed to be the first
non-Oriental illustrations of an etched
metal surface to have been published.

of which are reproduced in Figure 20. These have the distinction of being the first carefully drawn illustrations of an etched metal surface to appear in any technical publication, though it is probable that Persian or Japanese artists had much earlier depicted the patterns on their warriors' swords.

In the opening years of the nineteenth century, Clouet (professor of chemistry at the engineering school of Mézières) attacked the problem of welded Damascus steel with revolutionary vigor.[12] He showed how to build up faggots of mixed iron and steel shaped and arranged specially for forging into bars with a wide variety of internal structure, including complicated mosaic patterns and one which, like souvenir rock candy, would display the word *LIBERTÉ* on any section! So ingenious did French smiths become that one of them, Degrande-Gurgey of Marseilles, presented to the *Société d'Encouragement pour l'Industrie Nationale* a sword with the inscription *PREMIER ESSAI D'UN ART NOUVEAU D.N.G. 1819* legibly incorporated as an integral part of the blade.[13] This may have been done by local cementation or decarburization, but welding seems more probable in view of its background. The Italian Crivelli,[14] apparently under the misbelief that he too was reproducing true Damascus steel, had made by 1821 beautiful and serviceable blades by wrapping thick iron wire in a helix around steel strips, forging these flat, welding several together, twisting, and forging into a blade which was finished by polishing and etching. The swordsmiths of Solingen used a simpler laminar welding technique in many of their best nineteenth-century blades, one of which was

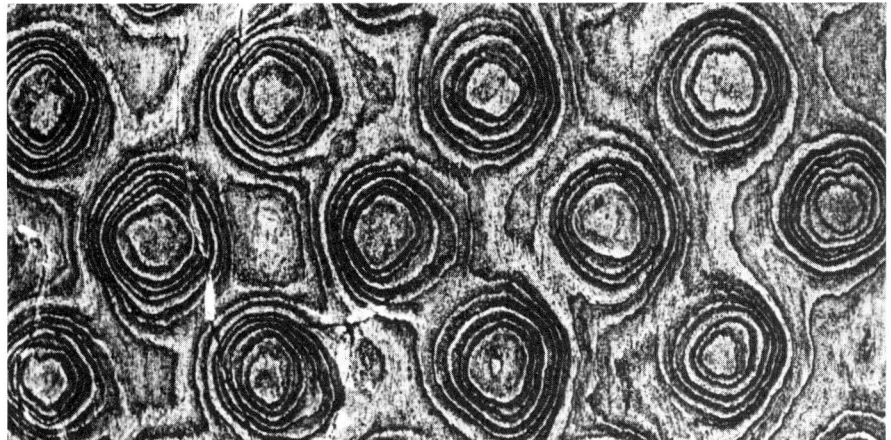

Fig. 21.—Modern Damascus steel. (J. A. Henckels, Solingen.) Natural size. Pattern known as ROSENDAMAST.

forging practices. Two of these are reproduced in Figure 20. These have the distinction of being the first carefully drawn illustrations of is a photograph of the surface of a steel plate with an average carbon content 0.4 per cent made by welding together different steels, the texture, known as *Rosendamast*, being produced by punching the surface into a waffle-like pattern and subsequent grinding and deep-etching. Other designs (Fig. 22) result from welded, twisted, and rewelded rods of duplex structure in a manner reminiscent of the Merovingian blades. This pattern is known as *Turkisch damaststahl* in Solingen. Still another Solingen damask is a true structural pattern of the kind suggested by Harnecker,[16] wherein a coarse grain-boundary network of cementite is produced in a steel with about 1.4 per cent carbon by long annealing and slow cooling. On appropriate subsequent working at a low temperature followed by a local indentation and grinding, textures as in Figure 23 are produced. Though it is interesting that the ancient art should be kept alive, all of these twentieth-century patterns have a contrived quality that leaves them aesthetically inferior to the older Oriental product.*

The most striking of all types of decorative designs made by mixed metal forging and etching is the kris of Java and other islands of the Malayan archipelago.[17] Significantly, the first textured blades in Indonesia occur under Hindu influence, during the Majapahit Empire (13th–15th centuries A.D.). A fine example in the Folk Art Mu-

---

* The writer is grateful to Drs. Carl Ebbefeld and Walter Wolff of the J. A. Henkels Zwillingswerk for these photographs and information on the present art.

Fig. 22.—*Textured steel made by welding together three twisted composite rods and sectioning near the surface or center, called* TURKISCH DAMASTSTAHL, ×1.8. *(J. A. Henckels.)*

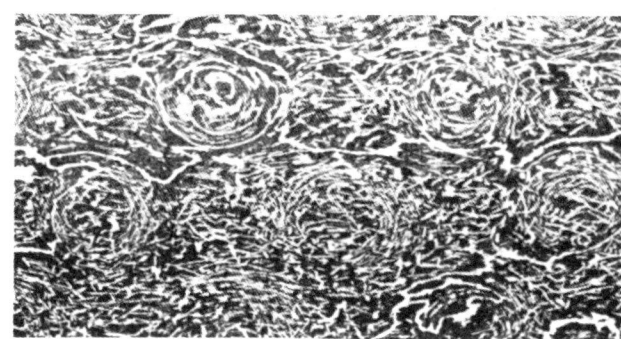

Fig. 23.—*Damask resulting from proeutectoid network of carbide in austenite grains. (J. A. Henckels.)*

seum in Santa Fe, New Mexico, is shown in Figure 24. The whole weapon, including the handle in the form of an abstract human figure (Fig. 25), was clearly shaped approximately to its final contours by forging and then finished in detail by filing or chiseling and deep-etching. The heterogeneity of this metal was probably no more than that of any steel of the period, which was, of course, made without melting. Once these textures came to be appreciated, smiths began to develop them intentionally and produced some extremely gaudy effects by combining metals of greatly different chemical properties. At the same time, the forged and carved solid metal handle gave way to a simple tang and a wooden grip, often made of wood with a spectacular grain to match that of the metal. An eighteenth-century kris from Bali with a fine pattern is shown in Figure 26. Details of five blades, each with a different type of texture, are shown in Figure 27, for which—as for so much else—the

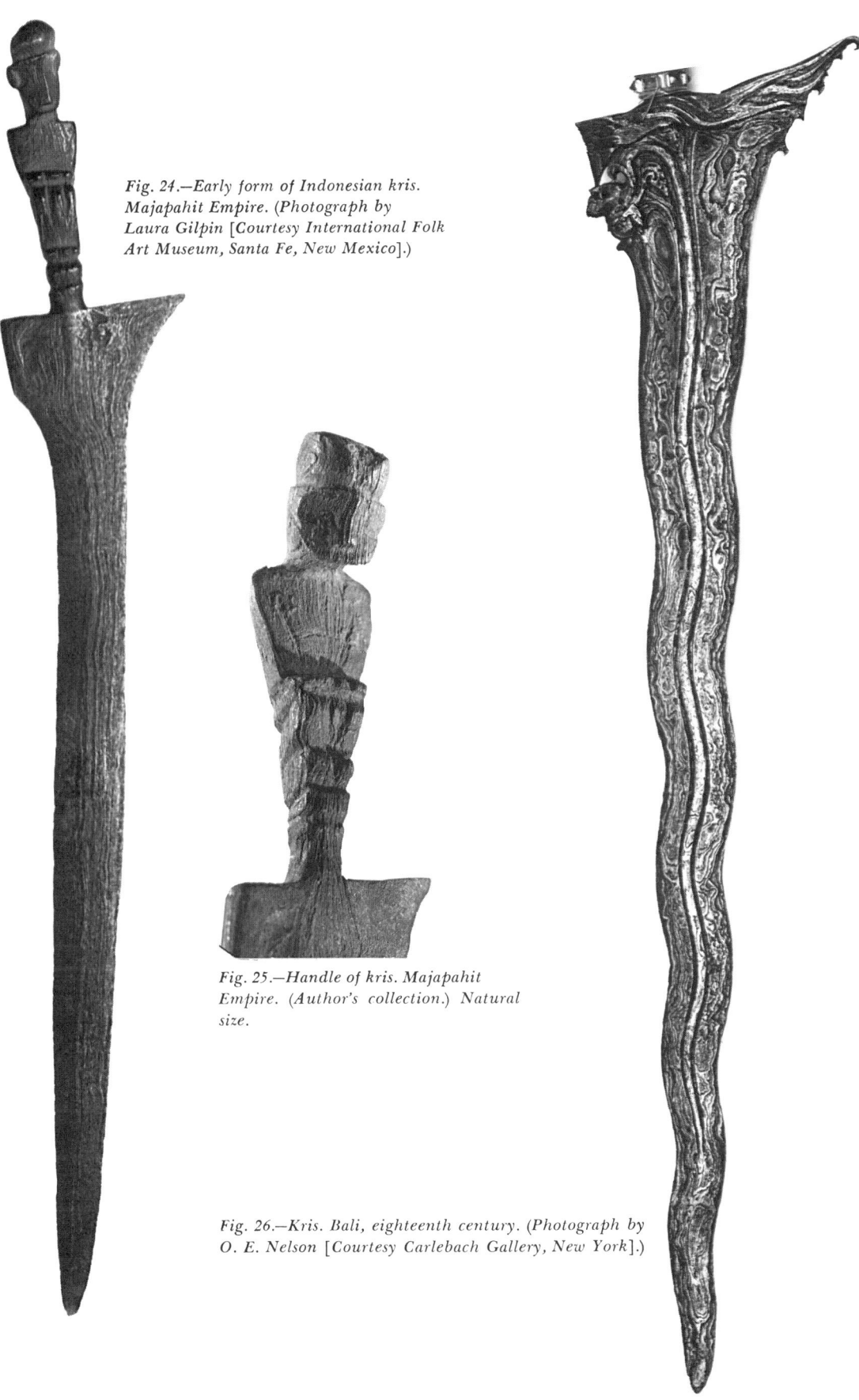

Fig. 24.—Early form of Indonesian kris. Majapahit Empire. (Photograph by Laura Gilpin [Courtesy International Folk Art Museum, Santa Fe, New Mexico].)

Fig. 25.—Handle of kris. Majapahit Empire. (Author's collection.) Natural size.

Fig. 26.—Kris. Bali, eighteenth century. (Photograph by O. E. Nelson [Courtesy Carlebach Gallery, New York].)

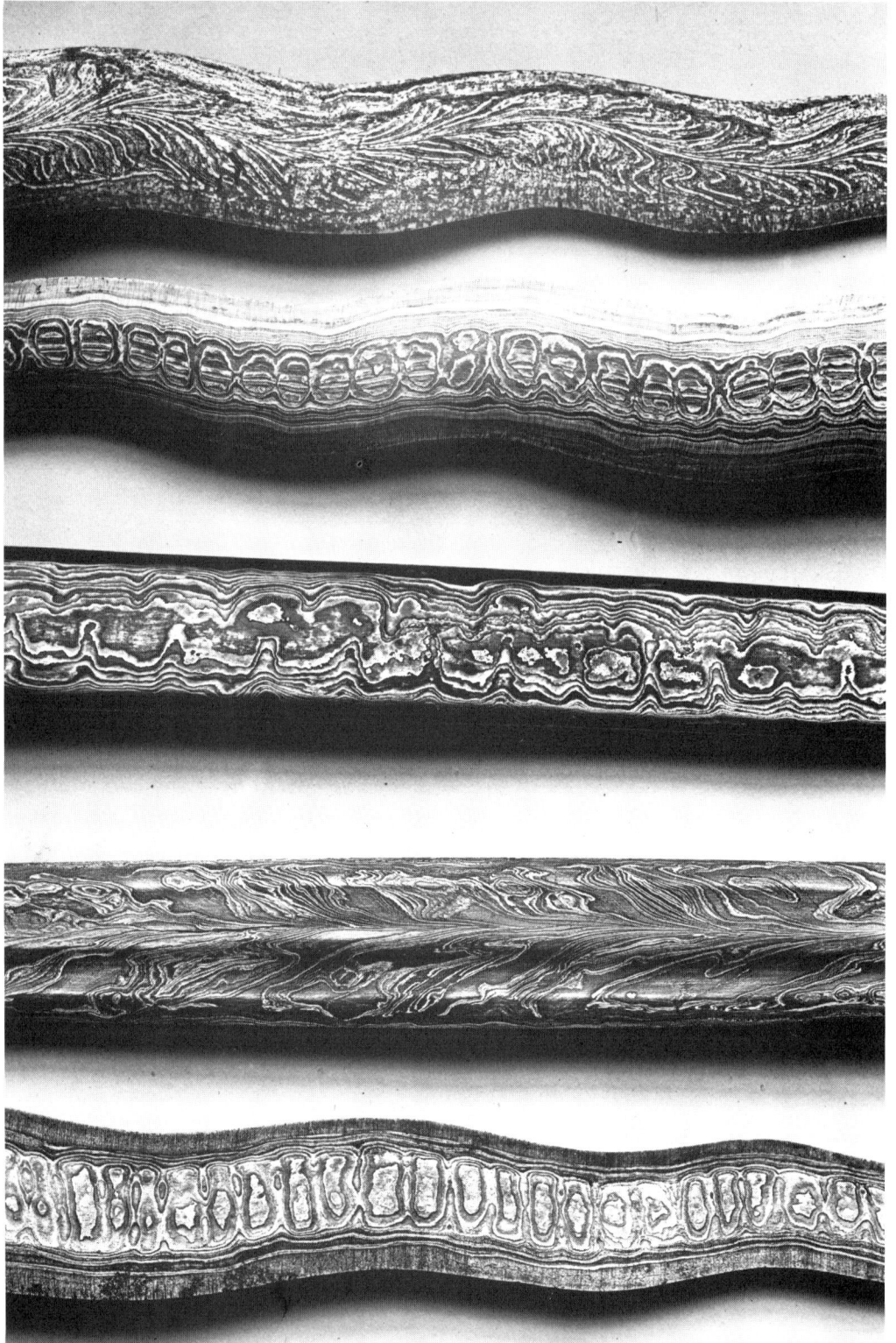

author is indebted to Mr. Herbert Maryon, O.B.E., a life-long student of textured metals.

The modern making of one type of kris was described by the great metallurgist Walter Rosenhain in a paper[18] written while he was still a student at Cambridge. Two different kinds of steel were forged together in a flat strip that was bent edgewise into a compact serpent-like form, and two such composite pieces were sandwich-welded between three strips of steel. After final forging, the blades were shaped by taper grinding to expose sections of the various layers and the blade etched in fruit acids, often containing white arsenic to give a better black.

A wide variety of patterns was produced by different smiths employing different methods of assembly and local contouring before finishing. Many blades contain bright, virtually unetched, silvery streaks in relief, and these are said to result from the use of high-nickel meteoric iron. Analyses by De Luynes showed[19] the presence of both nickel and copper in these layers, though his analyses are obviously unreliable. Experiment showed him that copper was more harmful than useful and he was able to duplicate the texture by the use of nickel alone. He packed nickel powder between sheets of iron which were stacked together, welded, and forged out. It is unlikely that the Indonesian smiths used pure nickel, though in later times they were reputed to have used European stainless steel. Whatever the composition, there has certainly been a great difference in the amount of chemical attack on the various layers. The sharpness of the interfaces, showing little interdiffusion, suggests that the welded metal had never been heated to a high temperature.

Fig. 27.—Blades of five Malayan kris. (British Museum [Photograph courtesy of Mr. Herbert Maryon].)

6. The Japanese Sword

The Japanese sword blade is the supreme metallurgical art. It involves unmatched skill in forging, in control of composition, in heat treatment and in finishing to produce surfaces that reflect qualitatively the underlying metallographic structures.* Figures 28 and 29 show the general appearances of three blades, and Figures 30 to 34† show various surface textures characterizing the work of different smiths or schools. The surface of the sword is finished by the use of a sequence of abrasive stones which reveal the structure both by slightly accentuating the slag particles and by leaving invisible scratches of slightly different quality on the hard and soft zones, so that there are subtle variations in the scattering of light from the slightly matte surfaces. Though the surfaces are geometrically defined, they are not finished with a smooth polish except for a zone at the back of the blade which is burnished. A good blade with its purity of outline and subtle variations in surface quality is a thing of great beauty and they have attracted connoisseurs of art in many countries. They are very difficult to photograph and the illustrations to this chapter give but a feeble reflection of the the quality of the blades, which can be appreciated only if the eye and the blade are moved in relation to each other and to source of light.

The swords were made by multiple folding, welding, and forging, to give a body of relatively soft iron with an edge of hardest steel, the combination allowing an extremely hard edge to be obtained without brittleness of the blade as a whole. Before the finished blade was heated for hardening, it was partly protected with an adherent insulating refractory layer to locally retard cooling when the blade was quenched, and so to prevent hardening except in the uncovered areas. By the introduc-

---

* There is a very extensive Japanese literature on the appraisal of swords. Their history and methods of manufacturing have been discussed in a number of European publications but here we need notice only Dobrée,[1] Joly and Inada,[2] Inami Hakusui,[3] and Yumoto.[4] The metallurgy of the swords has been treated by Professor Kuni-ichi Tawara in an article[5] and a fine book,[6] by Chikashige,[7] and by the present writer.[8]

† Figures 31 to 34 are photographs by Captain A. D. E. Craig, of blades in his own collection.

tion of layers of steel of different hardenability and combining this with careful shaping of the coating to give varying local gradients of cooling rate, very intricately shaped and shaded hard zones can be produced, and extraordinarily beautiful effects can be revealed under the final polish. Grains of martensite shine against a duller background and a fine network of pearlite changes in size and extent in zones of varying carbon content and cooling rate. All of this is superimposed upon a grosser texture coming from the manner in which the metal has been piled and forged, which gives a kind of damask pattern, either coarse- or fine-grained, from the sectioning of layers of metal or thin lines of interleaved slag in a distribution that depends upon the detail forging regimen used.

The method of making the swords, as many other things in Japan, was imported initially from China, but it was developed and refined beyond recognition. In China, steel was made perhaps as early as the first century B.C., by heating a stack of wrought-iron plates with cast iron until the latter melted.[9] Such material, if repeatedly folded, welded, and forged out in the manner of the Japanese smith, would give a satisfactory material of intermediate carbon content. The earliest Japanese swords that are known (from graves of the burial-mound period, before *ca.* A.D. 650) have a well-forged lamellar structure which almost certainly resulted from the intentional incorporation of many alternate layers of materials of two different carbon contents (either made in the Chinese method or by welding, followed by forging in a manner to prevent too much diffusion) and with the cutting edge hardened by local heat treatment. Later, a more complicated regimen came to be used, involving the incorporation of a low-carbon iron in the middle of the body of the blade and an extensively forged steel of about 0.6 per cent carbon on the outside. Although none of his swords have survived to the present day, some of the earlier books illustrate blades by a famous smith, Amakuni, who, in the eighth or ninth century A.D., started the custom of signing blades in certification that they had been made "in the proper way." A blade that has been preserved in the Shosoin Imperial Treasury at Nara (established on the death of the Emperor Shomu in A.D. 749) shows some of the structural features of more modern blades but, though it is reputed to have been untouched, it is not certain that the present surface finish is the original one. By the thirteenth century, and probably much earlier, blades were being polished in a way to bring out both the wood-grain damask in the metal composing the body of the blade, and the intricate interface between

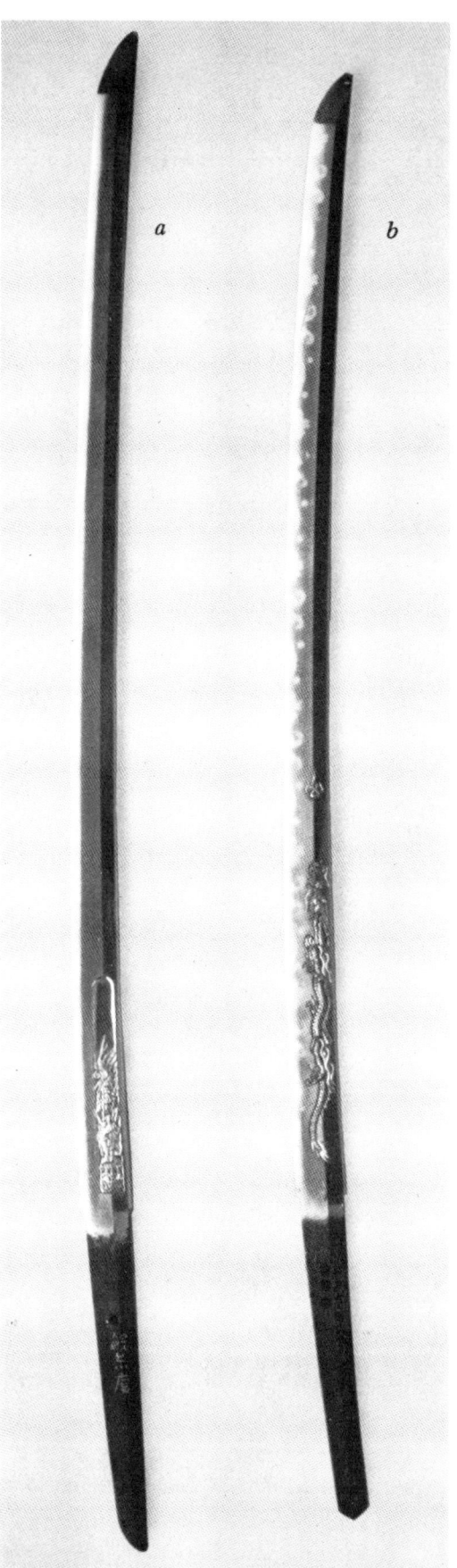

Fig. 28.—Two Japanese sword blades. (a) Signed Mioju Umetada, dated 1632 and (b) signed Hankei Noda, dated 1646. (Victoria and Albert Museum.)

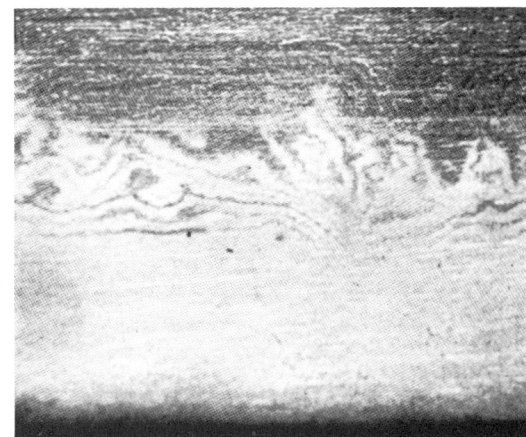

ig. 29.—*Blade by Kunihiro, 1357. From Junji Homma (see n. 17).*

Fig. 30.—*Blade by Amahide, twentieth century.
Showing grain due to layers of metal
of different hardness, ×5.*

the hardened edge (*yakiba*) and the softer steel of the rest of the blade. By the time of the Emperor Hideyoshi (1536–98) the appraisal of swords on the basis of their surface textures, shape, and other details had become a recognized profession of high standing.

The writer has argued[10] that the existence of decorative patterns that would not arise naturally from a straightforward forging or heat treating technique is a strong indication that the pattern was intentionally produced to be seen, and therefore that surface polishing to reveal structure must have been in vogue at the time the blade was made. Thus the late twelfth-century sword by Gwassan of Dewa (Fig. 33) has a sinusoidal pattern that probably resulted from a series of grooves cut regularly into the surface of the sword immediately prior to the last stage of forging, or perhaps from a forged undulating surface that was subsequently cut to a plane. However, the owner of this sword, Captain A. E. D. Craig, who has studied blades for many years and is a leading authority, does not accept this conclusion. Pointing out that the original surface of the sword had long been ground away, he believes that the pattern on all the early blades was an *unconscious* result of a forging technique that was used for no reason other than it had been found to give a serviceable blade. The dating of the first "metallographic" sur-

43

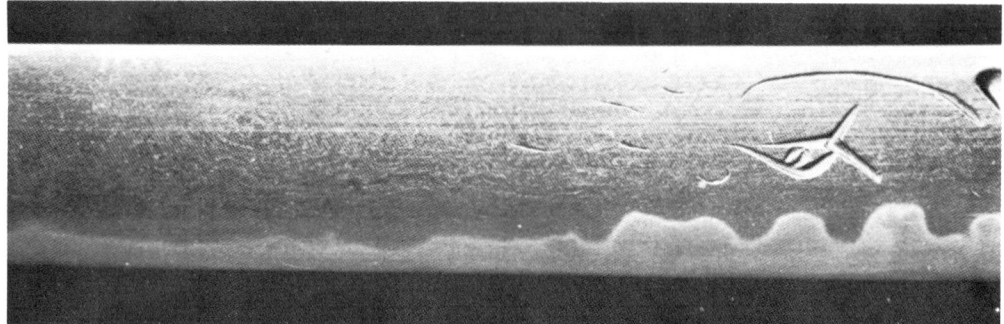

Fig. 31.—Blade by Hiromitsu, dated 1379. Showing JIHADA *with fine grain, due to well-distributed slag particles.*

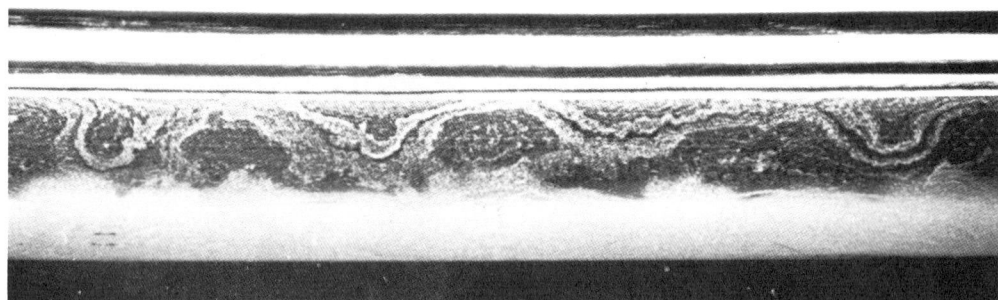

Fig. 32.—Blade by Akihiro of Sagami, dated 1369. Coarse grain results from layers of metals with different slag content.

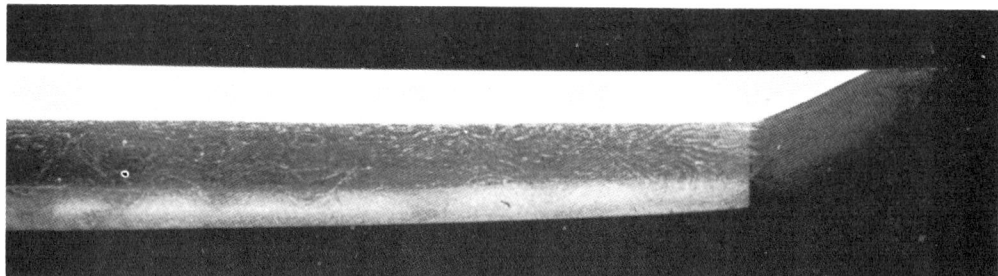

Fig. 33.—Blade by Gwassan of Dewa, late twelfth century. Grain mostly from interlamellar slag. Sword probably forged intentionally in manner to give sinusoidal pattern.

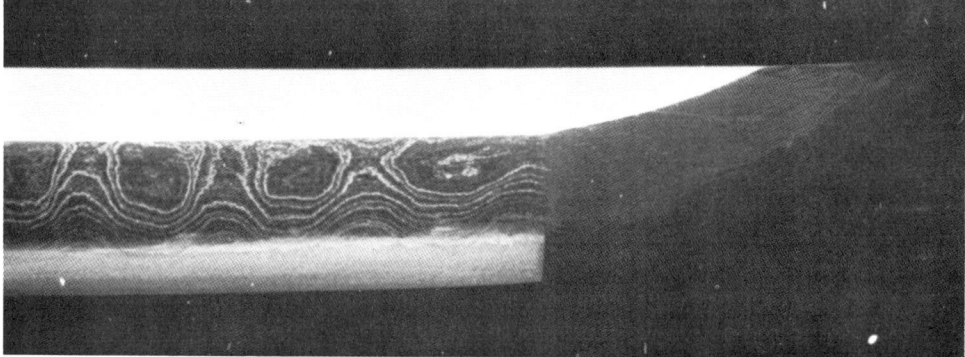

Fig. 34.—Blade by Gwassan Sadayoshi, dated 1865.

face treatment can only be done definitely on the basis of contemporary descriptions of the pattern, for mere mention of polishing is not enough.

The earliest reference to a visible pattern on finished swords known to the writer is actually a Chinese one. Needham quotes Shen Kua (*ca.* A.D. 1065) as saying in his *Dream Pool Essays:*

Ancient people use *chi kang* ("combined steel") for the edge, and a soft iron (*jou thieh*) for the back, otherwise it would often break. Too strong a weapon will cut and destroy its own edge; that is why it is advisable to use nothing but combined steel. As for the *yü-chhang* (fish intestines) effect, it is what is now called the "snake-coiling" steel sword, or alternatively, the "pine tree design." If you cook a fish fully and remove its bones, the shape of its guts will be seen to be like the lines on a "snake-coiling" sword.*

The forging technique is described in the various references cited above.[12] Perhaps the most interesting description from the standpoint of the metallurgist is that by Kawaba Suishinshi Masahida (1750–1825), a very celebrated practicing swordsmith, who was quoted by another Japanese writer and thence translated into English by Joly and Inada.[13] Chikashige[14] shows the sequence of steps in forging the metal for the cutting edge (Fig. 35). The carefully prepared steel thus forged was assembled in various ways with other steel to form the center and back of the blade. Some of the assemblies are shown in Figure 36, although there was almost endless variety possible in the disposition of the different components. The high-carbon cover metal was usually folded around the core to avoid a weld intersecting the cutting edge.

All of the descriptions agree on the essential forging operations. The center metal is made of a soft steel, indeed often of nearly pure iron, folded upon itself and forged out about a dozen times to give homogeneity but not enough to remove all slag, for it is found, according to Tawara,[15] that the wielding of a sword with slag in its center is less tiring than a solid one—an interesting practical result of internal friction. The metal that is to form the cutting edge and the adjacent sides is made by taking fragments of very high-carbon steel, or perhaps cast iron (previously quenched in water and broken into small pieces to free it from slag), stacking these on a previously made steel plate, forging the whole into a compact flat bar, then notching, folding, welding, reforging to the original size, notching and refolding at right angles to the previous direction, and so continuing for about fifteen to twenty foldings and

---

* This fish intestine analogy appears also in the early eighteenth-century "Sword Book" of Arai Hakuseki,[11] wherein it is attributed to Chin Sou Chu of So.

weldings. A mud of fine clay and charcoal powder, often mixed with
straw ashes, was applied before each heating to act as a flux and to control the carbon content. This was extremely important, for, with prolonged working operations, the opportunities for the incorporation of
scale or slag into the body of the metal and the loss of carbon by oxidation were very high indeed. Anvil and hammer were smooth and carefully cleaned. The pieces were shaped with convex surfaces so that welding proceeded outward from the center, squeezing out the flux without
entrapment, carrying with it the gangue in the crude metal as particles
were successively uncovered during the reduction. Beyond twenty foldings (giving 2^{20} or approximately 1,000,000 layers of the original steel),
the metal no longer improves. The final operation of the smith is to
weld the edge and cover metal in appropriate arrangement (varying
from smith to smith) around the softer center and to give the blade its
true shape. It is then scraped with a two-handled drawknife nearly to
the finished shape, and is ready for heat treatment.

A damask visible to the naked eye would probably disappear after
about eight or nine foldings, and many of the notable features in the
surface structure must arise in late stages of the forging operation. The
layers, in order to be visible to the naked eye, cannot be less than
about 0.1 millimeter thick (or, allowing for the increase in width due
to sectioning at an angle, perhaps as little as 0.01 mm. It follows that
metal with distinguishable characteristics must have been especially incorporated during the last few foldings. The visible difference in the
grain of blades by different makers arises from differences of carbon
content or from inclusions in layers of different thicknesses and distribution.

Several different types of grain are distinguishable, apparently resulting from different metallurgical procedures. The type known as *ko
mokumé* is similar to the welded "Damascus" patterns such as that
shown in Figure 21. It is a typical coarse wood-grain structure with
layers of steel forged nearly parallel to each other, but sufficiently irregular so that the sectioning plane surface discloses different layers
sequentially wherever the stratified mass is locally depressed or elevated.

The pattern is revealed by the polishing operation because of differences of either hardness or inclusion content. If the coarse *mokumé*
is due to a difference of hardenability alone, the pattern will be seen
most distinctly at the edge of the *yakiba*, for, at an intermediate cooling
rate, some layers of the metal will remain martensitic while adjacent
portions have, at the same cooling rate, transformed to pearlite (Fig. 30).

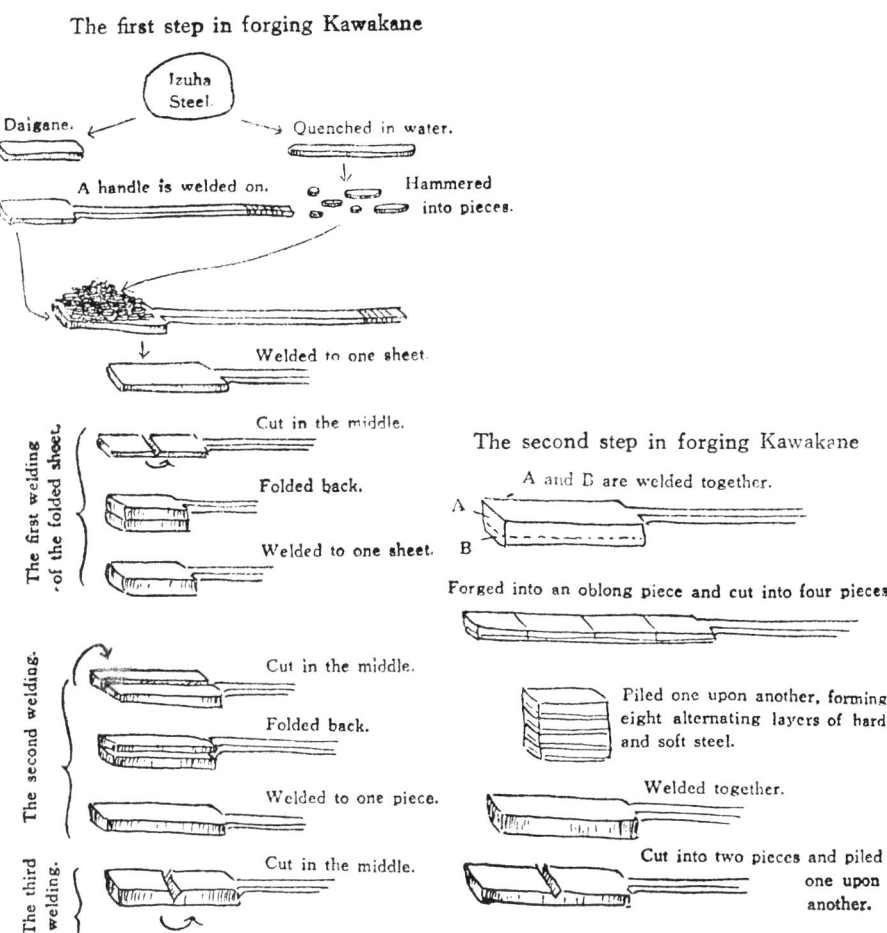

The first step in forging Kawakane

Izuha Steel.

Daigane.

Quenched in water.

A handle is welded on.

Hammered into pieces.

Welded to one sheet.

The first welding of the folded sheet.

Cut in the middle.

Folded back.

Welded to one sheet.

The second welding.

Cut in the middle.

Folded back.

Welded to one piece.

The third welding.

Cut in the middle.

The second step in forging Kawakane

A and B are welded together.

A

B

Forged into an oblong piece and cut into four pieces.

Piled one upon another, forming eight alternating layers of hard and soft steel.

Welded together.

Cut into two pieces and piled one upon another.

Fig. 35.—Stages in the forging of steel for the outer part of a sword (Chikashige).

Sometimes, however, the distinction between the different bands of steel that gives rise to the pattern is merely one of inclusion content, which confers a different quality to the polished surface without there being much effect on the depth of hardening (Fig. 32); and sometimes these two effects are combined.

The self-conscious grain in the sword shown in Figure 34 is of a late and decadent style rather reminiscent of the Indonesian kris. Under some magnification (Fig. 37) it can be seen that the particles have not elongated much, and from their distribution it seems likely that the pattern resulted from welding a stack of plates with refractory powder sprinkled between them, then cutting rounded transverse grooves into the nearly finished sword and finally forging flat and finishing.

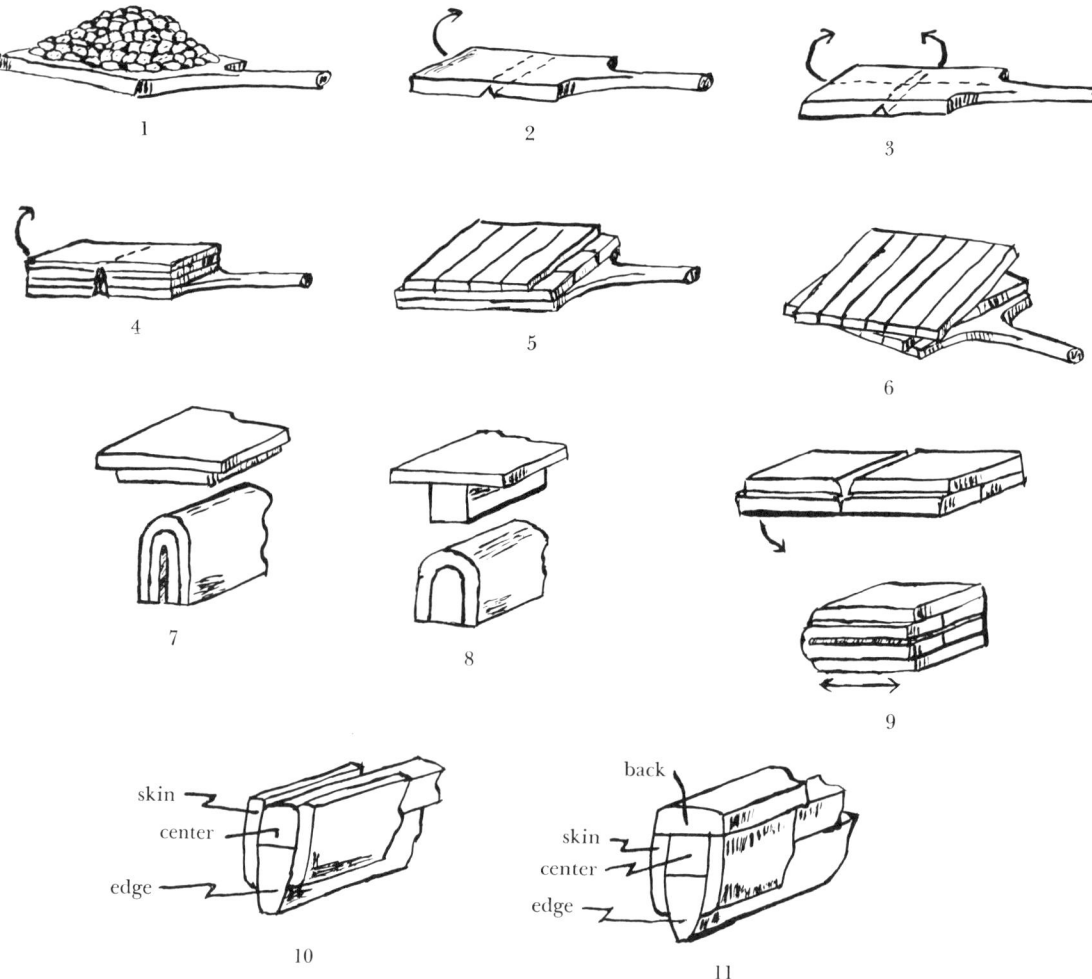

Fig. 36.—Manner of assembly of different steel pieces for final forging of the sword (Tawara [see n. 6], redrawn). Steels of different carbon content and quality were used for the different parts of the blade. (1) piling of steel pieces, (2) folding, (3) JUMONJI-KITAE, (4) TANZAKU-KITAE, (5) KONOHA-KITAE, (6) TSUZUREGI-KITAE, (7) MAKURI-TSUKURI, (8) KOBUSE-TSUKURI, (9) ORIKAKISANMAI-TSUKURI, (10) HONSANMAI-TSUKURI, (11) SHIHOTSUME-TSUKURI.

There is a type of grain with fine, often discontinuous, lines of no significant width. These seem to result from slag inclusions forming between layers of metal, themselves essentially slag-free, as they are welded together at a relatively late stage of forging: they are often quite irregularly distributed and may indeed be accidental in origin. There also is a random overall distribution of small spots not connected with each other to form lines or other patterns, and which seems to originate at an early stage of welding followed by so many foldings that the identity of the layers had disappeared. All of these various distributions will be modified by the refractoriness, plasticity, and surface tension of the inclusions, for some will spread out and disappear while others may remain indefinitely as discrete spots. A scientific study of the composition, distribution, shape, and structure of inclusions in Japanese swords would prove extremely rewarding and would undoubtedly disclose significant differences between the practices employed by various masters.

An even more interesting metallurgical feature of the blades is the *hamon*, or outline of the *yakiba*, the hard martensitic area produced by local quenching. To produce the desired local hardness of the cutting edge while leaving the body of the blade soft, the latter is coated with a thermally-insulating layer which remains in place during quenching and causes the metal beneath it to cool slowly while the exposed edge of the blade is rapidly cooled. For this purpose a mixture is used of clay, polishing-stone powder, charcoal, and sometimes some fusible salt. The coating at the edge is then cut away (or it was previously thinly applied) to give the desired contour to the *yakiba*. After the coating has dried, the blade is heated uniformly over a charcoal fire to the proper heat as judged by color, and quenched in water of such temperature that the edge of the blade has the correct hardness without tempering.

The most striking feature of a blade is the *nioi*, a line a millimeter or so wide marking the edge of the hardened part of the *yakiba*. The *nioi* appears after the polishing operation to be more matte than the adjacent areas, and shines as a clear or diffuse irregular line running the full length of the blade and outlining the *yakiba*. It is made up of mirror-bright spots, usually just visible to the naked eye although sometimes a lens is needed to resolve them. These spots are known as *niye* and are metallographically interesting for each spot is the martensitic central remnant of a single austenite grain, completely surrounded by softer pearlite. The body of the *yakiba* is entirely martensitic, and the main surface (*jigane*) of the blade pearlitic, while the *nioi* is of duplex structure with fine details arising from the gradual increase in the amount of pearlite which forms as the quenching rate decreases beyond the edge of the protective clay (Fig. 38). The structure, indeed, is precisely similar to the gradient in structure in a twentieth-century "Jominy test" for hardenability, somewhat complicated by composition differences superimposed on the thermal gradient. *Niye* also occur singly or in clusters (*tobiyaki*) far removed from the main *yakiba*, apparently as a result of

49

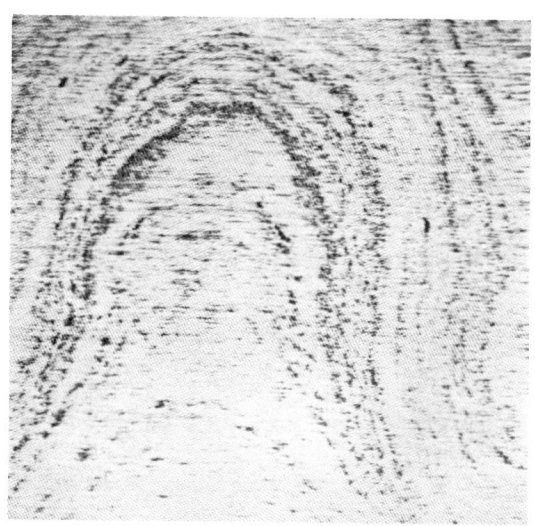

*Fig. 37.—Surface of
Gwassan blade (Fig. 34)
at higher
magnification, ×8.*

occasional isolated large grains in the steel and from locally thinned patches in the insulating clay coating.

The earliest swords had a straight edge to the *yakiba,* but even before the twelfth century irregular interfaces were produced which had both mechanical and decorative merit. Thereafter appeared a profusion of variants with indentations shaped like cedars, pill boxes, cloves, arrowheads, teeth, rats' feet, etc., some of which are shown in Figure 39. In the eighteenth century extremely intricate designs became common, with veritable pictures of mountains (sometimes with snowfields, rivers, and clouds), chrysanthemum flowers, breaking waves with spray, and a hundred others—all portrayed in the metallographer's martensite, in various grain sizes against a background of softer pearlite! Though the principal means of achieving the effect was the shaping of the clay coat to give local variations in the cooling rate during quenching, the incorporation during forging of streaks of steel of different hardenability in the zone of the temperature gradient was also important, for many fine details are on a scale finer than could be achieved by any possible local variations in cooling rate produced by variations in thickness or contour of the superficial insulating layer. Lines and streaks of *niye* away from the hard zone, such as those in Figure 40 clearly result from a forged-in layer of different metal.

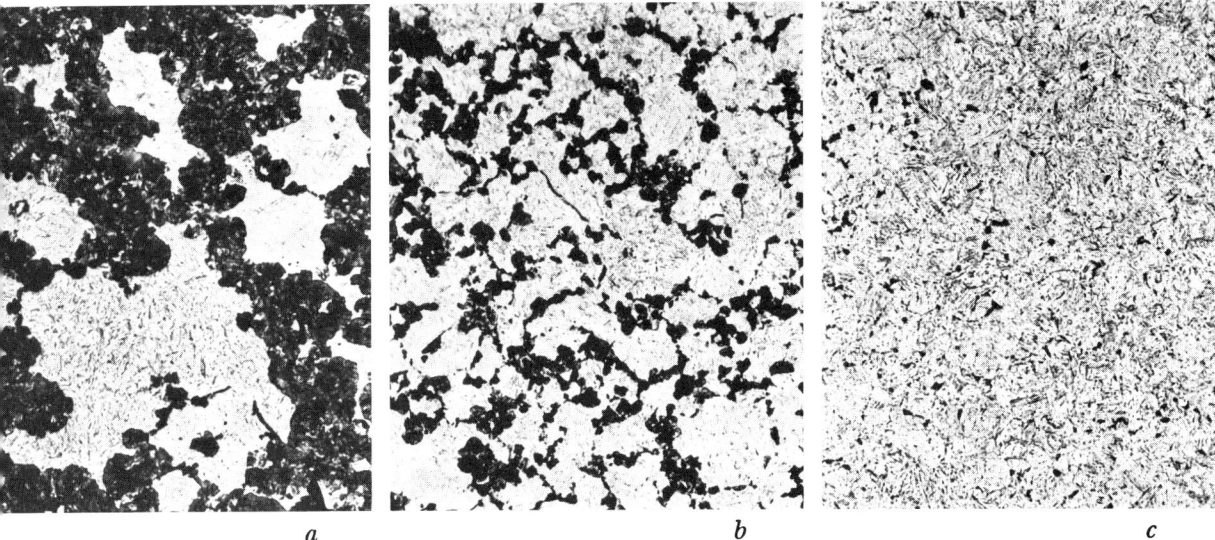

Fig. 38.—Surface of sword by Nobuyuki, nineteenth century. Metallographically repolished and etched, ×150. (a) NIYE *(spot of martensite in generally pearlitic area), (b)* NIOI *(intermediate zone of transition between martensite and pearlite), (c) edge of* YAKIBA.

Figure 41 is a photograph of an etched cross-section of a blade which clearly shows the bands of large grain size material incorporated in it in such a way that the intersection of the bands with the surface would give separate lines of *niye*. It should be noted that many of the structures require the final surface of the blade to be cut at an angle to the forged surface in order to expose underlying layers. The apparent width of a given body of metal would vary with the angle of cut, and the relationship of the forged shape to the contours of the finished sword would have to be carefully considered.

Some of this locally enhanced hardenability seems to be associated with a higher carbon content and with a larger austenitic grain size. Not only is the amount of transformed soft metal less than in the neighboring fine grained material, but the residual martensite is in large patches easily seen with the naked eye. Because transformation to the soft pearlitic structure commences at grain boundaries and moves in from these at an approximately uniform rate, a large grain is not able to proceed to completion while neighboring small grains have done so. This effect of grain size on hardenability is well known and can change the critical cooling rate by a factor of five or more. In modern alloy steels, alloying elements such as molybdenum or chromium are added to increase the hardenability but there is no evidence that any such were used in Japanese steel. Suishinshi Masahide (quoted in reference 2) mentions the use of copper by some Japanese swordsmiths. but this would have small effect on the hardenability and was added primarily to give a decorative grain. Masahide himself says it should only be used for ceremonial swords and not in serious weapons. The temperature used in forging the steel and the nature of the flux would certainly influence its final grain size characteristics and local differences would

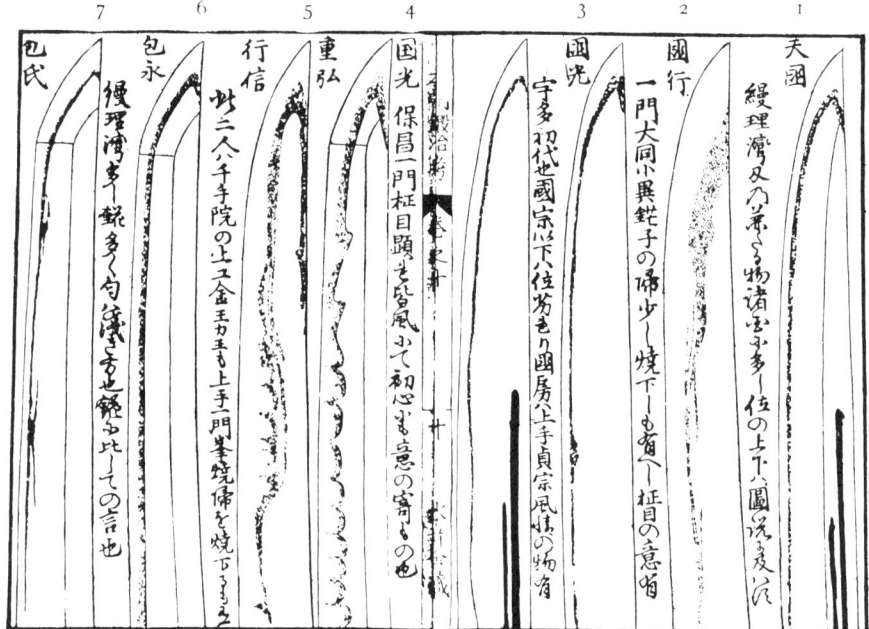

Fig. 39.—Yakiba *styles of various swordsmiths from the* honcho tanya biko (*from Joly* [*see n. 2*]). *Attributed to (1) Amakuni, (2) Kuniyuki, (3) Kunimitsu (Uda school), (4) Kunimitsu (Hosho school), (5) Shigehiro and Yukinobu, (6) Kanenaga, (7) Kaneuji. (From copy in Chicago Natural History Museum.)*

persist even through the final forging period. The existence of many intermediate structures, some with large areas in which partial transformation has rather uniformly occurred, means that the cooling rate must have been very critically controlled: the insulating clay cannot have been very thick (as it is almost always drawn in published descriptions) but only thick enough to prevent the hardening of the finest-grained material.

Some of the details of the *yakiba*, such as the "rats' feet" pattern, result from the forged-in curvature of a strip of high-hardenability steel which just meets the zone of fast cooling rate defined by a straight or nearly straight edge to the thin insulating layer. Some effects may pos-

Fig. 40.—*Surface of an unsigned sixteenth-century blade. Etched with nital,* ×2.

sibly result from preliminary annealing treatment in a different temperature gradient, by overheating and cooling before quenching, or by quenching while a temperature gradient still existed during rapid heating.*

None of these fine structural details could be seen without the final highly specialized polishing operation. Published descriptions[16] of polishing refer to the use of a graded series of blocks of abrasive stones, finishing with paper-backed slivers of a fine-grained limestone. Cinnabar powder ("mercury ore with some gold ore in it"[!]) is said to be used sometimes to bring out a difficult structure. There are probably some undisclosed materials or methods involved, for the final effect is totally unlike—and aesthetically far superior to—any abraded or polished surface produced in the West. The various surfaces are very nearly flat or are curved to intersect along clear-cut straight lines, in smooth geometric perfection which is one of the principal aspects of the beauty of the blades. Except for the back and the adjacent surface called the *shinogi-ji* which are burnished, a good blade has a finely matte surface, not a specular polish. The harder parts are perceptibly *more* matte than the softer ones and scatter light in a manner less affected by the direction of illumination. Photomicrographs of typical areas in the hard and soft zones on a thirteenth-century sword repolished in the twentieth

* It should be remarked that modern differential hardening produced by electrical induction heating or by local heating with intense flames results from quenching on a rising temperature with a steep temperature gradient, only the outer layers being above the critical point: it is quite different from Japanese sword hardening wherein the temperature at the moment of quenching is approximately uniform throughout but the cooling rate varies.

Fig. 41.—Blade by Nobuyuki (cf.
Fig. 38). Cross-section. Etched, ×12.

century are shown in Figure 42*a* and *b,* while Figure 42*c* shows the
intermediate zone or *nioi.* The local variations in the surface reflectivity visible to the naked eye are seen to result from the presence of much
more numerous scratches, often almost granular in appearance, in the
hard areas, with longer, deeper scratches on an otherwise smooth surface in the soft areas. These textures are reversed when small, soft zones
occur in generally hard areas, or vice versa.

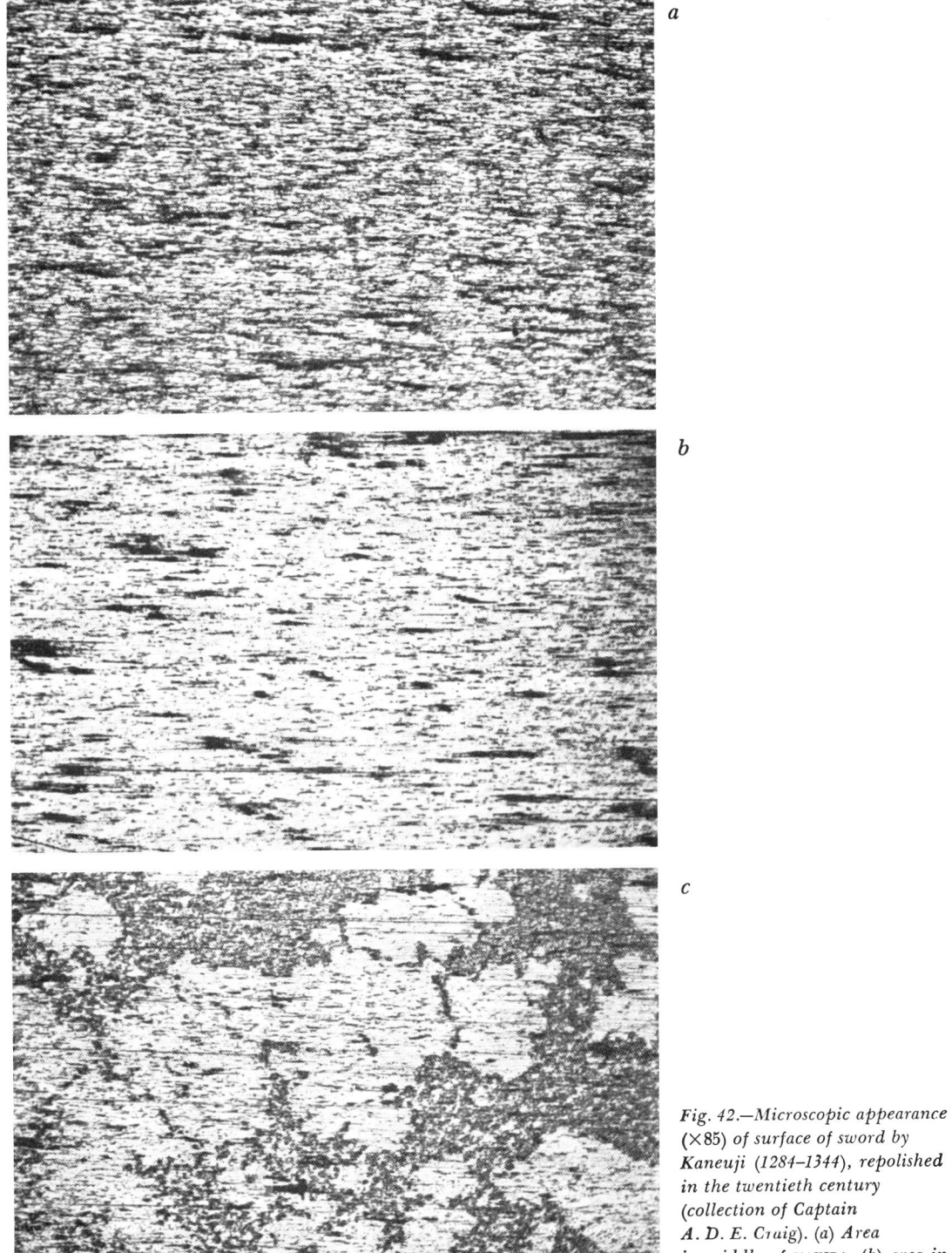

a

b

c

*Fig. 42.—Microscopic appearance
(×85) of surface of sword by
Kaneuji (1284–1344), repolished
in the twentieth century
(collection of Captain
A. D. E. Craig). (a) Area
in middle of* YAKIBA, *(b) area in*
JIGANE, *(c)* NIOI.

It is possible that some chemical action is involved in the Japanese finish—European connoisseurs who have attempted to duplicate the finish do use light etching immediately before the finishing treatment*— but the principal effects could perhaps result from the use, as a penultimate step, of a hard, medium-fine-grained abrasive on a hard backing, followed by a softer, finer abrasive on a yielding support. If used with discretion, this would bring large areas of the softer material to an intermediate state of polish more rapidly than the hard, while the reversal in the *nioi* would be a natural result of the partial protection of local soft spots and the prominence of hard, isolated *niye*.†

The reader will notice that two of the features—the large *niye* and the well-defined grain (not in themselves desirable, but nevertheless frequently found in treasured swords by the most famous makers)—are due to a large austenitic grain size and the presence of solid non-metallic inclusions, both of which features are regarded as being undesirable in modern steels intended for severe service. The extremely severe service tests applied at least to the older swords would surely rule out procedures that were harmful, and one must therefore conclude that there were compensating factors which made the high heat treatment temperature and the somewhat dirty forging practice actually beneficial. It must be remembered that the comparison was not with the best homogeneous modern steel, but with other steels of more variable quality. Steel with no inclusions would be impossible to make without fusion (or, for that matter, without melting and casting in a vacuum or inert atmosphere) and a good grain guarantees that extensive forging has been properly carried out. Similarly, the presence of *niye* implies that both the carbon content of the metal and its heat treatment was under a degree of control that can only be marveled at in the absence of chemical analysis and pyrometry.

Whatever may be the technique of the Japanese sword polisher, it is a direct antecedent of metallography—indeed it *is* metallography—for the surface textures depend on the details of the microstructure, and

---

* Mr. D. Tudor-Williams and Capt. A. D. E. Craig; private communications.

† It has now become fairly common to polish blades with a "decoration *yakiba*" (Japanese *kesho*). This results from manually localized treatment with an abrasive, and has only an approximate relation, or sometimes none at all, to the true metallurgical structure. It should be despised by a connoisseur, but even the "famous swords" in the atlas of fine photographs published by Junji Homma[17] have almost all had their beautifully delicate true *yakiba* hidden under a garish, locally abraded finish: the fine outlines of metallurgical detail depicted so boldly in the old books are now practically invisible.

the method of forging and heat-treating the blades must have been controlled in accordance with the indications of the final surface appearance. This is not to say that a knowledge of surface structure preceded the development of the essential steps in making a good blade: however, once blades were polished (in the eleventh century or earlier) a smith would certainly have related the different *yakiba* to the methods of manufacture that produced them, and it cannot have been much later when he began to develop types of *hamon* and grain with one eye on the aesthetic effect, provided it could be produced without sacrifice of serviceability.

JAPANESE SWORD FITTINGS

Everything to do with the sword was an object of admiration and reverence to the Japanese. The fittings were made as carefully as the blade. Extraordinary skill of design and workmanship was lavished on metal details at the ends of the scabbard and pommel, the handle of the little knife inserted in the scabbord and, above all, the guard or *tsuba*.* Unlike European metal work which usually utilizes only the fluidity and workability of metals, Japanese metal work often reveals a deep feeling for the structure of metals and their chemical properties. Several alloys were developed solely because of their ability to acquire a beautifully colored and textured pattern after pickling. Simple surface textures are proudly displayed that result from scaling, from corrosion, or from accidents in forging. A roughly forged or superficially pitted iron piece would be used alone or as a background for an exquisitely finished insect or leaf in gold and patinated copper; a heavily scaled or rusted surface would be inlaid with a few delicate lines of gold and silver. Often, particularly in later examples, purely artificial surfaces of great beauty would be produced on copper alloys by chiseling, punching, inlaying, and patinating, but the earlier workers prided themselves on working in steel or iron and showing its very "bones." For the present we are interested chiefly in the effect known as *mokumé* (wood-grain), which is related to the welded Damascus swords and gun barrels discussed in chapter 5. It was undoubtedly an outcome of the folding-welding-forging technique of the Japanese sword-maker, but in contrast to the deli-

---

* Good general discussions of *tsuba* are given by Tressan,[18] Joly,[19] Gunsaulus,[20] and Homma.[21] The last is a particularly fine collection of photographs but unfortunately includes no examples of *mokumé*. For Japanese metal work generally see also Pumpelly,[22] Rein,[23] Gowland,[24] Roberts-Austen,[25] Binkley,[26] and Wilson.[27]

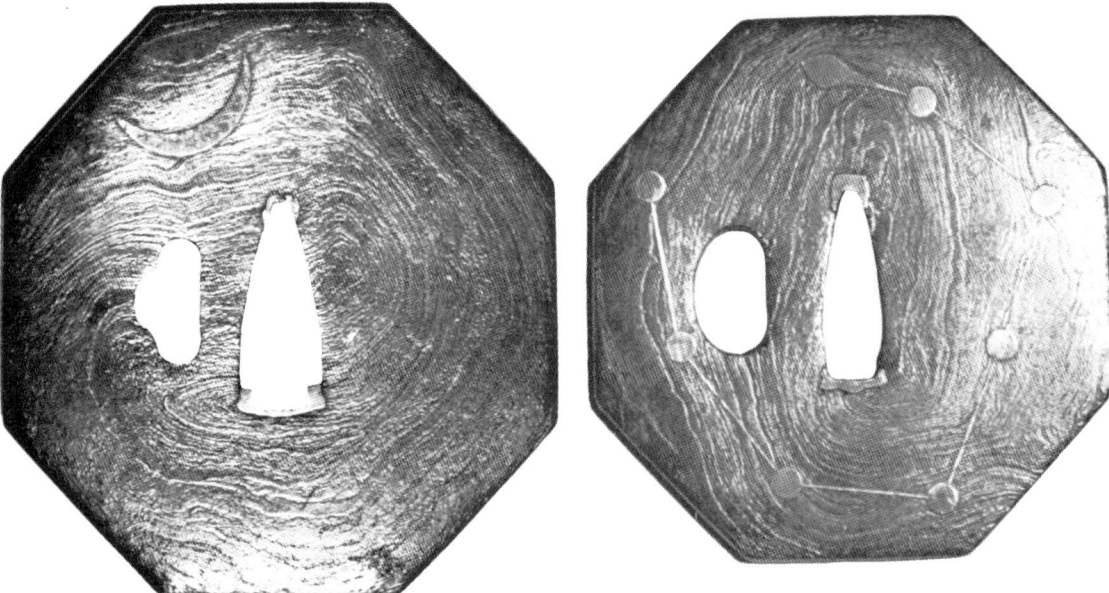

Fig. 43.—Pair of Japanese sword guards, iron MOKUMÉ, *inlaid with moon and constellations in silver. (Collection of Mr. G. E. Hearn.)*

cate polishing technique used on swords, the structure of the guards was developed by deep-etching to give a pronounced relief effect.

The most attractive of these *mokumé* surfaces depended for decorative effects purely on the natural flow-lines of a piece of metal in forging. Some guards were apparently made by welding a rolled-up iron or steel sheet to form a single solid cylinder, which was then flattened endwise into a disk. In others four or more rolls were welded side by side and compressed endwise. After being cut to shape, the metal was deeply etched to reveal the different layers of metals, and then sometimes inlaid (as in the guards shown in Fig. 43), incrusted (Fig. 44), or more often left without further treatment, the whole design being the macrostructure (Figs. 45, 46). Carving into a nearly flat welded composite of plates gave the structure in the chrysanthemum petals of Figure 47.

It is probable that the *mokumé* technique was first used by Miochin Nobuiye, who lived 1486 to 1564, one of the most outstanding of all *tsuba* makers.* There is no useful published description of the practical details of either forging or etching iron *mokumé*. The variable etching

* B. W. Robinson (Victoria and Albert Museum, London) private communication. The writer is deeply indebted to Mr. Robinson for introducing him to Japanese metal work in general and for many interesting conversations.

Fig. 44.—Iron sword guard, signéd Miochin Munetane, ca. 1850. MOKUMÉ *used as background for applied details in gold and* SHAKUDO. *(From H. Joly [Catalogue of the Hawkshaw collection].)*

rate may come from differences in carbon content due to surface carburization or decarburization in a single sheet before welding or to welding together pieces of two or more different metals. Perhaps even thin surface washes of copper may have been applied (as metal or from a cupriferous flux), which would diffuse into solid solution as the sheets were welded and forged but would confer locally enhanced resistance to the etchant.

More striking patterns (though less beautiful) were produced in nonferrous metals. Sheets of copper, silver, and special alloys selected for their color were soldered together and the resulting composite sheet, after thinning, was crumpled, locally dimpled, punched, or otherwise distorted before finishing with a plane surface cutting through and exposing the different layers. Figure 48 is a diagram from a paper by Roberts-Austen[28] which described the technique in some detail. An example of the finished work is shown in Figure 49.

The earliest use of this non-ferrous metal *mokumé* was in the eighteenth century, from which also dates the alloy *shibuichi,* one of its principal components. This is an alloy of copper with 25 to 50 per cent silver which, on proper pickling, develops a wonderfully attractive warm patination, between silver-gray and olive-brown in color, with a

59

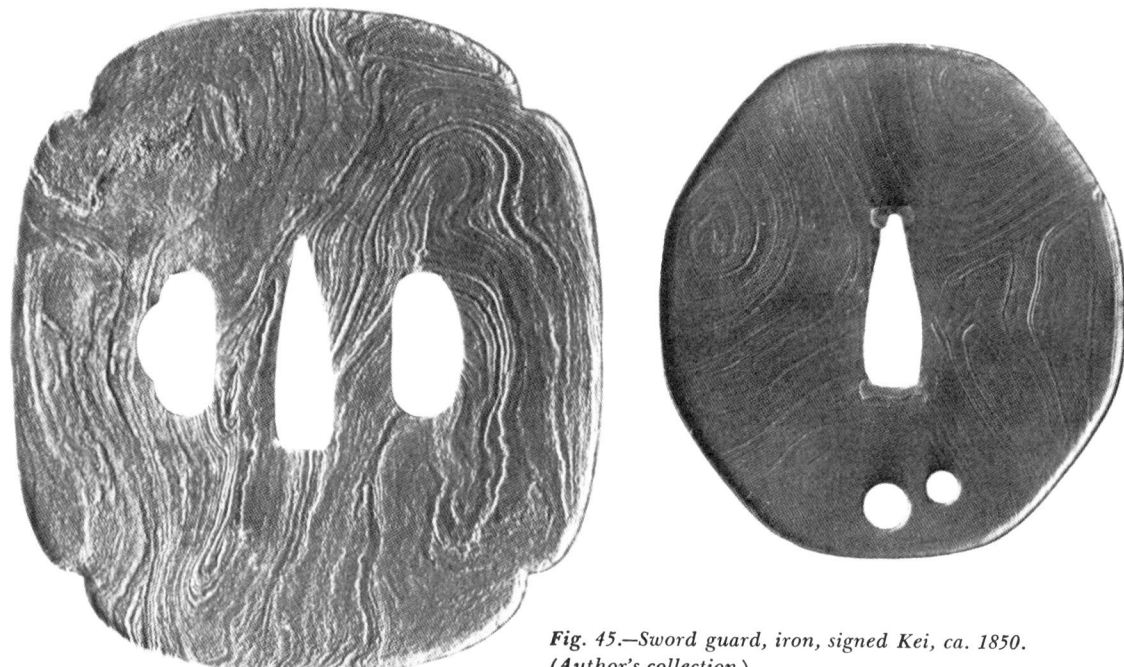

Fig. 45.—*Sword guard, iron, signed Kei, ca. 1850.*
(Author's collection.)

Fig. 46.—*Sword guard, iron* MOKUMÉ, *signed*
Miochin Munetane, ca. 1840. (Author's collection.)

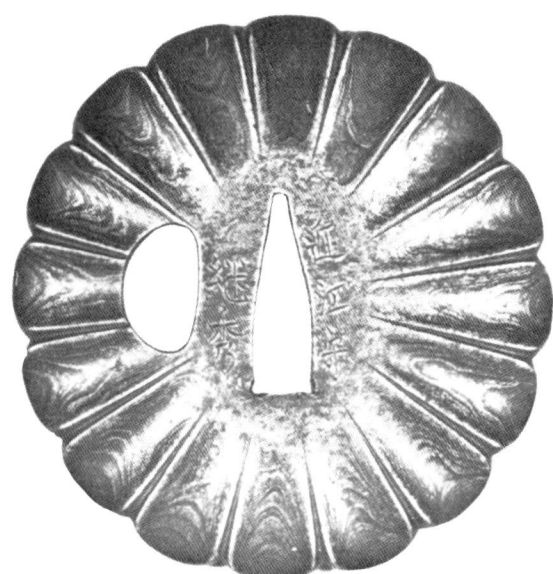

Fig. 47.—*Carved iron sword guard by Tsugihide,*
swordsmith, ca. 1800. (Victoria and Albert
Museum.)

Fig. 48.—*Sketch illustrating the principle of*
MOKUMÉ *manufacture. (Roberts-Austen.)*

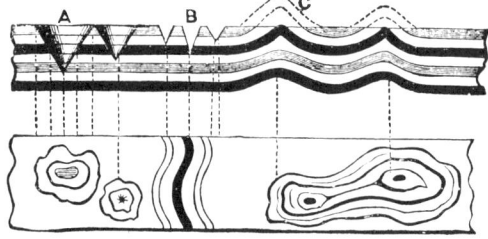

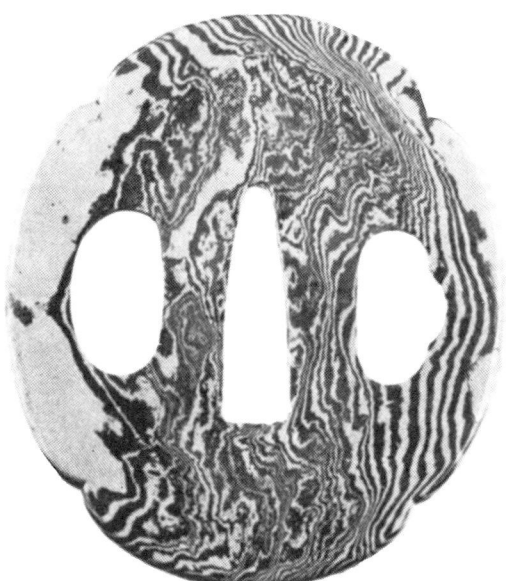

Fig. 49.—*Sword guard, mixed soft metal* MOKUMÉ, *unsigned, ca. 1800. (Victoria and Albert Museum [Crown Copyright].)*

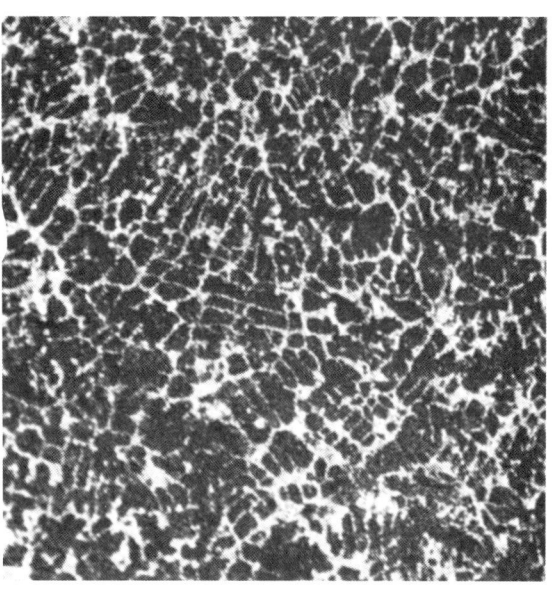

Fig. 50.—*Microstructure of* SHIBUICHI *guard, signed Masayuki, d. 1769. Original surface, not repolished or treated in any way,* ×85. *Oblique illumination.*

faint bloom which prevents harsh reflections. It is remarkably resistant to atmospheric corrosion and abrasion. A typical pickling solution, according to Roberts-Austen, contains 6.35 grams of verdigris (which may be replaced by sulphate of iron) and 4.25 grams of copper sulphate in a liter of water. Wilson prefers[29] a much more dilute solution. It was used boiling. When examined closely, the surface often shows a just visible speckled appearance which seems to depend upon a duplex microstructure, and microscopic examination confirms this. Figure 50 is a photomicrograph, at a magnification of 80, of an untouched surface of a *shibuichi* swordguard: this photograph was made in 1956 of a surface which had been finished by a *tsuba* maker who died in 1769 and which had been given no subsequent treatment during nearly two centuries. The undistorted rather fine white network of silver-rich eutectic shows that the alloy had been cooled rather rapidly from the liquid state and had not been much worked during the process of shaping the guard. It is this silver network, standing proud above the surface of the copper-rich portion of the alloy, which is responsible both for the characteristic sheen of *shibuichi* and for its resistance to abrasion. Another popular Japanese alloy *shakado* (copper with about 2 to 4 per cent gold) often has a faint microstructure due to coring, but its beautiful warm black color does not depend upon this.

Japanese pewter is reputed by Gowland[30] to acquire with time and gentle wiping a mottled surface, which may result from segregation of lead and tin during solidification of the casting.*

The finishing of a Japanese sword and its furniture is a metallographer's art par excellence. Yet, though it depends so intimately on an appreciation of the structure of the metal beautifully revealed to the naked eye, and has served to control practical forging and heat treating procedures, it has contributed nothing to the scientific understanding of the nature of metals or the manner of their solidification or transformation.† In Europe where both the microscope and the intellectual curiosity existed from the seventeenth century onward, the only metal surfaces available for study were either fractures or surfaces abraded or burnished so as to disguise the structure completely. Had either the Japanese been scientifically inclined or the Europeans better artists in metal, the history of metallography could hardly have failed to have been very different.

* The "spotty" surface on certain organ pipes made of a lead-tin alloy has long been associated with superior tone. The spotty appearance results from a network of dark rough depressions in the form of a two-dimensional foam pattern. It is a crystallization shrinkage effect, and occurs only on thin cast sheets of alloys containing between 35 and 65 per cent lead. Dendrites of lead evidently first crystallize uniformly throughout the metal on a fairly fine scale, but the eutectic colonies are nucleated infrequently—only about once in an area of 1–5 square centimeters—and grow radially outward. The shrinkage accompanying the solidification is concentrated at the places where the colonies meet, so exposing linear zones of lead dendrites with the interstitial eutectic drained away. The existence of the pattern is a guarantee that the pipes are made of cast metal that has not been rolled and to this day is considered to be a hallmark of the best pipes. This perhaps has some basis in the sound velocity and internal friction of the alloys, for acoustic tests have shown better reinforcement of the first seven harmonics in organ pipes of spotty metal than in those of other materials. The origin of this knowledge has not been traced, but it probably antedates the eighteenth century, when spotted metal was used by the famous English family of organ builders named Harris. (See W. L. Sumner, *The Organ* [London and New York, 1952], pp. 165 and 256–57.)

† It is interesting to note that some of the earliest microscopic studies of the shape of snow crystals that were made anywhere were done in Japan. European drawings date from the early seventeenth century but they seem to be merely naked eye impressions of hexagonal and star shapes. According to U. Nakaya (*Snow Crystals* [Cambridge, Mass., 1954]) eighty-six sketches of snowflakes made by Toshitsura Doi using the "Dutch glass" were published in 1832, with a supplement in 1836. Even before this snowflakes had been occasionally used as a decorative motif on sword guards but thereafter they become quite common. One, made by Goto Ichijo in 1854, is ilustrated in Plate LVII in the catalogue of the Gunsaulus Collection.[31] Some fine semicrystallographic details of snowflakes appear on an 1864 guard by the last of the great masters, Natsuo, reproduced in Plate 154 of the *Masterpieces of Japanese Sword Guards*, edited by Junji Homma.[32] In view of this early use of the microscope in Japan, it would be desirable to search pre-Meiji literature for descriptions or drawings of magnified sword details.

7. Moiré Métallique

All of the effects discussed previously in this section depend on local variations of composition or hardness to produce a difference in surface texture visible to the naked eye. As the metallographer well knows, differences in microcrystalline orientation can produce equally great contrast under proper chemical attack. In developing their patinated alloys, the Japanese must have used a wide variety of reagents on many different materials, including the brasses known as *shinshu* and *sentoku* which, in castings, would have a grain size easily made visible by etching. It does not seem, however, that this was ever used decoratively.

A cast brass door knob or railing subject to mild handling and weathering often develops an etched appearance clearly showing the shape and size of the crystal grains. The spangled appearance of galvanized iron and the etched crystals on the inside of a tin can that has contained acidic foods are commonplace. Gaudy effects are today commonly produced by organic crystal formation in lacquers, but crystals have apparently only once been used intentionally to produce decorative effects on metals. This was early in the nineteenth century when one Alard popularized for a time a material which he called *moiré métallique* for the manufacture of objects such as cabinet ornaments, dressing boxes, opera glasses, and the like.

The process was invented in 1814 and was promptly honored by a medal and a patent, though it consisted of nothing more than etching tinned iron plate (sometimes after a special treatment to give a particular pattern) and subsequently covering it with a clear or colored lacquer. The best description is in a paper by Baget.[1] Alard was proprietor of a factory in Paris for making lacquered tinware. He had advice from the famous scientists Thenard, Monge, and Gay-Lussac, the latter being so impressed as to state that the *moiré métallique* was the beginning of a new epoch in the history of the arts! However, Gay-Lussac also appreciated the scientific interest of it:

It is very interesting for science to see a sheet of tinplate, on which no crystallization can be distinguished, display, after the action of acid, a very pronounced crystalline appearance and a most agreeable sparkling effect. It is not the acid which produces the effect of crystallization, for this exists in the alloy from the moment of its formation, and the acid only uncovers it. The result is analogous to those obtained by Daniell, namely, that when a crystallized body whose surface is without form is submitted to the slow action of a solvent, the parts do not all dissolve equally: the regularly crystallized lamellae resist for a longer time than those which have been cut off and so exposed on their flanks to the action of the solvent.[2]

Peschel also considered[3] that the crystalline figures were formed during the tinning of the sheet, though Gilbert editorially suggested that the structure was a two-dimensional version of the lead tree, and resulted from tin being dissolved by the acid and reprecipitated by the iron in metallic form.

Translations of these papers and of several others on the subject were published in the *Annalen der Physik* in 1820. Engelbrecht[4] drew the analogy with the crystallization patterns produced by frost on a win-dowpane, discussed the effect of impurities, and described three differ-ent types of patterns: the natural one which the sheet takes in tinning; discontinuous patterns, which can take the form of letters, stars, etc., made by heating locally with a soldering iron, or granular effects pro-duced by heating and cooling locally with water; and finally actual scenic pictures "painted" on the sheet by locally heating with a blow torch. The patterns were developed by etching, usually with nitric acid, sometimes with nitric and hydrochloric acid mixed. Berry found[5] he could produce the effect on tin foil but not on rolled tin.

There is a report of Alard's invention in the English *Repertory of Arts* for July, 1818, in which very beautiful effects comparable with mother-of-pearl or malachite are glowingly ascribed to "crystallization operating by some new method on the surface of the metal," without any mention of etching.[6] Supposedly under the influence of some such rumor, Blakemore and James carried out work which culminated in a patent.[7] Their process did not involve etching but rather the treatment of the plate during tinning "to produce or leave on the surface of the sheet or plate . . . evident and visible marks or impressions of crystalliza-tion." To make this possible, the tin was alloyed with zinc, bismuth, copper, lead, or brass, the amounts being adjusted to the quality of the block tin used, "according to the depth of crystallization desired, and also according to the general character and crystalline figure or appear-ance which may from time to time be in vogue or called for or suited to the taste, caprice and opinion of the public by whom the manufactured goods are to be purchased or consumed."

Galvanizing of sheet-iron objects by immersion in molten zinc became common shortly after this time, with the surface spangle developing directly from crystallization, and has continued to the present day.

Though occasional references to *moiré métallique* (or *Atlas-Blech,* as it was known in Germany) can be found throughout the nineteenth century, there is no indication that any scientific study of crystallization was inspired by it. The process soon fell from favor. Figure C in Appendix C shows one of the few surviving objects.

A Baltimore artist, Bill Pond, is exploring the interesting effects produced by local distortion of the surface of large single crystals of aluminum, which are then annealed to produce recrystallization and etched to produce finely textured patterns against a background that reflects light only in a few crystallographic directions.

The examples discussed above show that metal workers have on occasion been sensitive to the inner structure of their materials and have empirically developed ways of revealing it. Empirical observation of the connection between changes in manufacturing methods and variations in the resultant surface structures which were desired for beauty or serviceability must have led to a kind of metallographic control of working or heat treating procedures, yet, except for the Damascus sword, there is no indication that artistic metal work ever gave rise to scientific speculation on the structure of metal. The duplex structure produced by welding is not a subtle one, but its very boldness made the discovery and improvement of etching reagents relatively easy. The more wonderful effects of crystallization and segregation remained too long a secret of the mysterious East and eventually the explanation of the Damascus blade fell out of work which it did not directly inspire. It seems, in fact, that ideas on structure developed slowly and largely in the minds of men who were ignorant of the suggestive practical techniques discussed above. This section, therefore, is a contribution to metallurgical science only in the negative way of showing to those who attempt to build a theoretical framework the value of contact with those who know matter intuitively and aesthetically from working directly with it.

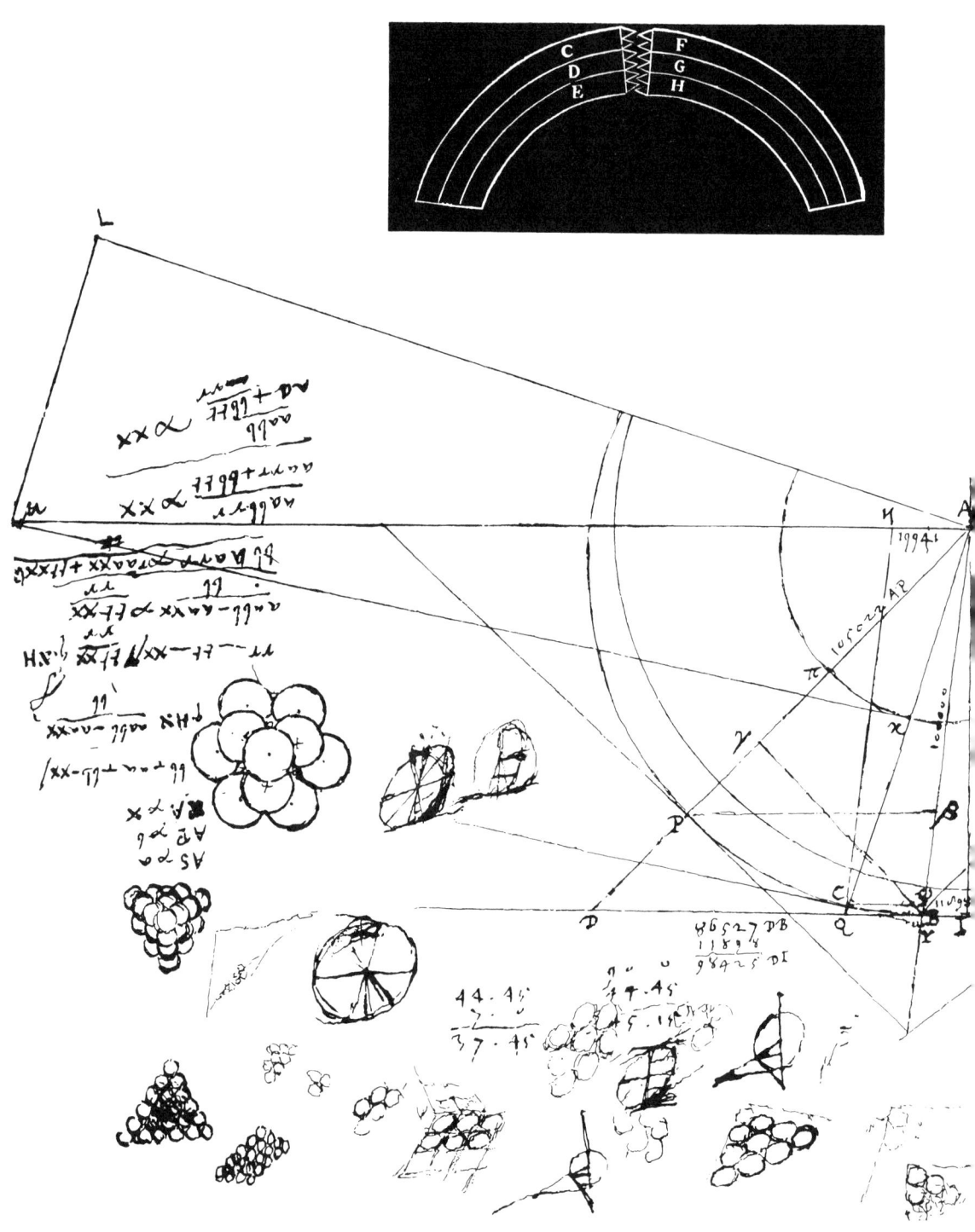

Hartsoeker's diagram of steel fracture (cf. p. 76) and a detail from Huygens' notebook (cf. p. 81).

THE PHILOSOPHIC
BACKGROUND

s ANY modern metallurgist knows, metals cannot be understood in isolation, but only in relation to the whole of matter. It is necessary, therefore, to trace briefly the history of ideas on the nature of matter with particular attention to those authorities who have attempted to explain properties in terms of structure—the atomists and their descendants—rather than in terms of the quality-dependent ideas of Aristotle and Paracelsus.*

In Aristotelian theory, the four elements, each associated with a tangible property (and themselves compounded from primordial matter) combined in varying proportions to give rise to all matter. Metals resulted from the condensation of the smoky exhalations (hot and dry) acting upon earth (cold and dry), but since they were fusible, they were also supposed to contain much moisture. Aristotle regarded metals as essentially structureless and homogeneous, although he invokes particles and penetrable pores in distinguishing between matter which can melt with heat and "melt" in water, and he knew that many natural materials were mixed in character and easily separable.

Metals have always had a central place in chemical theory, but until recently chemists have neither assumed nor explained any structure beyond the molecule. Paracelsus was making an important division of matter into classes of interatomic bonding when he selected salt, sulphur, and mercury as his principles, but he seems not to have felt the need for structural concepts. The obvious malleability and general fusibility of metals suggested a principle common to them. Their reversible conversion into calces on heating in air or with reducing materials was central to the rise of the phlogiston theory and to its vanquishing by modern chemistry, but the answer was in atomic or molecular terms, not in those of higher aggregates. Chemists have been indeed slow to realize the importance of crystallography, despite the possible beginning of atomism as an essentially physical concept and the fact that crystalline symmetry is most frequently responsible for the fixed combining proportions which provided the experimental basis for atomic chemistry early in the nineteenth century and with which chemists have

* Some of the material in this chapter was presented at the 1957 Institute of the History of Science, Madison, Wisconsin, and is published in Marshall Clagett (ed.), *Critical Problems in the History of Science* (Madison, 1959), pp. 467–98.

been overly preoccupied ever since. Only now are they beginning to realize that isomorphism and variable combining proportions are more common than the rigidly fixed stoichiometric ratios of the simple atomic theory.

Parallel with the chemical principles, however, there was another approach that was essentially structural in nature. Atomism began as a philosophic concept with the Greeks, passed through an exciting period in the seventeenth century when disputes between the atomists and Cartesians gave rise to mechanical corpuscular philosophy and eventually to the modern idea of the chemical atom. The writer believes that observation of the polycrystalline nature of matter, visible to the eye in the texture of a broken surface of steel or stone, played a large role in the beginnings of this viewpoint. Some form of corpuscular theory, the idea that matter is composed of aggregates of small parts progressing from the just visible through several stages of the invisible, could hardly fail to originate from an observation of the textures of broken pieces of steel, bronze, or stone or the grain of wood when splintered or polished. The early makers or users of metal would notice that fracture, sometimes without ductility and often after considerable obvious deformation, depends on the arrangement of "grains," which clearly suggests that strength and ductility depend upon the ability of parts of some kind to deform while maintaining contact with each other. Although a source of bitter disagreement among philosophers, the question of the ultimate divisibility or indivisibility of matter makes little difference in practice, for the effective particles for any given property may be either the primary ones themselves or aggregates of these into bigger units. That they were sometimes extremely small was early deduced from the prodigious extensibility of gold without change of character.*

The difficulty of building up a convincing and quantitative model

* Galileo (*Discorsi . . . intorno à due nuoue scienze* [Leiden, 1638; Eng. trans. Evanston and Chicago, 1930]) cites as proof that the ultimate particles are essentially small but indivisible the impalpable but not fluid powders of gold and silver obtained by the use of parting acid on gold-silver alloys and the fact that gold on a gilded silver rod could be drawn to extend its surface area, and hence thinned, by a factor of 200 without losing its characteristic appearance. Réaumur (*Mém. Acad. Sci.*, 1713, pp. 199–220) carefully dissolved away the silver core of such a wire in nitric acid, producing an intact gold tube which he believed to be no thicker than one millionth of a *ligne*, i.e., about 20 angstroms!

came from the fact that the only observations and experiments that were possible until the nineteenth century did not depend always on the behavior of the atom but usually on aggregates of atoms on many scales of organization. Some properties are due to the atoms themselves, some to molecules, some to microcrystals and not infrequently to statistically trivial imperfections in crystals. The philosophers attempted to explain everything in a qualitative sense; whereas the physicists and chemists, almost all of whom after the mid–seventeenth century took corpuscularity for granted, were far more restrained and invoked particles only when they could be specifically useful. The distinction is in fact almost that between the kind of speculation which passes through the mind of a scientist when he thinks what might be the explanation of his problem, and the more disciplined careful statements of his published work.

8. The Seventeenth-Century Corpuscular Philosophers

Marie Boas[1] has pointed out how the anti-Aristotelian trend of the early seventeenth century—partly precipitated by Francis Bacon—led people to study the earlier atomistic philosophers whom Aristotle's opposition had relegated to oblivion for so long. Democritus, Lucretius, Plato, and Epicurus were revived by Sennert,[2] Magnen,[3] Gassendi,[4] Hill,[5] Basso,[6] and others. Though metaphysical in tone, the writings of these men established a background which deeply affected the more experimental philosophers like Boyle and Newton. Although Basso expressed the idea that atoms associate with each other in corpuscles of varying degrees of complexity, the writings of the atomists do not reflect any concern with crystallinity. Sennert does cite a number of examples in support of atomism which are taken from metallurgical operations, but these all involve the identity of the metallic atom through operations such as solution, precipitation, melting, hardening, or amalgamation. Gassendi explained the hardening of iron (i.e. steel) on heat treatment as due to the fact that heat opened the structure of iron by the penetration of atoms of fire, while quenching caused water particles to "sub ingress into the amplified pores of the iron . . . . remaining imprisoned in the small incontiguities or inane spaces which otherwise would have been empty, making the body of the iron somewhat more solid or hard than otherwise it would have been." Annealing by the smith relaxes the structure and evaporates the water. Soft bodies generally have parts "so separated from each other in many points as if more and larger inane spaces be intercepted among them," while hard bodies are reduced to close order. The great ductility of gold is due to the compactness of the structure, the great tenuity of the component particles, and "the multitude of small hooks and claws whereby those particles reciprocally implicate each other and maintain the continuity as a whole mass. . . . No sooner does one particle dissociate from its neighbors but instantly it lays hold of and fastens upon another, and as firmly cohereth thereonto as to its former hold, so that mutual coherence is maintained even above the highest degree of extension or attenuation which any imaginable art can promise." Brittleness, conversely, is due to the fact that the percussional pressure is transmitted from point to point successively throughout the structure.

This kind of explanation of properties of bodies in terms of the shape and interaction of particles was developed much more extensively by René Descartes, and particularly by the Cartesian physicists who flourished toward the end of the seventeenth century. An essential part of Descartes' theory of the universe was the existence of three different particulate forms into which primary matter was divided. Descartes was interested in metals and must have observed fairly closely the operations of some iron smelters and toolmakers. Thus in 1639 he corresponded with Mersenne on the reasons for the hardening of iron on quenching,[7] which he explained in terms of particles of subtile matter and the smallest parts of earthy matter, of which a great number are always present in fire and air and, being agitated by heat, constantly issue from the pores at great speed. The particles are continually replaced by others as long as the iron is red hot, but if it is plunged into water nothing can re-enter the pores except the subtile matter contained in the water itself, consequently the pores contract, and the parts draw together making the iron hard, but it is brittle because the parts were changing position so fast they have broken some of their bonds with each other.

In his *Principia philosophiae*[8] (first published in 1644), Descartes returns to the hardening of steel, and explains it with a rather confused picture of little drops or clumps (*guttulas sive grumulos*, translated in the French edition simply as *"gouttes"*) which are agitated by the fire and the little branches of the particles that compose the surface of the drops are pushed back inside by the action of other drops, thus making the drops themselves more compact, but less tightly joined to neighboring ones. This structure, maintained in position by rapid cooling, makes the metal hard but brittle. Hardened steel is stiff and springy because the arrangement of its parts cannot be changed by bending it, but only the shape of the pores. (Elsewhere Descartes had pointed out that ductility depends upon the ability of parts to change their relationship with their neighbors.) Slow cooling, conversely, allows the small branches of the particles to come out of the surface into their natural positions, where they become interlocked or interlaced with similar ones projecting from the surfaces of other drops.

It is apparent from Descartes' description[9] of the smelting of iron and steel that he has watched the operation and has observed the granular nature of wrought iron during its production from cast iron in a finery hearth:

The entire bath of molten metal becomes divided into many small lumps or drops, the surfaces of which become smooth. For all the particles of metal that are somehow joined together make up one of these drops, and because this drop is pressed upon all sides by other drops which surround it and which move in different directions from its own, none of the points or branches of the particles could come out of its surface ever so little more than the others without being continually pushed back toward its core by the other drops. This causes the surface to become smooth and also causes the particles that compose each drop to draw together and be even more closely joined.

It takes only a little imagination to see in this the concept of surface tension and particularly the crystallization of the metal into numerous grains with atoms at the interfaces belonging to none. Of great interest is Descartes' approach to the idea of crystalline arrangement or order of the parts composing each "drop." He says that the aggregation of matter into drops results from the adjustment against each other of the channels in the smaller particles, which when achieved allows the circulation of cosmic subtle matter to proceed with less interference, and so the particles which are properly arranged are held more closely together than they are to others. This aggregation occurs during the smelting of iron ore as the metal is separated from other matter:

Concerning iron ore as it is smelted to convert it into iron or steel, it must be assumed that the particles of metal, while being agitated by the heat, first separate from the other matters with which they are mixed, and afterwards continue to move separately from each other, until those of their surfaces where the halves of the here-to-fore described ducts are embedded, happen to be so adjusted toward each other that these ducts become complete. When this has happened the channeled particles, which are present in the fire in as great a number as in other earthly substances, flow continuously through these ducts, and by their movement prevent the small surfaces (through the joining of which the ducts were made) from changing position as easily as they did before. In addition the mutual attachment of the particles of metal for each other as well as the force of gravity helps to keep them joined. In the meantime, because these particles continue to be agitated by the fire the result is that several become attuned to following the same movement, and thus the entire bath of molten metal becomes divided into many small lumps or drops, the surface of which becomes smooth.

It is obvious from the above that Descartes saw quite clearly that matter was composed of particles which, although they could be of exactly the same kind, could aggregate into clumps the particles of which were more closely associated with each other than with those of neighboring clumps. This idea permeates most subsequent thinking although Descartes' brilliant idea that the orientation of the particles was responsible for the difference was not followed up. The molecule displaced the crystal in man's mind for two centuries.

Descartes' ideas on matter were developed by both philosophers and physicists, the most notable addition perhaps being a willingness to hypothecate a wide variety of special shapes, which permitted almost any property to be accounted for by packing with more or less order and more or less space between them. Le Grand,[10] for example, combined (as did Descartes himself but in less degree) a mixture of iatrochemistry with corpuscularism. He believed that metals were composed of salt, oil, and earth mingled in the deep-seated parts of the earth and there linked together into particles of characteristic shape. "Substances become distinct from each other because of differences in the contexture of their parts, as when by the accession of new parts others are thrust out of their places, or they become otherwise ranged." Iron is composed of thick and branchy particles whose surfaces lie close together and whose pores are penetrated only by striate matter (i.e., salt particles). "The reason for this ductility of metals is because the particles of metals are of a longish figure and are so disposed that they lie upon one another according to their whole surface; which makes, that when they are present under the hammer or in drawing, they fall down sideling and joyn side to side without any separation. Thus it comes to pass the metal under the hammer may be extended into length and breadth, still retaining the firm cohesion of their parts." On the hardening of steel Le Grand reproduces Descartes' account almost verbatim.

But it was the French corpuscular physicists who pushed to extremes the explanation of the properties of matter in terms of particular structure. Claude Perrault[11] (who disagreed with Descartes on biology, but nevertheless approximated the philosopher's ideas on the structure of matter) carried out experiments on the density of alloys of tin and copper from which he concluded that the particles of tin penetrated the pores of copper and rendered the alloy harder because the flat and polished surfaces on the composite particles which resulted could join to each other more firmly than the original particles of copper.

In general, Perrault believed that hard bodies are those in which the greater part of the faces of the corpuscles were perfectly flat and applied directly one to the other, being held together, as was all matter, by the pressure of subtile "air" (much like Descartes' ethereal third element). Soft bodies are those which have only a few faces joined, and friable ones, those that are unevenly applied to each other. Forging or cold working hardens metal because strong compression causes a greater number of faces to join together and presses out sliding particles which had been interposed, although there are some metals like lead and tin which are softened by cold working, which mixes up the particles of a fluid nature with the flat-faced ones. Fire introduces its fluid and slippery parts even into iron, but if the metal is cooled slowly they are able to escape. Steel is hardened because the fire separates its parts, and on sudden cooling the hot particles, which are still soft and mobile, readjust themselves into the best possible arrangement, expressing the slippery particles which had been rendered mobile by the fire. However, since steel when quenched occupies a greater volume than in its annealed condition (Perrault was himself the first to observe this) and its grain is changed, it was necessary to introduce the idea that there are pores, frozen in on cooling, of such a size that the subtile part of the air can act although the coarser parts of the air, which compress the whole together and which are responsible for cohesion, cannot penetrate. Tempering or annealing allows these to return easily to their first state.

Throughout the whole book there is emphasis on the laying together of shaped units with intermeshing flat surfaces: It is hardly surprising to find that the author of this treatise was connected with masonry—he was, indeed, none other than the famous architect of the Louvre.

In a book[12] whose title (*Principes de physique*) promises physics but which is mostly uncontrolled speculation, Hartsoeker develops even more specifically shaped building blocks. Starting from the Cartesian belief that cohesion depends on the pressure of the column of "subtile air" extending to the surface of the earth, he shows that it will be more difficult to move parts with large plane surfaces than with small, from

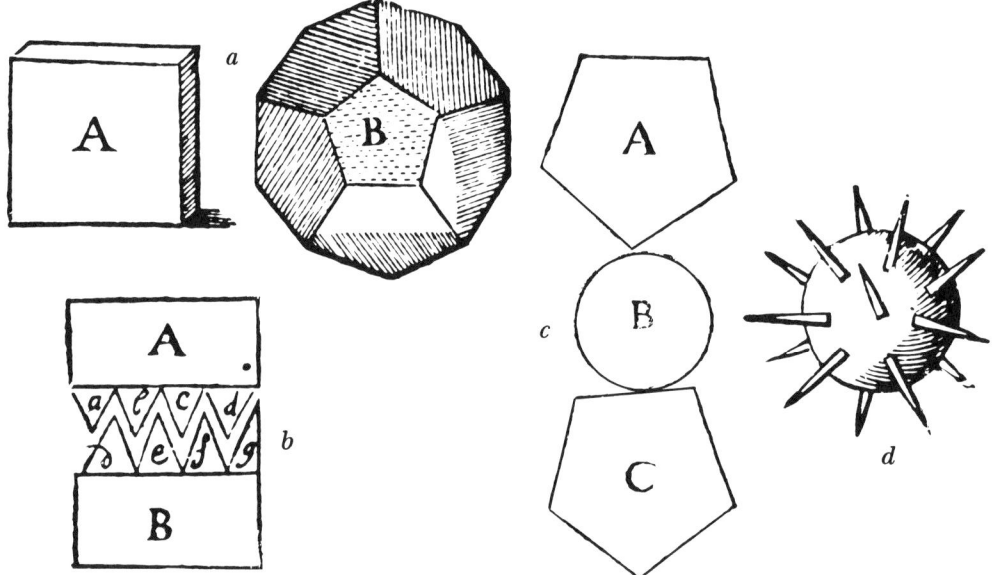

Fig. 51.—*Conjectural shapes of the particles of matter. (Hartsoeker, 1696.) (a) A is a refractory, B a fusible material. (b) The interlocking of particles of bodies like iron which are hard when cold but malleable when heat separates the parts. (c) Particles of mercury penetrating between polyhedral gold particles. (d) A composite particle of corrosive sublimate.*

which he deduces that the particles of refractory materials are composed of nearly cubic particles, whereas the parts of easily fusible ones are dodecahedral, which he illustrates (Fig. 51). By varying the shapes and areas of contacting surfaces, and with an additional variable in their roughness, he is able to explain all properties, or rather from known properties he deduces the configuration of the parts "without which it is impossible to make any progress in physics." Bodies which are fairly hard but which extend easily when hot, like iron, do so "because the parts like A and B, of which these bodies are composed . . . have very ample and very extended plane surfaces, with eminences as shown in the following figure [Fig. 51*b*]. In this manner these parts are not able to slide easily over each other unless they are somewhat separated from each other by the action of fire, up to the point that, for example, the asperities *a, b, c, d,* of body A are no longer in a condition to snag those *e, f, g, h* of body B." Hartsoeker proceeds in a similar fashion to explain the properties of all kinds of substances. The particles of gold (which may themselves be compounded from indistinguishable smaller ones) must have parts with very large flat surfaces, but, since gold is ductile, these surfaces must be very smooth and polished to allow the particles to slide easily over each other. Chemists, incidentally, waste their time trying to duplicate such things. Mercury is, naturally, composed of smooth and polished balls, hence it is easy for mercury to penetrate between the parts of gold, and the gold polyhedra (A and C, Fig. 51c) can roll over each other by means of the balls of mercury (B), whereupon gold becomes fragile and brittle. Lead is rather like mercury, but the balls are not quite spherical, being polyhedral with an infinity of little flat surfaces, and the little bodies which form the polyhedra themselves are rather easily separated.

Although some of Hartsoeker's molecules approach the modern concept of a close association of different elementary parts, some of them are amazing contraptions: Corrosive sublimate is a ball of mercury with, stuck all over it, the particles of salt and vitriol which are needles and cutting blades (Fig. 51d). Air is a hollow ball built of wirelike rings to give it the necessary elasticity to allow music to be heard.

One could in this way seek the configuration of the parts of most of the bodies that come before our senses and could see that salts are perhaps nothing other than an infinity of little longish parts of which some are pointed at one of the two ends, others at both; some are large and others minute; some are simply pointed like needles without being cutting and others with cutting edges like knife blades without being pointed; some are both pointed and cutting. But I do not wish to deprive the reader of the pleasure of himself making the search following the principles that have been established above.

His imagination reaches its apex in the structure that he deduces to explain the varied properties of iron. He postulates that there are formed in the entrails of the earth hollow parallelepipeds or prisms each composed of many tiny bodies and with faces of considerable area which are more or less jagged. Forging renders crude iron soft and malleable, because the heat allows the particles to arrange themselves and the hammering makes the jagged surfaces become smooth and polished so that they slide more easily over each other. The hardening of steel is due principally to the fact that the little bodies composing the prisms have been put in movement and deranged by the fire and do not have time to rearrange themselves if quickly cooled; the surface of prisms thus remain jagged, and the steel is hard and brittle.

The last Cartesian whom we will quote is better disciplined, and displays better physical intuition. He is Jacques Rohault, whose *Traité de physique*[13] was published in Paris in 1671. Rohault was the son-in-law of Claude Clerselier, who had been a friend of Descartes and devoted most of his life to promoting the philosopher's ideas. With true

Cartesian breadth, Rohault covers all material science in his book, including astronomy, medicine, and meteorology. It had a great influence on the development of French science, and, by means of a critically annotated Latin translation by Samuel Clarke, was instrumental in popularizing its rival Newtonian physics in England.

Following Descartes, Rohault believed that all properties of matter depend upon the shape, arrangement, and motion of its particles. The particles were essentially all of the same composition, although chemical differences are sometimes explained in terms of a kind of segregation. In its motion, the elementary matter gets broken up into three kinds of particles, a very fine dust produced by the attrition, particles which are rounded in form, and polyhedral particles which are more or less the first fragments. The particles, forming the Cartesian elements, are in principle transformable into each other. Although in their origin they are not assumed to be uniform, yet in their use to explain the properties of matter they seem always to be essentially of a single size and shape in any given substance.

Although Rohault is much concerned with the manner of packing the particles, he is not specific as to the type of force that holds them together, whether their own attraction or under the influence of the external pressure of subtile matter.

There are pores between the third kind of particles, through some of which the second matter can move and through others only the first. The pores will be enlarged under a tensile stress, contracted under compression: In a bent spring the subtile matter in circulating through the pores may either wear the particles or restore them to the state they were in before bending, thus causing springback.

Cold working a metal increases its power of springing back, because the beating does nothing else but make the parts approach nearer to one another, thus straightening the pores. Bodies that bend plastically without breaking are conceived as being made of parts with complicated textures intermixed with each other and hooked together like the rings of a chain or the threads of a cord, whereas brittle bodies are of simple texture with particles touching one another at only a few places, so that they cannot be separated even a little without the whole continuity being destroyed.

In Part III Rohault talks of the engendering of metals in restricted cavities in the earth from rising particles of water, salt, and oily substances mixed together and combined to form little hard bodies which are supposed to be the component parts of metals. Such a model would allow rearranging the small component parts to produce transmutation of the metals, though it is incredibly improbable that it can ever be done by man.

The portion of Rohault dealing with ductility and hardness is particularly revealing and will be quoted *in extenso:*

But however this be, we cannot but think that the component Parts of Metals are long; otherwise we cannot understand how Metals should be so ductile as they are, whether they be forged upon an Anvil, or drawn through a wire-drawing Iron; whereas, if we suppose them to be somewhat long, it is easy to conceive, that when they are pressed on one Side, they will slip sideways of each other without quite separating.

Further, it is not possible to conceive, that when a Piece of Metal is continually pressed upon one Way the Parts of it should be able to lye cross; on the contrary, we cannot but think, that they must necessarily so order themselves as to place themselves by each other's Side, and correspond Length-ways to the Length of the whole Piece, which will make it easier to bend that Way than any other; And this agrees with Experience; for Metals which are beaten into Rods upon an Anvil, or drawn into Wire through a Wire-Iron, are very strong Length-ways, but Breadth-ways they are many Times easier to break than Workmen would have them. And we observe Strings in them, as in the Slip of an Ozier.

These Strings ought not to be in Metal that is cast and has not been forged: And so we find that cast Metal is as easy to break one Way as another.

Steel, which is nothing else but fine Iron* is capable of being made the hardest of all Metals: The Way of making it so is this; only to heat it red-hot in the Fire, and then throw it all at once into cold Water; and this manner of hardening is what they call *tempering* it, and this makes it capable of cutting or at least of breaking all Sorts of Bodies without Exception, even Diamonds themselves: For it is certain they will break in Pieces with a small Stroke with a Hammer if it hits right.

In order to account for this Effect (which perhaps is one of the most admirable, and doubtless one of the most useful Properties that we know) we must suppose that the Heat of the Fire, which makes the Steel almost ready to melt, puts the small Particles, which each component Part is made up of, into Motion, and thereby causes the Particles of the two nearest component Parts, (whose Distance from each other was very small, though far enough) to approach a little nearer one another, so that the Metal becomes more uniform than it was before; after this, being cast on a sudden into the cold Water, the metallick Parts lose the Motion they were in, before they have time to gather

* The concept that steel is nothing but refined iron was almost universally believed prior to Réaumur. It can be traced to Aristotle (*Meteorologica* iv. 6) and is perhaps a reasonable supposition since steel resulted from the further heating of iron in the fire, and fire is known to purify. Robert Hooke took the opposite view, believing that vitreous matter was present in steel and responsible for its properties.

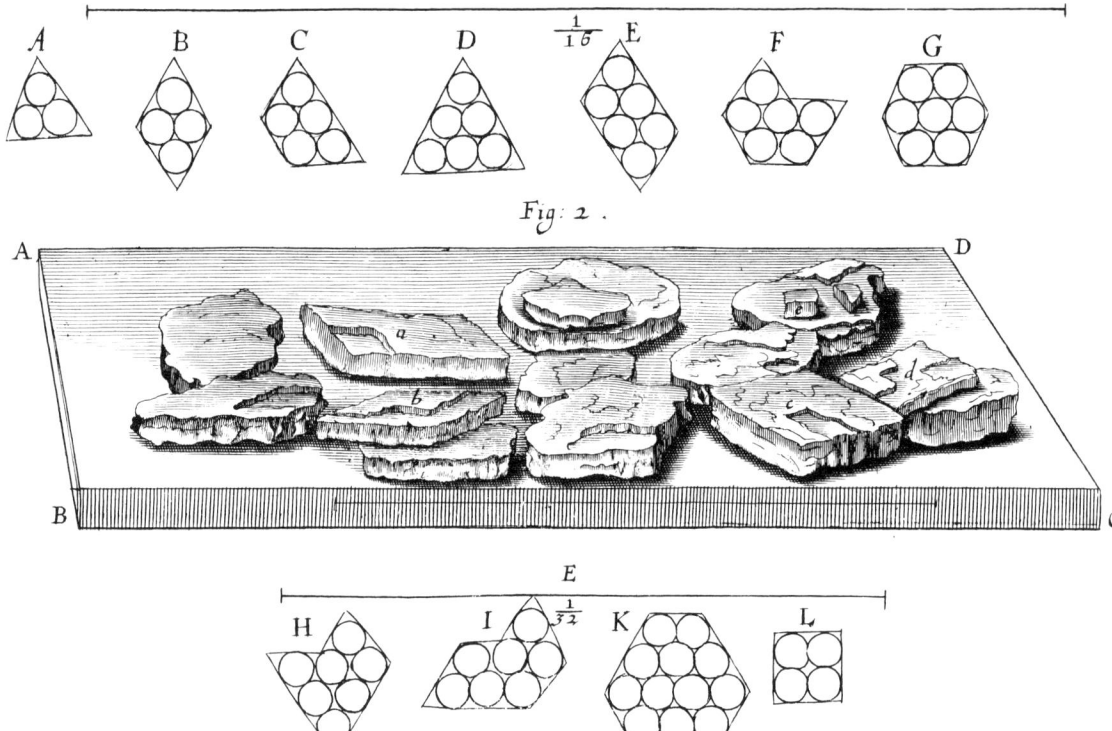

Fig. 52.—Packing of spheres to match the natural shapes observed in alum crystals. (Hooke, 1665.)

together again into gross component Parts, with considerable Intervals between them: Whence it follows, that the Points or Edges of Gravers and the Teeth of Files can only slip over them without entering into them.[14]

On the hardening of steel Rohault had earlier remarked, "Thus we can perceive by our Senses that the Parts of Steel which are not tempered [i.e., hardened], are larger, and consequently the Pores wider, than those of tempered Steel" (I, 134). He was referring to the visible fracture, which becomes finer on quenching, and it seems certain therefore that the parts he is talking about verge on the visible—in other words, are commensurate with the modern metallurgist's grains.

In discussing the formation of little cubes on evaporating a concentrated solution of salt, Rohault remarks on the ordering of the particles as they lie against each other during crystallization, but he is farther than was Descartes from the concept that crystallinity consists just in order. Nevertheless, Rohault's "gross component parts" (French *gros grumeaux*, the common word for "clot" or "clump," obviously descended from Descartes' *gouttes* or *grumulos*) carry clearly the idea of an aggregation of particles and were analogous to the microcrystalline grains, except for the emphasis on crystalline order.

Crystalline order was meanwhile being considered by scientists of another cast of mind. The possibility of building up crystalline shapes by packing spheres in various arrays was first explicitly stated in 1611 by Johannes Kepler, in a little book on hexagonal snow.[15] Robert Hooke[16] outlined a rather complete experimental approach to chemical crystallography but seems not to have pursued it. With brilliant insight he realized that simple close-packed arrays of globular bodies

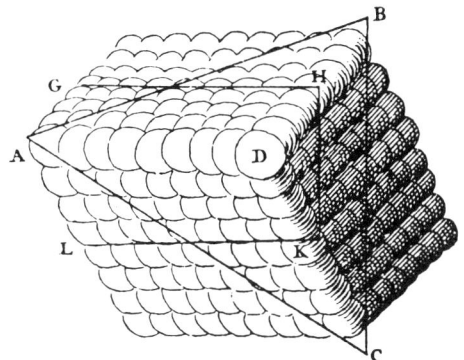

Fig. 53.—Iceland spar crystal composed of stacked spheroids. (Huygens, 1690.)

(Fig. 52) would duplicate all of the shapes observed in alum crystals, in both two and three dimensions. He even postulated a simple cubic array (*L* in Fig. 52) for rock salt, with other crystals "compounded of these two textures but modulated." His experiments with spherical bullets must have shown him many grain boundaries, but he records only single crystalline forms.

This idea is developed by Christian Huygens in a work[17] which was written in 1678 but published in 1690. After mentioning the frequency of polygonal-faced bodies in nature (including a variety of rock crystal in small stones "piled up directly upon one another, which are all of pentagonal figure with rounded edges, and the sides a little folded inwards"—perhaps just a random aggregate in which pentagonal faces are the most common) Huygens suggests that Iceland spar may be composed of an array of slightly flattened spheroids* (Fig. 53). In one

* The original page of notes in Huygen's handwriting in which he records the concept of stacked spheroids to explain double refraction still exists. Headed "*Eureka* 6 Aug. 1677," it is covered with calculations, geometric constructions, and sketches of aggregates of spheres and spheroids. It is reproduced in *Oeuvres completes de Christiaan Huygens,* Vol. **XIX** (The Hague, 1937), between pp. 426 and 427. See illustration on p. 66 of this book.

stroke he had simultaneously explained the optical birefringence, the cleavage properties, and the growth shape of the crystal! But such things were not much in men's minds and the concept lay unused to be independently conceived by Wollaston in 1812, some years after the stacking of unit polyhedra had been popularized by Grignon and Haüy (see chap. 14). In any case this viewpoint focused attention upon perfection and could not for a time have helped the metallurgist. Another approach that was to bring delayed benefits was Isaac Newton's succinct and restrained summary of corpuscular philosophy, stripped, as one would expect, of most of its fanciful aspects but still virtually ignoring crystallinity. To quote once more a much-quoted passage from the *Opticks:*

> The Parts of all homogeneal hard Bodies which fully touch one another, stick together very strongly. And for explaining how this may be, some have invented hooked Atoms, which is begging the Question, and others tell us that Bodies are glued together by rest, that is, by an occult Quality, or rather by nothing; and others, that they stick together by conspiring Motions, that is, by relative rest amongst themselves. I had rather infer from their Cohesion, that their Particles attract one another by some Force, which in immediate Contact is exceeding strong, at small distances performs the chymical Operations above-mention'd, and reaches not far from the Particles with any sensible Effect.[18]

He believed, however, in aggregates of particles on successive scales:

> Now if we conceive these Particles of Bodies to be so disposed amongst themselves, that the Intervals or empty Spaces between them may be equal in magnitude to them all; and that these Particles may be composed of other Particles much smaller, which have as much empty Space between them as equals all the Magnitudes of these smaller Particles: And that in like manner these smaller Particles are again composed of others much smaller, all which together are equal to all the Pores or empty Spaces between them; and so on perpetually till you come to solid Particles, such as have no Pores or empty Spaces within them. . . .[19]

And later, more specifically:

> There are therefore Agents in Nature able to make the Particles of Bodies stick together by very strong Attractions. And it is the Business of experimental Philosophy to find them out.

> Now the smallest Particles of Matter may cohere by the strongest Attractions, and compose bigger Particles of weaker Virtue and many of these may cohere and compose bigger Particles whose Virtue is still weaker, and so on for divers Successions, until the Progression end in the biggest Particles on which the Operations in Chymistry, and the Colours of natural Bodies depend, and which by cohering compose Bodies of a sensible Magnitude. If the

Body is compact, and bends or yields inward to Pression without any sliding of its Parts, it is hard and elastick, returning to its Figure with a Force arising from the mutual Attraction of its Parts. If the Parts slide upon one another, the Body is malleable or soft. If they slip easily, and are of a fit Size to be agitated by Heat, and the Heat is big enough to keep them in Agitation, the Body is fluid....[20]

These statements of Newton's are obviously based on the earlier philosophical speculations—indeed (except for the assumption of attractional forces between the particles, which is an inspiration of genius even if it is quite as occult as the pressures due to imaginary motions against which he is inveighing), it represents a fair summary of them, stripped, however, to essentials and without the ridiculous extensions into casually adjusted detail to fit special cases. A. R. Hall[21] has remarked that "Cartesian mechanism, the dinosaur of seventeenth-century scientific thought, could not adapt itself to a new environment. It was committed dogmatically in too many points of detail where it proved to be false." Yet at least from the viewpoint of present studies, the passing of the dinosaur was lamentable, for its fanciers showed a greater interest in the diversity of real matter than did the mathematical physicists who followed them. The subject was too complicated for a rigorous approach. It was Rohault's Cartesian speculations that inspired Réamur's important practical work, not the more scientific Newtonianism. Indeed, as science became more mathematical under the influence of Newton (the *Principia* more than the *Opticks*), speculation on the nature of matter was perforce excluded, and even interest in it vanished. Following the *Principia* and Galileo's *Due nuove scienze*—both superb works—mechanics became purely theoretical, and theory for a time was *only* mathematics. The most important tool of science came to be regarded as its body.

Many great contributions to nineteenth-century mathematics were inspired by problems in elasticity theory. At the other extreme there were outstanding practical engineers who were testing metals and building fine bridges and other structures. Between the two, however, there arose a strangely formalized subject taught to engineers as "Elasticity and Strength of Materials" which decayed into exercises in applied mathematics with almost complete disregard of the intricate irregularities in the structure and behavior of real materials. The field points up well both the advantages of the mathematical approach and its dangers. The elegance and beauty of mathematics rightly make it the queen of the sciences; yet when mathematics is applied to physical

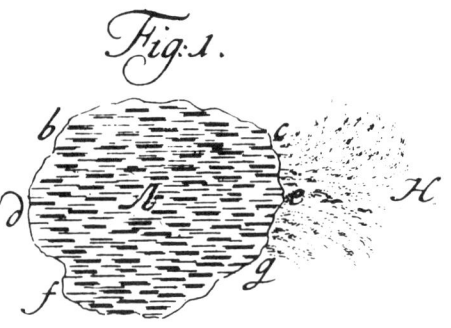

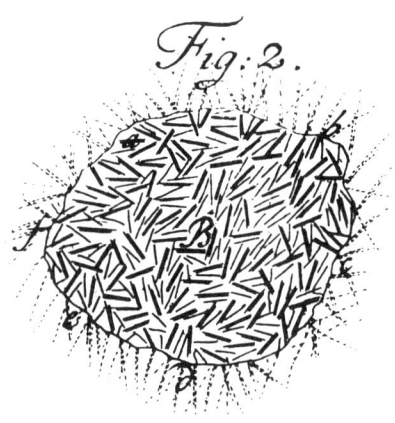

Fig. 54.—*The arrangement of parts in a
ferromagnetic material when magnetized
(Fig. 1), and demagnetized (Fig. 2).
(Swedenborg, 1734.)*

or engineering problems, the model has to be a simple one if it is to
be computed, and the simplifications are often later forgotten. It is
questionable whether present mathematical methods are capable of
dealing with interacting systems as complex as in an ordinary piece of
steel, yet the complexity does not reach the point where purely statistical methods can be used with precision. Most of nature is of intermediate complexity and the artisan—or artist—sensing rather than
specifying the balance of innumerable factors will continue to be important in many fields. The interplay of practical requirements and
fundamental concepts with the idealized mathematics of "strength of
materials" would be a most illuminating study for a historian of science.
An extensive critical bibliography already exists.[22]

On the whole, the active speculation on the structure of matter in
the seventeenth century was followed by a period of remarkably little
interest in the eighteenth. Swedenborg is almost alone among eighteenth-century philosophers in contributing new ideas to this field. In
his *Principia rerum naturalium* (1734),[23] he has some quite remarkable
orbital motions in his structural units which one modern historian[24]

regards as a forerunner of the Bohr atom. From our viewpoint his most interesting section is his brillant anticipation of ferromagnetic domains. After describing the loss of magnetism in iron on heating (which had been reported by William Gilbert in 1600), Swedenborg attributes this to the disorder of the parts produced by heat. Polarity comes from alignment of the unit magnets. He has two fine, though symbolic, drawings, which are reproduced in Figure 54.

In general, however, as science became more mathematical, the role of speculation that could not be experimentally checked was rightly diminished. Chemists virtually ignored the structure of their materials, although the introductions to chemical textbooks would usually have some superficial statement on the "parts" of which matter was composed. Despite the supremely important work of Boscovich, who laid the basis for the calculation of interatomic forces in 1758, most physicists, though taking the existence of corpuscular structure for granted, were completely uninterested in it and proceeded with the study of inertia and elasticity as if matter were homogeneous. In the field of metallurgy, the great French scientist R. A. F. de Réaumur alone proceeded to develop corpuscular theory into something useful. Almost a century before the particles of corpuscular theory had developed into the chemical atom on one hand, and the stacking of polyhedra had led to mathematical crystallography on the other, Réaumur was using the idea of parts aggregated on different scales to explain the structures that he saw on fractured metals and to develop theories of the nature of steel and iron on the basis of which he made very practical innovations of great industrial importance. This will be discussed in the next section.

6) From these few experiments, but made with all possible accuracy, I think we may draw the following conclusions:

differences of Irons & steels.

1. That though aqua-fortis is found mostly to have a stronger action on steel than iron, in virtue of its great attraction to the inflammable matter, of which steel contains something more than iron does; yet at the same time it sooner loses its power on the steel, and deposites thereon a sediment, consisting partly of inflamable matter, & partly of iron calx, which covers the steel and hinders the action of the etching water, and likewise gives the steel a more or less black surface, according as it contains more or less of the inflammable matter, or as it is more or less hard, insomuch that we are in condition to judge in some degree of the steel's hardness from the degrees of lighter or darker grey colours which it takes in the etching. On the contrary this sedi= ment never fastens at all on iron, whereby the etching water

Persian coin (obverse of Fig. 57) and detail from Chisholm's translation of Rinman's paper on etching (cf. pp. 145 and 249).

SECTION **III**

OBSERVATIONS

OF STRUCTURE

BEFORE SORBY

IN THIS section is summarized a miscellany of structural observations made by craftsmen and scientists in the period before structures were revealed unmistakably by metallographic techniques. Some information was gleaned by casual observations through the microscope, while more important contributions came from the careful scientific consideration of the appearance of fracture. Chemists found that they could grow good crystals of metals, and etching began to be used by metallurgists to show chemical and structural heterogeneity, many decades before the true significance of the observations was realized.

9. The Early Microscopists

Early in the seventeenth century the microscope first began to reveal the life and structure that lay unseen in all kinds of natural and artificial things.* The opening up of the world of minute things served to enlarge man's mental horizons quite as much as did the telescope's penetration into space. The impact was far greater on biology than on the inorganic sciences because the parts of living structures were more clearly visible without any treatment to differentiate or develop them, and because the resulting knowledge radically changed the basis of the science. In the beginning, however, many crystalline materials were studied for their beauty and interest, and it was not until the nineteenth century that treatises on microscopy for both professional and amateur became exclusively biological. For example, the first edition of Henry Baker's popular *Employment for the Microscope*[2] published in 1753, devotes almost half its space to salts, minerals and inorganic experiments generally, including the growth of dendritic crystals or trees of many metals.

When the microscope was first turned on the world, metals were as eagerly examined as other things. Interest in them was, however, soon exhausted because the only features visible on the surfaces that were available at the time depended more on accidental circumstances than on significant internal structure.

The first microscopist to mention metals was Henry Power,[3] in 1664. He has observations on mercury, gold, silver, iron, copper, tin, and lead. On the solids he remarks: "Look at a polished piece of any of these Metals and you shall see them all full of fissures, cavities, and asperities, and irregularities; but least of all in Lead, which is the closest and most compact Body probably in the world." On mercury

* For the invention of the microscope and its early development, see *The History of the Microscope* by Clay and Court.[1]

$\frac{1}{16}$ poll: Ang: or of an inch

Fig: 2:

Fig. 55.—*Edge of razor as seen under the microscope.* (Hooke, 1665.) *The scale is indicated by the line representing 1/16 of an inch. In the original, this measures 3.03 inches, corresponding to the magnification of 49.*

he reports that even the tiniest sphere seems like a globular looking glass reflecting the surroundings in fine detail and "whereas in most other metals you may perceive holes, pores and cavities; yet in mercury none at all are discernible."

Power observed the fused globular nature of sparks struck by flint from steel but credits the explanation of their origin to Robert Hooke, who was to give illustrations and amplify the explanation the following year in his famous *Micrographia*.[4]

The *Micrographia*, a superbly illustrated folio volume published in London in 1665, summarizes the microscopical observations shown by Robert Hooke to the members of the Royal Society at various meetings in 1663 and 1664. Although Hooke was by no means the first to use the microscope, and was not the inventor of the compound microscope, which has often been attributed to him, his reports of its use far surpass all others of the seventeenth century in meticulous observation, careful draftsmanship, and particularly in scientific insight. His microscopic observations are illuminated by discourses on a host of diverse physical, chemical, and biological topics ranging from meteorology, interference colors, and capillarity, to the meteorite hypothesis of lunar craters, the fluidization of finely divided solids, and the crystallization of alum, vitriol, and other salts. His metallurgical observations, however, are disappointing. Promisingly enough, his first subject is the point of a needle, which he regards as very gross compared with naturally occurring points, both organic and inorganic. Proceeding from point to line, he next examines a razor, which the minute book of the Royal Society[5] records as having been shown to its members on July 8, 1663, at one of the weekly demonstrations.

A Razor doth appear to be a Body of a very neat and curious aspect, till more closely viewed by the *Microscope,* and there we may observe its very Edge to be of all kind of shapes, except what it should be. For examining that of a very sharp one, I could not find that any part of it had anything of sharpness in it; but it appear'd a rough surface of a very considerable bredth from side to side, the narrowest part not seeming thinner than the back of a pretty thick Knife.

The razor when set between lens and light so that reflections occurred from its very edge appeared as in Figure 55. The scratches *gh, ik* were

caus'd perhaps by some small Dust casually falling on the Hone, or some harder or more flinty part of the Hone it self.

The parts of the razor that had been finished on a grinding stone appeared rougher,

looking almost like a plow'd field, with many parallels, ridges, and furrows,
and a cloddy, as 'twere, or an uneven surface: *

Had the edge really been

such as it appear'd through the *Microscope,* [it] would scarcely have serv'd
to cleave wood, much less to cut off the hair of beards, unless it were after
the manner that *Lucian* merrily relates *Charon* to have made use of, when
with a Carpenters Axe he chop'd off the beard of a sage Philosopher, whose
gravity he very cautiously fear'd would indanger the oversetting of his
Wherry.

Hooke observed (following Power) the melting and partial or complete
vitrification of sparks struck from steel by flint—and explained this as
due to the violence of the concussion working on a very small part of
the steel, which became heated.

. . . the filings or small parts of Steel are very apt, as it were, to take fire, and
are presently red hot, that is, there seems to be a very *combustible sulphureous*
Body in Iron or Steel, which the Air very readily preys upon, as soon as the
body is a little violently heated.

The globular form arose from the common property of all fluids to
acquire this form.

Led into the subject by observations on interference colors, Hooke
has a section on the hardening and tempering of steel. Unlike most of
his contemporaries who thought steel is the purest, most refined iron,
Hooke believed steel to be iron in which certain salts had been incorporated and

that Steel is a substance made out of Iron, by means of a certain proportionate
Vitrification of several parts, which are so curiously and proportionately
mixt with the more tough and unalter'd parts of the Iron, that when by the
great heat of the fire this vitrify'd substance is melted, and consequently
rarify'd, and thereby the pores of the iron are more open, if then by means
of dipping it in cold water it be suddenly cold, and the parts hardned, that
is, stay'd in that same degree of *Expansion* they were in when hot, the parts
become very hard and brittle,

Tempering the steel in a convenient heat till the colors appear on it
causes the hardness to relax by degrees

* Finding similar scratches even on the best finished glass for optical use, Hooke says
that "it seems impossible by art to cut the surface of any hard and brittle body smooth, since
Putte or even the most curious powder that can be made use of, to polish such a body, must
consist of little hard rough particles, and each of them must cut its way and subsequently
leave some kind of gutter or furrow behind it."

namely, the action of the heat does by degrees loosen the parts of the Steel that were before stretched or set *atilt* as it were, and stayed open by each other, whereby they became relaxed and set at liberty, whence some of the more brittle interjacent parts are thrust out and melted into a thin skin on the surface of the Steel, which from no colour increases to a deep Purple, and so onward by these *gradations* or consecutions, *White, Yellow, Orange, Minium, Scarlet, Purple, Blew, Watchet* &c. and the parts within . . . . gradually subside into a texture which is much better proportion'd and closer joyn'd whence that rigidnesse of parts ceases and the parts begin to acquire their former ductilness.

Correctly relating the colors to those possessed by thin films of mica and other substances, he incorrectly identified their origin. His "parts" are obviously the hypothetical units of the corpuscular philosophers, not grains or other structure that he might have seen with the microscope. By vitreous he meant non-crystalline material, an antecedent of the amorphous metal of the early twentieth century. This is not a bad theory of the hardening of steel, and his explanation of tempering would have been correct had he chosen to have the hardening substance rejected to intergranular spaces instead of to the surface. The increase of hardness on cold working he believed to be due to the putting and restraining of the parts in bended postures, while annealing occurred by the agitation of the heat loosening the parts and suffering them to unbend themselves.

After observing beautiful hexagonal dendritic crystals of ice forming on urine, Hooke comments that he had several times observed similar structures on the starred regulus of antimony where all the stems and branchings were "bended in a most excellent and regular order." Similar but much smaller figures appeared on the surface of lead containing arsenic and other impurities.

Hooke also comments on the green transparency of gold leaf either to the naked eye or to the microscope. He attributed this to the transparency of the ultimate particles of metals and believed that other metals would be transparent and have their particular colors were it possible to produce them in thin enough foils.

Though it is perhaps not surprising that Hooke did not examine any etched material, it is strange that he never examined microscopically the fracture of a piece of metal, particularly since he made a careful naked-eye study of fractures in relation to solidification mechanisms (see chap. 10).

The French physicist Rohault,[6] about whom much more was said in chapter 8, observed (1671), the surface of gold after it had been burnished with a blood stone: if such gold be looked on with a micro-scope, "it will appear very rugged and uneven and like a great number of little mountain ranges on each side with the valleys betwixt them" —which may be a description either of scratches or of the orange-peel effect resulting from deformation of a material of large grain size.

In 1672 Isaac Newton wrote to the secretary of the Royal Society referring to the small pores in speculum metal, visible only under the microscope, which interfere with its polishing to correct figure. A dec-ade later Newton was still interested in the structure of metals, for he records[7] the appearance of the fracture of a number of alloys with which he was experimenting. Alloys of antimony, copper, bismuth, and zinc showed fractures variously described as glassy, with glittering granulae, or steel-like, though Newton seems not to have deduced anything significant from their observation.

Grindel van Ach[8] published in 1687 a book of observations made with his compound microscope. He includes engravings illustrating the appearance of gold wire, of the point of a needle, and of the head of a pin, all at a magnification of 100. Beyond die scratches and chatter marks on the wire, these show no metallurgical detail whatever, and the author indulges in no speculations on structure.

Next follow the remarkable microscopic observations by Antoni van Leeuwenhoek, who used his simple microscope to observe the fine structure of so much of nature. He was unable to see much in metals and his comments are perhaps of greater interest in showing a mental picture of structure than in revealing the true state of affairs, for he usually misinterpreted surface features as true parts of the metal. Leeuwenhoek's first observations[9] were in 1702. He describes the notches and holes at the edge of a razor, although nothing was revealed that even approximated the true structure of the metal. He mentions a charlatan who "bragg'd mightily of a magnifying glass that he had whereby he could see into Metals and Minerals. I laughed at what he said and made answer that the Pores of Metals were so close and im-pervious that it was impossible to see through them."

Next, Leeuwenhoek goes on to observe a tree of silver[10] grown by placing a fragment of copper in silver nitrate solution.

... then viewing my silver water with a microscope, I observed with a great deal of pleasure, how the Silver in this clear Water was coagulated into such bodies as are described by the above mentioned trees;* which coagulation we may call an Inclination; & that I may give you a better Idea of this kind

* This refers to some animalculae found in canal water.

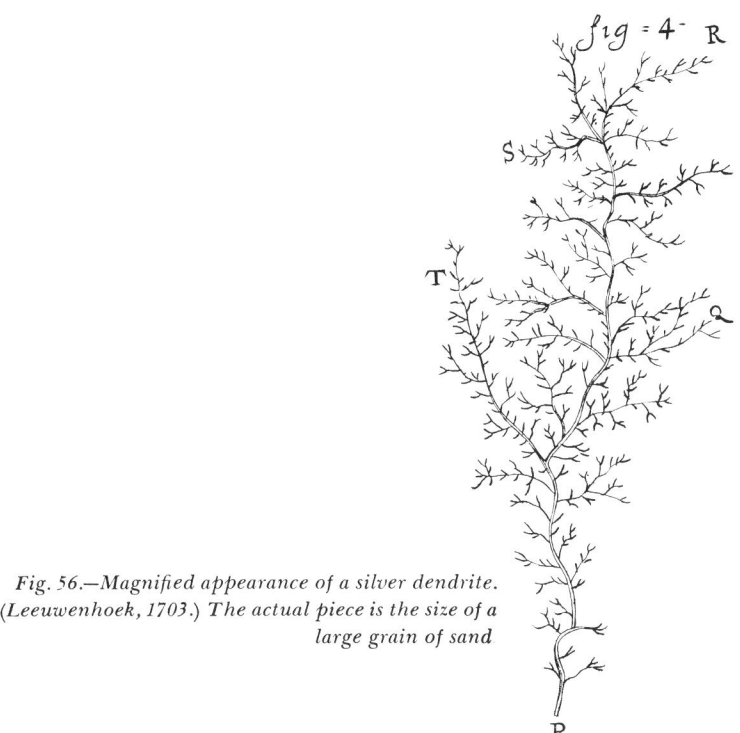

Fig. 56.—*Magnified appearance of a silver dendrite.*
(*Leeuwenhoek, 1703.*) *The actual piece is the size of a
large grain of sand*

of Coagulation I caus'd the painter to describe a small Particle of the Silver
no bigger than a large sand, vid. Fig. [56].

I was amazed to see in a few seconds of a minute, such coagulation of
branched particles (which even through a good microscope are invisible)
arising from a seeming clear Diaphanous and Liquid Matter.

Leeuwenhoek seems not to have recognized the crystallinity of his tree,
although this had been suggested a few years before by Homberg.[11]
The growth of the *Arbor Dianae* from mercury or silver amalgam
placed in silver nitrate solution is a very old chemical trick, which
aroused many speculations as to the vegetability of metals. Its electrolytic nature was recognized early in the nineteenth century. Homberg,
who was one of the first to study it scientifically concluded that it is
"not at all a true vegetation but is only a simple crystallization." He
believed that the tree was a congelation of mercury and silver together,
but that the mercury, being accompanied by nitrous needles in the
solvent, followed the direction of the latter in forming the dendritic
branches—an interesting, if incorrect, foreshadowing of the idea of
epitaxy.

By 1709, however, Leeuwenhoek was more confident of his ability
to see some structure in metals.[12]

One may indeed by the help of a good microscope, just discover the exceeding small Particles of Gold and Silver but one cannot perceive of what figure they are; and who can tell of what Multitude of parts these little particles, which we see by the help of a microscope, are again composed.

And again:

[In steel] we can only discover the broken gaps or Notches of a Razor, for instance, and the coarser the parts are, of which those metals are composed, as we may see in Cast Iron, the less valuable are the said metals; but the finer the particles are, the more valuable in my opinion will be the steel and iron which they compose.

Forged metals, he says, are

consolidated by strokes and pressure of the Smith's hammer that they seem to us to be but one body, tho' they do consist of a great many small particles, the coarsest of which are always obvious when we come to break the mettals: and how often soever you melt any of these mettals and break them again after they are cold, you will always be able to discover the grainy particles thereof; but you will find them so strongly joyned and riveted in one another that they appear to be one body.

Magnification was used by Lewis[13] in his studies of gold-platinum alloys. "On viewing the fracture with a magnifying glass," he says, "the gold and platina appeared unequally mixed and several small particles of the latter were seen distinct; nor was the mixture entirely uniform after it had been again and again returned to the fire and suffered many many hours of strong fusion."

In the eighteenth century microscopes became common, yet there were astonishingly few references to metallic structure, and, until Sorby, no consistent observation. Since they are mostly incidental to other studies, they will be referred to in later chapters.

10. The Study of Fracture

The character of the fracture of defective metals must have been noticed at the very beginning of the Bronze Age and an observant artisan would soon have related the appearance of specific fracture characteristics to variables in smelting or melting procedures. It seems highly unlikely that either the proper composition of a bronze for casting or the correct way of working a hearth to give a malleable iron or hardenable steel could have been found by anything other than a series of fracture tests, for the surface ordinarily seen is grossly insensitive to the factors that determine utility.

As in the case of etching, some use of fractured surfaces for decorative purposes preceded their scientific study. *Orekuchi-ishime,* one of the many beautiful irregularly textured surfaces given to Japanese sword guards in the seventeenth century and later, was achieved by repeatedly striking the surface with the fractured end of a steel bar used as a punch. A broken punch was sometimes used by English printers to give texture to type faces. A much earlier use, however, was in the striking of coins, best exemplified by the Persian *siglos.** This was a small silver coin, apparently made by melting a preadjusted mass of silver (about 5.4 gm.) on a flat or slightly grooved surface and allowing it to solidify under surface tension forces alone. The resulting sessile drop was then flattened and finally struck (perhaps hot) on an engraved die, using a rectangular punch, smaller than the coin, to drive the metal into the deepest-cut parts of the die. The deeply formed incuses are a fine record of the shape of the end of the punch, and show considerable detail, although this is usually distorted by wear of both punch and product. Figure 57 shows the reverse of a *siglos* of the fifth century B.C., the rectangular impression probably being that of a fractured cast bronze punch, with what seem to be traces of the columnar and granular

* The writer is indebted to Miss Dorothy H. Cox of Yale University for bringing these interesting coins to his attention. A metallurgically interesting discussion of early coining methods is given by Hill.[1]

Fig. 57.—Persian SIGLOS, *fifth
century B.C. Showing impression of
fractured bronze (?) punch,* ×6.5.

structure of the casting still visible despite some wear. A number of such coins are illustrated by Noe.[2] The impressions show that all the punches had irregular ends that were unmistakably shaped by the fracture of a brittle material. The punches were always convex and some of them may have been shaped by chipping or flaking with multiple hammer blows. Wear precludes precise metallographic conclusions. The use of a high-tin bronze (18–23 per cent tin) is suggested by other evidence, but it cannot be definitely excluded that the punches were of a textured crystalline stone, and occasionally the impressions suggest a broken poorly forged faggot of wrought iron and steel.

It is not until Biringuccio (1540) that there is a specific description of the use of fracture to control a metallurgical process. Biringuccio[3] remarks that tin

is known to be pure the more it shows its whiteness, or if when broken it shows itself granular like steel inside, or if when some thin part of it is bent or squeezed by the teeth it gives its natural crackling noise, like water when frozen by cold.

In making steel by immersing blooms of sponge iron in molten cast iron, the process is controlled by taking out a sample and forging it into a bar which is quenched as hot as possible in water and broken. The master then observes "whether every little part has changed its nature and it is entirely free inside from every layer of iron. . . . The kind that has a white very fine and fixed grain is the best." Similarly, in alloying tin with copper for casting various kinds of objects, the skilled workman recognizes the composition of bronze "by the whiteness and brittleness: changing from red which is the color of copper, it becomes white; from soft and flexible, it becomes hard and brittle like glass." Biringuccio instructs his readers to withdraw test samples as necessary before casting bells and guns. Ercker[4] describes the gray fracture of brass made with calamine from Goslar which was undoubtedly due to its lead content.

In 1627 Louis Savot[5] even more explicitly describes the use of the fracture test as a method of controlling bell metal:

Founders judge the quantity of tin which they should put [in bell metal] by breaking a piece of the material before they cast it to make a bell: because if they find the grain too large they put in more tin; if it is too fine [delié] they increase the copper. They put more tin to render the grain finer and, by the same means, the sound better. They augment the copper in order to render the material less subject to breaking, and little bells can carry more tin than big ones because they are not struck with such heavy blows. Bismuth is added to clock gongs to give them a better sound and a finer, smaller grain, but they are also very subject to breaking.

It was in the same year that Mathurin Jousse[6] defined in detail the selection of iron and steel for blacksmiths' work on the basis of its fracture. Jousse was one of the builders of the Jesuit College at La Flêche (famed as Descartes' school) and in his book describes both blacksmith and fine locksmith work as they pertain to building. He describes the fracture of iron that is suitable for various purposes in phrases such as the following:

Soft iron can be recognized by the color of its fracture. When this is black over the entire cross-section of the bar, it is a sure sign that you have good iron which can be formed cold and worked with the file; for the blacker the fracture, the softer will be the iron to the file and the more ductile.

Still another kind of iron has a mixed fracture—part of it being white and the rest black or gray—and a somewhat coarser grain than the iron previously described. This is often the better iron. It forges more easily, is not likely to have cindery spots, has no [imbedded hard] grains, and polishes more easily.

Another kind of iron has a coarse, bright grain in its fracture like bismuth or talc. This iron is practically worthless, for it is cold-short and weak in the fire, being unable to withstand great heat without burning.

Steel also is examined for its fracture, which, according to Jousse, should have a fine, distinct, small, white grain with no black veins or streaks of iron. A steel that was much valued in France was one imported from Germany or Carinthia under the name *Acier du Carme* or *à la rose* which was characterized by a colored spot in the center of the fracture:

Right in the center there is an almost black area verging on violet, with a very fine grain, and without any yellow spots or signs of iron, and if this area is found in the cross-section of almost the entire bar from one end to the other, these are sure indications that the steel is good.

Réaumur gave some good illustrations of this rose in 1722. It was correctly explained by Perret in 1771 as due to internal fracture on quenching and the subsequent formation of temper colors. Though its existence would guarantee a high carbon content and a uniform steel, it is indeed amazing that it was for so long regarded as a sign of quality for it would have been necessary in use to weld up the fracture.

Félibien[7] in his well-known *Principes de l'architecture* (1676) has a long section on the blacksmith's selection of his material, but he is quoting Jousse almost verbation without acknowledgment. Two paragraphs on the use of mineral coal are his only original comments on the blacksmith's craft. Joseph Moxon[8] is far more rewarding. When advising the blacksmith in his *Mechanick Exercises* (1677) he says,

Therefore when you chuse Iron, chuse such as bows oft'nest before it break, which is an Argument of Toughness; and see it break sound within, be grey of Colour like broken Lead, and free of such glistering Specks you see in broken *Antimony,* no flaws or divisions in it; for these are arguments that it is sound and well wrought at the Mill.

and later:

The Rule to know good Steel by. Break a little piece of the end of the Rod and observe how it breaks: for good Steel breaks short off, all Gray, like frost work Silver. But in the breaking of the bad, you will find some grains of Iron shining and doubling in the Steel.

It is so common today to think of the various varieties of cast iron as being classified by their fracture that it is rather surprising to find that the distinction was first recorded only in 1665, by Dud Dudley.[9] In his unsubstantiated claim to have successfully smelted iron with "sea-cole or pit-cole" he remarks that there be three sorts of Cast Iron:

1. The first sort is Gray Iron.
2. The second sort is Motley Iron, of which one part of the Sowes or **Piggs** is gray, the other part is white intermixt.
3. The third sort is called white Iron, this is almost as white as Bell-Mettle, but in the Furnace is least fined, and the most Terrestrial; of the three, the Motley Iron is somewhat more fined, but the Gray Iron, is most fined, and more sufficient to make Bar-Iron with, and tough Iron to make Ordnance . . . being . . . more malliable and tough then [*sic*] the other two sorts before mentioned.

Although all of these material-selection tests based on fracture presuppose the existence of some kind of a structure and its relation to serviceability, they do not by themselves contribute much to the understanding of the nature of the component parts of a metal. It was Robert Boyle in his *Essay about the Origine and Virtues of Gems* (1672) who first recorded experiments on fracture done with a scientific motivation.[10] A rational corpuscularian, he speculated on how the corpuscles of the body may be so shaped that on their coalition they would form "glossy planes" and that, however distorted they may be, if kept in a state of fluidity they may better rearrange themselves into smooth and shiny planes.

The situation of those planes in reference to one another will be more uniform and regular than almost anyone would expect in a concretion so hastily made; notwithstanding which their internal contexture will be much diversified by circumstances, as particularly the figure of the vessel or mold wherein fluid matter concretes.

He had a remarkably clear idea of the ordering of the parts that occurs during solidification and the orientating effects of temperature gradients. Had he been a little more specific on the abutment of adjacent crystals of differing orientations, he would have prevented the centuries-long confusion of a grain with some kind of an enveloped entity like a grain of wheat. He clearly distinguishes between the random orientation of a slowly cooled ingot and the directional columnar grains in a chilled one and recognizes the origins of these structures in the manner in which parts congeal on the advancing solid surface. Bismuth, when fractured,

will discover a great many smooth and bright planes, larger or lesser according to the bigness of the lump: which sometimes meet and sometimes **cross** one another at different angles: Considering this, I say, I thought it **probable** that a body that had already melted, and was apt to convene in such **places,** not only would do so on another fusion but might have the order and **bigness** of these planes diversified by the figure and capacity of the vessel I should think fit for my purpose. Wherefore having beaten a sufficient quantity of it [bismuth] to powder, and when it was well melted, cast it into a good pair of iron moulds, whose cavity was an inch in diameter, we had a bullet, which being warily broken, did as we expected, seemed to be, as it were, made up of a multitude of little shining planes, so shaped and placed that they seemed orderly to decrease more and more as they were further and further removed from the superficies of the globe; and they were so ranked, that they seemed to consist of a multitude of these rows of planes reaching every way, almost like so many radiuses of a sphere from the center or middle part; whereas if we melt tinglass in a crucible, and let it cool there, the matter being taken out and broken, will appear indeed full of smooth planes, but (as was lately intimated) very irregularly and confusedly associated or placed.

Boyle goes on to explain this as being due to the fact that

the coagulation being thus begun, the parts of the remaining fluid, as they happen to pass by this already cooled matter, with a motion which . . . was now slackened, they were easily fastened on the already stable parts . . . and the refrigeration still reaching further inwards, till it came last of all to the middle of the globe, that being the remotest part from the refrigerating agents; and the apposition was successively and orderly made till the whole matter was concreted.

Similar structures were observed in cast antimony and other bodies but these did not show as clear a structure as bismuth.

Réaumur, whose interest in fracture was aroused by his studies of the changes accompanying the conversion of iron into steel, published in 1724 some remarkable studies of cast metal fracture,[11] which will be reviewed at some length, for it illustrates so well the conceptual difficulties associated with the problem. He saw that the configuration of the elementary parts or "molecules" could not be deduced from the fracture because successive ingots of antimony (i.e., antimony sulphide) had different arrangement of the needles characterizing their fracture. He properly related this to the manner of cooling, supposing that as the fiery particles which separated and agitated the molecules were removed they allowed the molecules to settle in a certain order. The molecules that would settle first would be those that were nearest to the ones already fixed, and they would fall successively end to end to give rise to fibers, the direction of which would show the manner in which cooling had taken place. By varying the position and manner of cooling he obtained the structures shown in Figure 58. Note particularly No. *6*, where the direction of solidification was reversed by putting a cold poker into the middle of the melt.

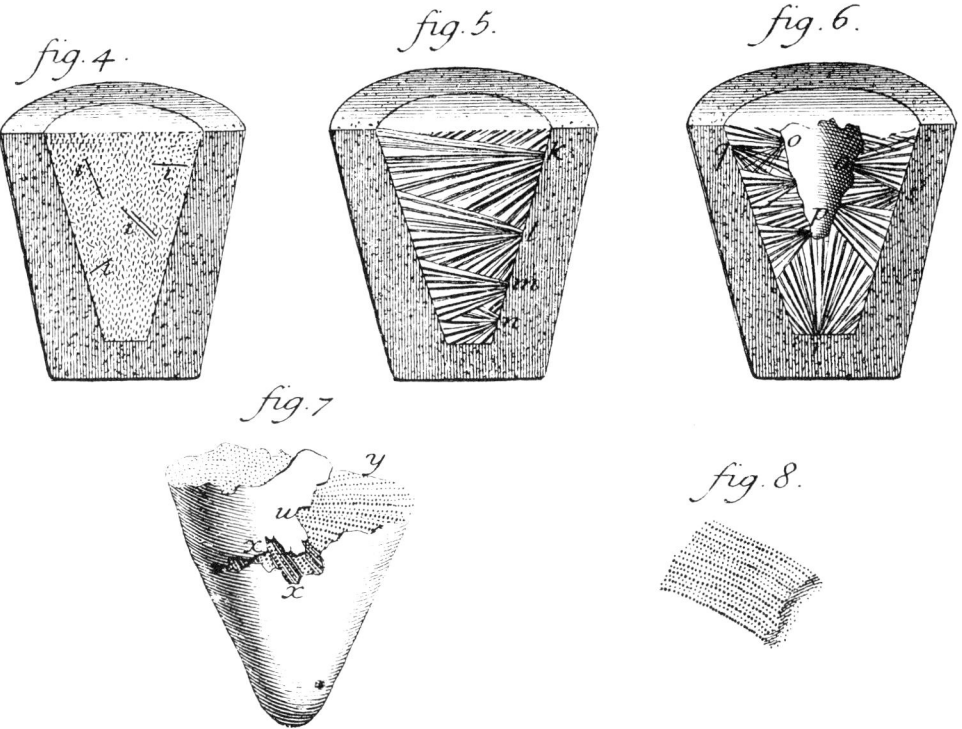

Fig. 58.—Ingots of antimony solidified under various conditions and subsequently broken; also a lead ingot broken near the melting point, showing intergranular fracture. (Réaumur, 1724.)

Rightly believing that a temperature gradient was necessary, Réaumur misinterpreted the structureless fracture of a single crystal obtained by extremely slow cooling as being due to the molecules remaining fixed in the places where the fire had left them, lacking the ordering tendency of previously solidified surface. He remarks that physicists had attempted to cool metals extremely slowly in order to show the arrangement of their parts but had been unsuccessful because of the deformation that occurs when the samples are broken and consequently are not disposed as they originally were but are in a quite different state. He brilliantly gets around this difficulty in the case of copper, zinc and, best of all, in lead, by using an old trick used by gun-founders to break up large cannon into handleable pieces—by heating them to the point where they become brittle. When the metal is hot,

the molecules are too much driven away from each other. They hold poorly together, and can be entirely separated by the first somewhat sharp blow. . . . They are then like brittle bodies; their fractures can show to us under these circumstances the arrangement of their interior parts.

(This effect is actually due to the beginning of melting at the grain boundaries, and is greatly facilitated by the presence of impurities to give a low-melting constituent.) A piece of lead so broken looked like a piece of quenched steel, but, slowly cooled, the same piece showed no grains whatever when broken.

By casting lead into the crucible and, when just solid, removing it and striking it with a hammer, Réaumur was able to show the same general appearance of fracture that he had seen in antimony, with packets of fibers perpendicular to the side, Figure 58, No. 7. However, there was a very important difference between the fibers in lead and the needles of antimony, for the latter were brilliant and shiny like little mirrors adjusted side by side, whereas the fibers of lead were less brilliant, not flat but visibly rounded. Unfortunately the figure referred to in the memoir was not reproduced, but the grains are described as appearing to the naked eye or with a low-power lens as a row of little balls, arranged like grains of a string of beads. A higher-power lens or a microscope does not show in each one of the groups of fibers the very rounded figures but always the fiber appears to be formed of grains applied one to the other by only part of their terminations. While the sides of the needles of antimony are straight, those of the fibers of lead are dentelated. Réaumur realized that the shapes of lead and antimony fractures were different, though he failed to see that in antimony the fracture was principally through the grains, whereas in lead, fracture near its melting point was intergranular with details due to dendritic solidification. For the first time there was a specific statement of the connection between grain shape and properties: "Perhaps it will be found that it is on this shape of the grains and their arrangement that the ductility of metals and of some other materials depends."

Réaumur believed that all soft bodies, including soft waxes and fats, probably had a similar arrangement of fibers, but it will not be perceived as they are not brittle. However, all brittle bodies do not necessarily show anything on their fracture. As salts cannot form crystals if they are cooled too quickly or agitated, so cast bodies do not take a regular arrangement if they are cooled quickly or agitated during cooling. Another cause may interfere with this arrangement or perhaps completely prevent it, that is, when the melted body is not a uniform fluid but is composed of parts which have the disposition to separate from the others and which cannot bear the same degree of heat. The formation of fibers, plates, or needles is the result of successive cooling or, more exactly, of the parts becoming rigid only in sequence. If some elongated parts want to solidify before others which are nearer have lost their fluidity, there is no reason for them to form a straight row continuous with the others.

It is obvious from all of this that Réaumur had a fairly good idea of the nature of solidification in both pure metals and alloys. He relates the process to crystallization of salt. Although his crystals are directional in their growth, the use of the term needle, molecule, or grain, does not seem to necessitate internal order and there is no discussion whatever of the grain boundary or nature of the division between one grain and another. The concept that they differed *only* in orientation was a century away.

Réaumur is better known for his work on iron and steel, published in his famous *Memoirs* of 1722.[12] This is an almost unmatched example of combination of observation, planned experiment, and theory in the best tradition of the applied scientist. At the moment we are particularly concerned with his concepts of the structure of iron and steel. Not satisfied with mere selection of iron for carburizing on the basis of an arbitrarily defined fracture, and with stopping cementation when the desired steel fractures are produced, he speculates on the nature of the underlying invisible structure, and comes remarkably close to modern ideas except for the one all-important realization of orientation difference between adjacent grains. As a Cartesian in spirit, he avoided the concept of indivisible atoms but he clearly realized that "the arrangement and shape of the visible parts are usually the effect of certain natural tendencies and qualities of the invisible parts." He understood that, behind the grain which was visible in the fracture, there was sometimes an aggregation of smaller parts, and beneath that still another or perhaps others. It is by the changes of the clustering of these particles on various scales and of the diffusion in and between the clusters of a material called "sulphurs and salts" (i.e., modern carbon) that he explains all of the properties of iron and steel. He discusses structures in the following words:

When I speak of the structure of wrought irons, I mean the shape, the size, and the arrangement of their molecules; and it is by means of their fracture, of the surface at the location where they have been broken, that one can judge how these molecules differ. When different kinds of stone are broken, the fractures show the difference in the grains of each one. When different kinds of wood are broken, fibers of different size, which are sometimes differently arranged, are seen in their fractures. When bars made of different irons are broken, such obvious diversity will be noticed in their fractures that to the unaided eye the fractures of these bars sometimes seem to differ more among themselves than they differ from other metals such as lead, tin, and silver. There is not only as much and more difference in color, but the difference is even greater in the shape and the arrangement of the parts. By paying only casual attention to the fractures of these different kinds of iron, it is seen at once that they can be divided into two groups: into irons with only grains or platelets and irons with fibers in their fracture. The fracture of irons of the first group resembles either the fracture of stones or that of bismuth; and the fracture of irons of the second group resembles that of wood. (The artisans say that the latter are fibrous.) But this classification is still too general. The detailed study we intend to make requires finer subdivisions.

He then proceeds to list and illustrate seven classes of iron, the first having a fracture which

shows very brilliant white platelets, which appear to be as many little mirrors but of irregular shape and arrangement; they rather resemble those in the fracture of bismuth.

Some of the plates are as much as two-twelfths of an inch in diameter, and there is often space between them occupied by much smaller parts, which Réaumur termed grains. The next three classes are composed of increasing areas of finer and finer grained iron, the color progressively darkening. These grains are compared with those in steel broken in the hot short range (where a true intergranular fracture would occur), but they are somewhat less rounded. Finally, the very softest irons and the best for making steel are composed of fibers alone. Typical illustrations of some of these fractured irons are given in Réaumur's plates, one of which is reproduced in Figure 59. This includes some smaller sketches made under the microscope, seemingly at a magnification of about 4 and not very helpful.

Réaumur specifically assumes that the structure of the iron is that shown by the fracture, i.e., that an iron showing platelets is itself composed of platelets, and similarly a granular or fibrous fracturing iron is composed of differently shaped parts, to be called grains or fibers. Although he realizes that a cutting tool may sometimes remove clumps of particles, he does not realize (neither did any other scientist for over a century) that the manner of fracture was of overwhelming importance and that the distinction between a fibrous and a platelet-filled fracture could be no more than the manner in which a crack crosses the grains. Observing that in the process of carburization an iron with a fibrous fracture developed platelets, he remarks that the "sulphurs and salts" have to cut the fibers, and the most fibrous irons require the longest carburization time because "the fire has more work to do: before it can arrange the platelets it has to break up the fibers and reassemble the

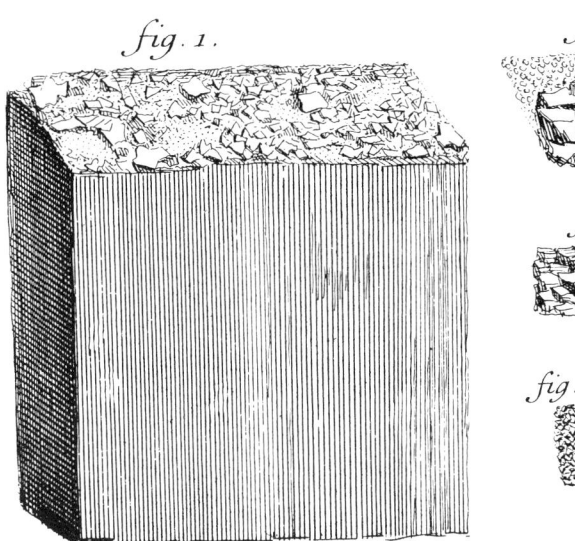

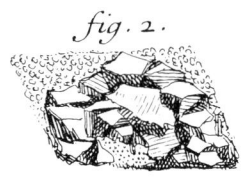

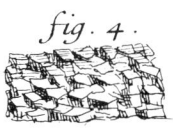

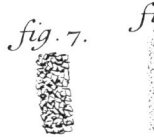

Fig. 59.—The appearance
of the fracture of a
wrought-iron bar. 1 is
natural size; 2 and 4 show
the appearance under a
magnifying glass.
(Réaumur, 1722.)

pieces." It seems that "the fire has less work to do in combining grains
or slender platelets together in order to make new platelets than in
breaking up fibers and changing them to similar platelets." The fracture of steel, he observed, is duller than that of iron and the platelets
become smaller and smaller until they appear to be only grains. This he
interprets as being due to the fact that "the shape and arrangement of
the parts of iron are changed in proportion as our sulphurous and
saline substances penetrate it."

Regardless of whether the iron had fibers or grain, everything is re-
arranged by cementation, so that the fracture will have platelets which
are duller than those in the fracture of bad irons and arranged more
uniformly. The weight and volume of the iron increase and consequently the parts have moved farther from each other, being separated
and subdivided by the substances that have diffused in. The longer the
process is continued the more the parts of the metal are subdivided. He
uses the curious analogy of a tangled or matted mass of fibers, and concludes that steel is better "carded," by which he apparently means more
thoroughly subdivided, rather than more ordered in its arrangement.

When testing steel after it has been made, Réaumur not only describes mechanical tests for body, but also reduces grain size determination to a quantitative basis. This he did by comparison with a set of
standards which permit a study and measurement of the variation of
fracture grain size with both temperature of heat treatment and the
nature of the steel. These tests were later resurrected and are known today under the names of Metcalf and Shepard. Réaumur clearly realized

that different steels had different inherent grain sizes, and that the size of grain would increase with increasing temperature above the point at which any change occurred at all. Figure 60 shows the entire range of fracture of a bar of steel, the upper end of which was heated to a high temperature while the lower end was cold, the whole being quenched in water. The area marked *4* is of unchanged steel terminated by an abrupt line, *3*, which we now know to correspond to the critical point. Between this and *2* the grain is too small to be resolved with a magnifying glass (though Réaumur supposed that it would show platelets at a higher magnification), while beyond *2* the platelets appear and grow in size to the hottest end of the bar. An extremely significant experiment!

In considering the reason for the hardening of steel Réaumur introduces his famous drawing of the microstructure* (Fig. 61). This is said to be a magnified view of the grain visible to the naked eye at the spot *G* (which is 1.17 mm. dia., so that the enlargement is about 40), while the adjacent little cluster, *pp*, is a still-further-enlarged view of one of the grains *M*. There is a hierarchy of parts, without regularity of shape or arrangement and with voids in between them. To quote:

This grain [Fig. 61 *G*] which is easily seen by the eye, is itself an accumulation of an infinite number of other grains, which we shall call the "molecules" of this grain (*M*, *M*). The microscope brings these molecules into the field of our vision. But these molecules are themselves composed of other parts (*p*, *p*). It is possible, if it seems desirable, to suppose that the latter are the elementary parts, although in reality we may have to continue the division vastly much farther before we reach them; however, we can stop here. We thus have to consider a grain, the molecules of which it is composed, and the elementary parts of the molecules. As the salts and sulfurs intimately penetrate the iron, we can at least assume that those by which steel outnumbers iron penetrate the molecules of the grain. If I expose to the fire a soft steel containing the grain on which we have concentrated our attention, the fire will melt the sulfurs and the salts of the molecules of this grain before it melts the molecules themselves. It will drive part of the sulfurs and salts by which steel outnumbers iron out of the molecules in which they were wedged. Whereas, be-

---

* This figure has a remarkable similarity in appearance to the microstructure of a polished and etched medium-carbon steel in the normalized condition, the grains *M* being pearlite and the spaces *V* ferrite (or vice versa). It has indeed been so interpreted by some readers of Réaumur (for example, More, *Trans. AIME*, 1936, **120**, 14, who also misconstrued Réaumur's device for hot-straightening iron as a creep test), but there can be little doubt that, though inspired by observations of fracture, the drawing is mainly symbolic to indicate the hierarchy of aggregations in many stages down to the elementary parts. He did not think of the parts themselves in geometric order and therefore could not see that the difference between the various stages of aggregation (which today we know as atom, molecule, subgrain, and crystal grain) depends quite as much upon the geometry of order and orientation as it does on scale.

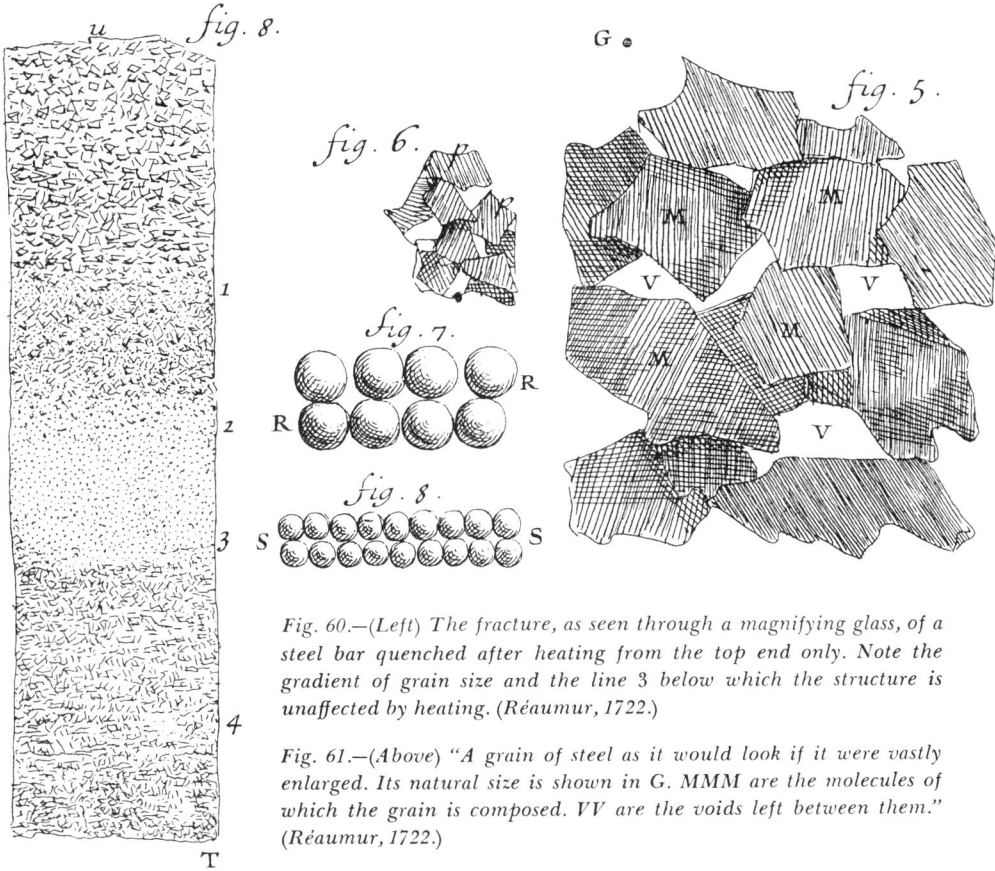

Fig. 60.—(Left) *The fracture, as seen through a magnifying glass, of a steel bar quenched after heating from the top end only. Note the gradient of grain size and the line 3 below which the structure is unaffected by heating.* (Réaumur, 1722.)

Fig. 61.—(Above) "*A grain of steel as it would look if it were vastly enlarged. Its natural size is shown in G. MMM are the molecules of which the grain is composed. VV are the voids left between them.*" (Réaumur, 1722.)

fore this, they had penetrated these molecules, they will now, as a first step, go into the gaps between them. This will be all the effect caused by a moderate fire. . . . Let us therefore not hesitate to admit, that, when our grain has reached a certain degree of heat, the empty spaces between the molecules of which it is composed will be partly filled by a sulfurous matter which was not there previously and of which the molecules have been deprived; that part of this sulfurous matter, which the fire has started on its way out of the iron, has passed from the molecules themselves to the intervals left between them. In this state let us plunge the bar of steel with the grain we are studying into cold water. We shall instantaneously fix the sulfurs and the salts which float around together. We shall deprive them of their fluidity; they will no longer be in condition to re-enter the molecules. However, the small intervals between these molecules of the grain will now be more completely filled, and filled by a substance which we can suppose to be almost as hard as we want it to be. The molecules of the grain will therefore be more firmly bound to each other. For this reason our grain of steel will be more difficult to divide or to break; in other words, our grain of steel has now become harder. The same thing has happened to all the other grains of steel that had acquired the same degree of heat. Consequently, our steel has now been hardened; or, to be more precise and in order to keep in mind what we actually wish to explain, we should say all the grains of our bar of steel have now become hardened.

This is a remarkably acute structural explanation of the hardening of steel, although, in common with Rohault, Perrault, and all of the Cartesian philosophers who clearly influenced Réaumur, it is in too simple mechanical terms. Nevertheless, when today we say that carbon atoms diffuse into the interstices between the atoms in the austenitic iron crystal, and is retained approximately in the same relation on rapid cooling, while slow cooling allows it to diffuse out of the iron and collect in interferritic spaces as cementite, this picture though inverted is of precisely the same kind. Réaumur had combined the vague Cartesian picture of the universe with careful real observation of the structure of metals as shown by their fractures and produced a model which was an extremely satisfactory guide for subsequent experiment; not until the microscope permitted the various segregates and solutions to be actually seen could it be improved upon.

Réaumur also used the fracture test to select cast iron for various purposes, such as for making castings for subsequent malleablizing, and particularly for studying the process of malleablization, of which he was the originator. He describes how, during the annealing of a white cast iron, the initial fracture—compact and free from grains, though with platelets—is replaced around the edge with a band of grains which slowly press inward. At the same time the color of the cast iron in the middle of the bar becomes darker as the material is softened, a fact which he explains as due to the evaporation of sulphurs and salts leaving behind cavities in a spongy mass of iron grains. Prolonged annealing recompacts the iron because the particles adhere to each other, since they are in a state approaching that of fusion.

He has a fine section on the fracture of cast iron in which, after observing the appearance of closely compacted platelets radiating from the center in a white cast iron (distinguished from wrought iron in which the platelets are randomly arranged, with spaces between them) and showing both mottled and gray cast iron, he proceeds to examine them under the microscope, commenting as follows:

If both classes of cast iron are examined under the microscope, the very white irons will always show a compact texture. There may be some interspersed flat platelets, which are, however, much smaller than those of steel. They cannot be seen through the same lens that brings out those of which the grains of a steel are composed which has been quenched at a low degree of heat.

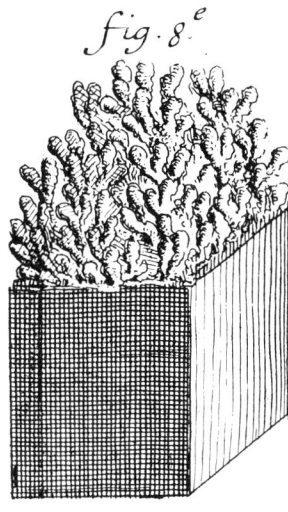

Fig. 62.—*Fracture of a gray cast iron, as seen under the microscope. "It seems to be composed of an infinite number of small branches, . . . each of which is made entirely of small plates placed one upon another."* (Réaumur, 1722.)

Gray irons when viewed under the microscope seem to have such a spongy texture that everything seems to be a mass of some kind of crystallization or, if you wish, a thicket, some sort of chemical vegetation made up of an infinite number of interlaced branches but each composed of little platelets arranged one on top of another [Fig. 62]. If one places under the focus of the microscope grains of both kinds [of iron] as small as the grains of an extremely fine sand, they appear more transparent than the most crystalline sand. [Réaumur here confuses specular with internal reflection.] Their transparence, and especially the vividness of their color, resembles more closely the transparence and brilliance of the diamond. In spite of the vividness of the color which the grains of the different cast irons thus have, the color of the gray irons can be distinguished from that of the white ones. The gray ones are more like polished steel and the white ones like polished silver. [Eng. trans., p. 262.]

This description of cast iron as composed of some kind of chemical vegetation is a delightful and precise description of a dendrite such as can be separated from the sink head of a casting. Yet Réaumur's illustration like the *Arbor Dianae* of Leeuwenhoek lacks the regularity of a real dendrite or crystal, which was left for Grignon to delineate accurately fifty years later.

Réaumur studied the white fracture of rapidly cooled cast iron and the relation of fracture to thickness of section, an effect which he first thought due to the escape of vitreous material to the surface, but later, in a paper[13] written before 1726, although not published until 1762, he correctly explained as due to cooling rate: "White cast iron is simply cast iron that is quenched."

It will be obvious in reading these excerpts that Réaumur was able to deduce much information from an examination of simple fractures and that, despite his missing the central point of crystalline order, he nevertheless was able to construct a model of the structure of matter in which features of different scales were interrelated with each other and this he used very astutely, not only to plan experiments of philosophic value, but also to lead him rapidly to processes of industrial importance. He was ahead of his time, however, both in knowledge and approach, and although the fracture test continued to be used, there was little if any advance of scientific understanding of fractures until near the end of the nineteenth century.

FRACTURES OF NON-FERROUS ALLOYS

The relationship of fracture type to the tin content of bronzes was long ago recognized (see pp. 98–99). As soon as corpuscular philosophy began to develop early in the seventeenth century, it was realized that the "parts" of one metal could mix with and partially interpenetrate those of another. The concept gave rise to Glauber's original work[14] on densities of alloys, followed by a series of investigations by the Royal Society and others. It was shown that the parts packed somewhat differently when mixed together than with their own kind, hence that the Archimedean assumption (although it was not an assumption made by Archimedes himself) that the volume of an alloy is exactly equal to the volume of its component metals is not correct. This is a point of considerable theoretical interest. The early experimental methods were incapable of resolving it satisfactorily since the density of a casting is likely to depend more on shrinkage cavities and blowholes than on the ideal density of the crystals, which only recently has become measurable as a result of X-ray lattice parameter measurements.

Geoffroy, in a paper on the alloys of copper and zinc presented in 1725,[15] uses the nature of the fracture of the material as an important part of his study. He comments on the striae visible under the microscope in a certain commercial alloy and on the variegated yellow and white striae visible in another. On alloys that he himself made up, he found that equal portions of brass and zinc produced a brilliant metal breaking like glass. Two parts of brass to three of zinc are similar, but of a duller grain, and at this composition the brilliant mirror in the fracture ceases. Augmenting the zinc beyond this, the metal, though still brittle and sonorous, becomes dull and bit by bit develops the grains with the little facets which are peculiar to zinc.

Three parts of brass with two ounces of zinc produce a white, sonorous, brittle metal, mirror-bright on its broken surfaces and with a grain with some tendency toward striations. On then augmenting the weight of brass, ounce by ounce, one sees the birth of metallic striae. . . . If one continues in other tests to augment the amount of brass, it developed a more enhanced yellow color, then becomes variegated with yellow and white with an appreciable grain.

With more brass, the grain becomes larger, golden, though a little dull and with still more becomes ashy brown with no fibers and the metal is ductile. Geoffroy's descriptions leave little doubt that he is referring to the gamma, beta, and alpha phases of the modern metallurgist and he relates them definitely to composition and to serviceability. The fracture of brass changes rapidly as the zinc content passes the point where beta is the principal constituent on solidification, a fact that was commonly used in the nineteenth century for checking the composition of muntz metal. A fracture test on a 60–40 composition was in use even as late as 1923 for testing spelter for antimony content. The nineteenth-century studies of Kalisher, Calvert and Johnson, and Thurston, though important in determining the range of composition of useful alloys and reporting the fractures of large numbers of brasses and bronzes, added relatively little to structural understanding. Thurston indeed retrogresses in thinking that the pentagonal facets on an intercrystalline fracture indicate a dodecahedral form of a cubic crystal.

The typographer Fournier,[16] writing in 1764 on the preparation of type metal, deduced the fact of segregation in type metal from its fracture:

The broken surfaces show a small sparkling grain, and the harder the metal [i.e., the greater the proportion of antimony in it] the more uneven does the grain appear, for the reason that the regulus being lighter than the lead will rise toward the top before the ingot sets hard, but it mixes evenly again in the metal pot, where it is melted for casting.

He then went on to mention that the alloy, being more porous, melted more easily than either of its components. Fournier illustrates the crystalline pattern on the top of a cast lump of regulus of antimony, but he does not depict the fracture of type metal showing the segregation.

The German physicist and chemist, Karl Franz Achard,[17] in carrying out his monumental studies on the properties of alloys in 1788 realized the importance of the appearance of the fracture, and meticulously notes the appearance of the broken surfaces of nearly every one of his

896 alloys, which represent virtually every possible combination of all the metals then known. He makes comments such as:

The fracture was gray, reddish without being brilliant with little very serrated grains. . . . The fracture of the upper part of the cylinder was white, not very brilliant and in grains and little leaflets with a very serrated texture; the lower portion was crystallized in small very brilliant leaflets. . . . The fracture was gray, crystallized in brilliant plates, irregularly disposed. . . . The fracture was reddish-gray in brilliant grains intermixed with little platelets, the platelets being more abundant and more red in the lower part of the cylinder than the upper. . . . The fracture was gray without brilliance in large grains intermixed with little platelike crystals.

Some of his descriptions undoubtedly refer to a two-phase crystalline structure, but his book is only a record of a great mass of experimental data, and there is no attempt to interpret it. This is unfortunate, for Achard was a scientist of the first rank, responsible *inter alia* for the observation of the relationship between conductivity for heat and for electricity. His interest, however, became diverted into the development of the German beet-sugar industry, and he did not continue work in metals or physics. His book on alloys passed almost entirely without comment by contemporaries and it had no influence on the development of metallurgy.

NINETEENTH-CENTURY WORK ON IRON
AND STEEL FRACTURE

When engineers began to use the tensile test on materials with a view to determining their proper employment in structures, it became common to make notes on the appearance of the fracture and there must be many thousands of qualitative descriptions. Though these were empirically related to quality and utility, the engineer's purpose did not require any further explanation of them. An important exception to this was Mallet,[18] who published a paper in 1855 relating the fracture of guns to the manner of solidification, and clearly pointing out the dangers of salient and re-entrant angles in producing

an equally sudden change in the arrangement of the crystals of the metal, and that every such change is accompanied with one or more planes of weakness in the mass [Fig. 63].

From these experiments he concluded that

when the particles consolidated under the influence of heat in motion, their crystals arrange and group themselves with their principal axes in lines perpendicular to the cooling or heating surfaces of the solid; that is, in the lines

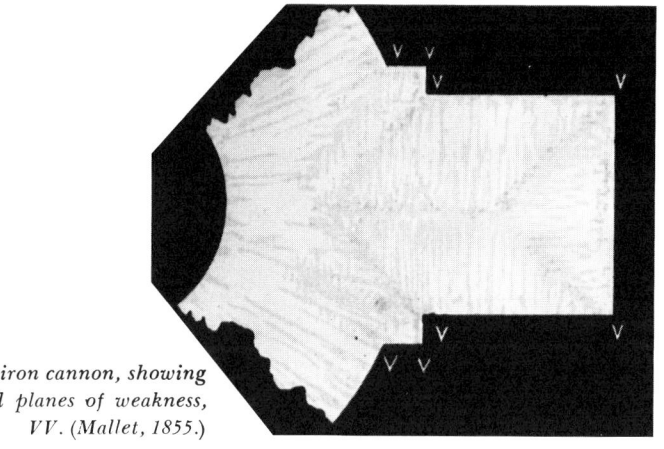

Fig. 63.—*Fracture of a cast-iron cannon, showing direction of crystallization and planes of weakness, VV. (Mallet, 1855.)*

of direction of the heat wave in motion, which is the direction of least pressure within the mass.

He later applied the same principle to the change in fracture of wrought iron upon forging, for he confused crystallization with fiber due to slag stringers. His work is nevertheless of value as an early attempt at a scientific study of the problems of ordnance manufacture and was widely read.

The *Philosophical Magazine* for 1839 and 1840 contains a long paper by C. Schafhaeutl[19] on the different species of iron, which contains some tantalizing observations on fracture, which he had obviously studied with great care. He believed that the visible molecules (i.e., grains) of all kinds of iron were composed of elementary parts, belonging to the cubic system, of approximately equal size, never exceeding 0.0000633 of an inch(!); their form appears to be always the same, and their different modes of aggregation in the various sorts of iron is all that can mechanically form the difference between cast and malleable iron, as far as ocular inspection, assisted by magnifying glasses, can penetrate. Dark gray iron has the molecules aggregated in the most perfect crystalline form, with all of their faces in one plane; forged, annealed, and cast steels have a tendency to nodular aggregation; hardened steel has the molecules equally distributed through the mass, the facets never appearing in one plane so that the fracture appears under the microscope "to be composed of innumerable bright points . . . the whole having a somewhat similar appearance to the full moon when first viewed through a common telescope." The difference between hard and soft steel he attributed to the effect of heat increasing the distance between the molecules, and destroying the stability of the equilibrium of their crystalline positions. Much of Schafhaeutl's paper is concerned with the solution in acid of cast iron, though he was unable to observe flakes of graphite in them; and in general his imagination was fertile but unrestrained and his observations erroneous or trivial.

From the earliest times the different appearance of the fracture of cast iron has been related to its serviceability in a rather empirical way. Had cast iron not been such a highly complex substance chemically, it could have led the way to metallography quite directly.

The doctoral dissertation of E. F. Dürre in 1868[20] contains an extensive description of the many different textures and details to be seen in the fracture of cast irons and a good summary of the literature of the time. Dürre was an advocate of the use of a good lens to study the fracture of castings, but he believed the use of the microscope with high magnification to be impractical. Although it is hard to relate his findings definitely to modern crystallographic structure, he describes clearly both dendrites and cubic crystals growing in shrinkage cavities, transcrystalline and intergranular fractures, and various forms of crystals of both carbide and graphite. Although he discusses at length the luster of cast iron and the appearance of graphite on polished sections, he does not etch. Nevertheless, Dürre's work did serve to set the tone of structural work by engineers in Germany, and influenced both Schott and Martens (see chap. 15).

There are two papers on steel by the Russian metallurgist Tschernoff which are remarkable for showing the amount of information that can be obtained on structure and structural changes without the aid of modern metallographic methods. The second of these deals with crystallization of steel ingots and is discussed in chapter 11. The first, published in 1868,[21] discussed fracture grain size in relation to heat treatment and carbon content. Starting with a purely practical observation during the forging of guns, he deduced both the existence of a critical point below which steel would not harden and the existence of another temperature, varying with carbon content, which marked the onset of the formation of large grains. This critical-point observation was a direct result of his noticing a zone of a dull texture on the machined surface of a gun forging in a lathe. "The ribbon of dead tint was so clearly marked that it was easy to trace its boundaries with a pencil." It was sharp on the low-temperature side and faded off in the zones that had been heated hotter (Fig. 64). This feature he found corresponded

Fig. 64.—*The line of metal having a different surface texture after machining, observed by Tschernoff (1868) on a gun forging heated at only one end.*

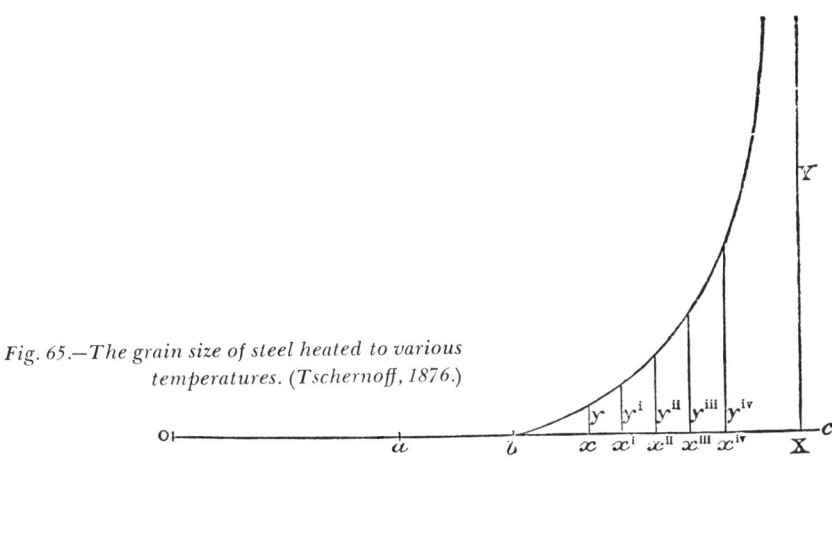

Fig. 65.—*The grain size of steel heated to various temperatures. (Tschernoff, 1876.)*

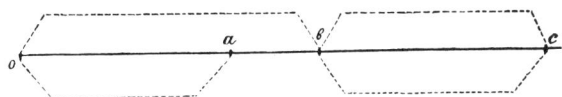

Fig. 66.—*Tschernoff's diagram showing the critical points in steel (1868). Line ob is designated as "No change of structure"; line oa, "Will not harden"; bc (upper), "Rising from b to c, amorphous structure"; and bc (lower), "Falling from c to b, crystallisation."*

with the point up to which the large ingots had been pushed in the re-heating furnace! This is an extraordinarily interesting example of the kind of practical observation that in an astute mind will lead to important theoretical advances.

On the basis of his observation of fractures, Tschernoff drew the parabolic curve of Figure 65 to show the increase of grain size with temperature. He thought (following Jullien) that steel in the high-temperature form was amorphous and that crystallization occurred during cooling in a manner analogous to alum melting and crystallizing in its own water of crystallization, and he drew a diagram that was later to become famous (Fig. 66). Carbon, he supposed, began to dissolve iron above the critical point *b*. On cooling, crystallization occurred, the larger crystals growing only if the steel were allowed to cool without disturbance; otherwise a multitudinous mass of small crystals would form. To obtain fine grains in forging it was necessary to work quickly, leaving no spot untouched with the hammer because "the heated piece of steel must be considered in an analogous condition to a saturated solution of a strongly crystallizing salt which, the moment it is allowed to cool quietly, develops strong crystals."

Tschernoff studied the fracture of a piece of cast steel under the microscope and found that there were considerable amounts of interstitial matter between the groups of grains. He noticed that when a cast ingot of steel not heated above a critical point was forged, the crystals or grains would be pressed together and become elongated, which did not happen above the critical temperature.

The fracture of such steel has a silky lustre, and under the microscope it is very difficult to trace the limits of the individual grains; they present the appearance of whole groups of waxy little balls squeezed together under a powerful press. If you cut off and polish the surface of a piece of steel so treated, and then immerse it in weak sulphuric acid, after a time a pattern will form on the surface, which presents the appearance of an irregular interlacing of crooked lines, the size of the network depending on the original size of the crystals, the manner of forging, and so on. I have already stated that the tendency to crystallisation, as well as the form of the crystals and their relative positions, depends on the purity of the steel, and the conditions under which the cast ingots are poured and cooled. In the higher qualities of 'boulat,' the tracery developed by acid is of remarkable beauty and regularity.

The cause of the patterns appearing is the various groupings of the crystals during their formation. These crystals have not the same chemical composition; the lighter parts of the tracery contain much more carbon* than the darker parts—a fact which I have demonstrated—and consequently, simultaneously with the grouping of the crystals or grains, there is a segregation of like chemical compounds. If you heat the piece of steel thus marked (damascened) up to the temperature *b*, or a little higher, and allow it to cool again, you will no longer be able to obtain any pattern by the action of acids. From what has been already said, the cause of this must now be quite plain, and I need not dwell on it any longer.

In his later paper[22] Tschernoff illustrates the fracture of large-grained steel, and for the first time in history gives good illustrations of the real

* In the original Russian (1868) the origin of the tracery is attributed to differences of crystalline compactness instead of segregation of carbon content. The change supposedly occurred during Tschernoff's correspondence with Anderson in 1875. The only other significant difference between the two versions is the addition of the grain size curve and its discussion, pp. 293–94 of the English version of 1880. See note, p. 140.

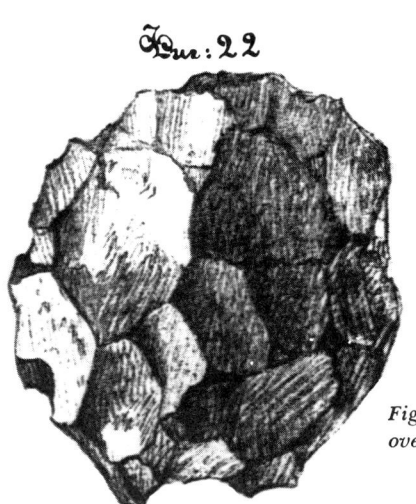

Fig. 67.—A single grain of steel removed from an overheated bar. (Tschernoff, 1878.)

shape of metal grains (Fig. 67). He remarks: "It is evident that these grains have only a slight likeness to regular crystals: there is no regularity of form or angle and the edges are more or less curved and crooked." Since he believed that these crystals grew in a matrix of more or less amorphous steel, it was not surprising to him that they were similar in shape to the crystals formed on the solidification of molten metal in a casting (see chap. 11).

Another important paper on fracture is that of Brinell published in 1885,[23] which was a critical year in the development of metallurgy. Brinell's work was destined to be completely overshadowed by that of Osmond published in the same year, yet it is a monument of imaginative and careful work and shows how much can be learned about steel without knowledge of its microstructure. Brinell was writing after chemists had found the distinction between hardening carbon and cement carbon in heat treated steel and he related such chemical findings directly with changes of fracture. His paper was translated into German, but in English was known for a time only from the abstract in the *Journal of the Iron and Steel Institute* (1886 [i]). However, this did provide a good summary of the important parts of the paper. It reads:

The Changes in the Texture of Steel on Heating and on Cooling.—From a large number of experiments, over eighty of which were with samples of steel from the same Bessemer ingot, having the following percentage composition—

Carbon	Manganese	Silicon	Phosphorus	Sulphur
0.52	0.48	0.13	0.026	trace

J. A. Brinell (*Jern-kontorets Annaler,* vol. xl., pp. 9–38.) draws the following conclusions:—

1. If steel loses its coarsely crystalline texture without mechanical treatment, it is always due to the change of state of the carbon, either from cement-carbon to hardening-carbon, or *vice versa.*

2. The change of texture is only complete when the temperature to which the steel has been subjected is sufficiently high to convert the whole of the carbon into another form. In accordance with this is the complete change in the texture of coarsely crystalline steel, hardened or unhardened, when the exact temperature is reached which is sufficient for the conversion of cement-carbon into hardening-carbon.

3. In order to convert the carbon into cement-carbon, steel raised to a white heat must cool down slowly to a temperature below that to which unhardened steel must be raised in order to produce the same result.

4. The conversion of the cement-carbon into hardening-carbon occurs rapidly, as soon as the temperature is sufficiently high to permit of that result taking place. The change of state from hardening-carbon into cement-carbon, on the other hand, takes place much more gradually.

5. The change from hardening-carbon into cement-carbon is always accompanied by the evolution of heat, and it is therefore probable that, when cement-carbon is converted into hardening-carbon heat is absorbed, although the author was unable to substantiate this view in his experiments.

6. At the moment when the hardening-carbon changes into cement-carbon, either on heating or on cooling, crystallisation instantaneously sets in, and the fracture will be the more coarsely crystalline the more it was so originally.

7. Rapid cooling never produces an amorphous or finely crystalline fracture in the case of any steel which, at the moment before cooling was effected, was largely crystalline, but it prevents a steel which at the commencement of the cooling operation was either amorphous or finely crystalline from assuming a crystalline structure during the progress of the cooling. That is to say, rapid cooling only fixes the structure the steel possessed at the moment when it was acted on.

8. In order to convert hardening-carbon into cement-carbon, not only is the right temperature required, but also time; while in opposition to this, the conversion of cement-carbon into hardening-carbon only depends on the temperature. A consequence of this is that rapid cooling can prevent the change from hardening-carbon into cement-carbon.

9. For the crystallisation of the steel, too, not only is a certain degree of heat required but also a certain period of time; and as regards this latter, if the time during which the cooling is effected is diminished, either by plunging the steel into water, or by any other means, the formation of the crystals is either diminished or totally prevented.

The author observed the manner in which different varieties of iron and steel cooled down from high temperatures by noting the colour phenomena, and he found that in the first period of the cooling, steel cooled down more rapidly than iron, while in the later stages the opposite was the case, and he considers that this was due to the formation of crystals; crystallisation, however, being a form of work, heat is absorbed, and as steel crystallises more than iron does, more heat is absorbed in the case of steel than in that of iron, and consequently during the crystallising period the cooling is the more rapid in the case of steel.

The reason for the steel becoming hotter than the iron towards the end of the cooling is, in the author's opinion, due to the heat evolved on the conversion of the hardening into cement carbon, and in part also to the tendency of the steel to give up its crystalline form during the progress of the change in the state of the carbon. Hardening and annealing experiments were also made, as well as others in which the steel was attacked with nitric acid.

Brinell not only gave good illustrations of the different types of fractures, but displayed in clear graphical form the sequence of heat treatments producing them. Had it not been for the sudden rise of the simpler, more definite indications of microscopic metallography, Brinell's work would have been regarded as one of the most important metallurgical papers ever to be published. There is a fine critical analysis given by Howe in his *Metallurgy of Steel*.[24] It is probable that the confirmation of the new findings of the microscopists by the age-old fracture test did much to make them acceptable to practical steelmen.

It will be noticed that almost all of the above discussion of fracture has related to ferrous materials, and that with the exception of the appearance of graphite in cast-iron fractures little has been said about the heterogeneity of chemical constituents and our concern has been only with crystal size. Although it is easy to recognize the separation of two liquids as in copper-lead alloys in liquation and of mattes, speisses and slags in smelting, the observation of two or more crystal types existing in solid alloys is difficult. Graphite, of course, was seen very early in cast-iron fractures, but the earliest suggestion that iron carbide could be seen in a steel fracture seems to be that by Lampadius,[25] somewhat after it had been recognized by chemical extractive methods. Usually the assumption seems to have been made that the grains of the solid would have the same average composition as the liquid from which they formed, and that the substructure would be on a scale too small to see. In the absence of the notion of crystalline order and before the law of simple combining proportions, there was no difficulty whatever in the concept of solid solutions with any components.

The study of fractures reached its apex with Tschernoff and Brinell and, though observations were frequent, there was thereafter little addition to the understanding of structure derived from them until the development of "fractography" by Carl Zapffe in the middle of the twentieth century, and the very fertile studies of fracture (often made with the electron microscope) that have been done under the inspiration of dislocation theory. The mechanical aspects of fracture and the propagation of cracks in steel plate has become of immense importance because of the failure by "brittle fracture" of ships, bridges and other structures, particularly welded ones, and has attracted a great deal of research attention in the last fifteen years, but it is more of mechanical than structural interest and must be excluded from our present study.

"CRYSTALLIZATION" OF METALS BY FATIGUE AND IMPACT

Nothing can indicate more clearly the difficulty with the concept of a crystal without a geometric shape than the acceptance in popular circles (and even in advanced engineering circles for a time) of the idea that metals would crystallize under repeated vibration or sharp impact. The beginnings of the idea I have been unable to trace. Many broken water-wheel axles, and parts of hammers and stamps must have been replaced with thicker ones without thought. During the early days of railroad engineering, however, serious accidents from the failure of wrought-iron axles focused much attention on the manifestly crystalline appearance of the fractured surfaces, in marked contrast to the fibrous torn ends of a similar axle bent when tested before service and which gave no visible clue to the underlying crystallinity. To this day the idea that vibration crystallizes iron is still fairly widespread, even among educated people with some contact with engineering matters, yet the correct explanation of the phenomenon was clearly expressed almost in the first discussions of it. The "brittle fracture" of steel is still a source of heated technical discussion, but at least the argument revolves around initiation or propagation of fracture and not about the fundamental structure of the material that is failing!

The spectacular growth of the railroads in the 1830's was accompanied by an equally spectacular growth in the number of failures of wrought-iron axles, and the failure of some of the iron bridges that were being built in large numbers. This aroused considerable suspicion as to the reliability of cast and wrought iron, particularly hot-blast iron, the use of which was rapidly growing though it had not become standard. The matter became of such importance that Queen Victoria appointed a high-level commission to consider the feasibility of iron for railroad use, particularly instructing its members to

endeavour to ascertain such principles and form such rules as may enable the Engineer and Mechanic, in their respective spheres, to apply the Metal with confidence, and shall illustrate by theory and experiment the action which takes place under varying circumstances in Iron Railway Bridges which have been constructed.

In an appendix to the commission's report[26] which was published in 1849, Eaton Hodgkinson described an extensive series of experiments on the behavior of iron subject to oscillating loads and repeated impact which are the first systematic studies of fatigue.

Although it was not central to their topic, the commissioners asked most witnesses their opinion as to whether the internal structure of an iron beam becomes altered by being subjected to a succession of light blows at a low temperature, as in rails, axles, or springs. Of the ten witnesses who replied three were sure that there was a change, five were equally sure that there was not, while two were doubtful. The commission concluded:

A great difference of opinion exists among practical men with respect to the first of these questions. [Whether the substance of metal which has been exposed for a long period to percussions and vibrations undergoes any change in the arrangements of its particles by which it becomes weakened.] Many curious facts have been elicited by us in evidence, which show that pieces of wrought iron which have been exposed to vibration, such as the axles of railway-carriages, the chains of cranes, &c. employed in raising heavy weights, frequently break after long use, and exhibit a peculiar crystalline fracture and loss of tenacity, which is considered by some engineers to be the result of a gradual change produced in the internal structure of the metal by the vibrations. In confirmation of this, various facts are adduced, as, for instance, that if a piece of good fibrous iron have the thread of a screw cut upon one end of it by the usual process of tapping, which is always accompanied by much vibratory action, and if the bar be then broken across, it will be found that the tapped part is a good deal more crystalline than the other portion of the bar. Others contend that this peculiar structure is the result of an original fault in the process of manufacture, and deny this effect of vibration altogether, whilst some alledge that the crystalline structure can be imparted to fibrous iron in various ways, as by repeatedly heating a bar red-hot, and plunging it into cold water, or by continually hammering it, when cold, for half an hour or more.

Mr. Brunel, however, thinks the various appearances of the fracture depend much upon the mode in which the iron is broken. The same piece of iron may be made to exhibit a fibrous fracture when broken by a slow heavy blow, and a crystalline fracture when broken by a sharp short blow. Temperature alone has also a decided effect upon the fracture; iron broken in a cold state shows a more crystalline fracture than the same iron warmed a little.

The same effects are by some supposed to be extended to cast iron.

It is indeed surprising that with an essentially correct viewpoint expressed by authoritative engineers such as Robert Stephenson, Brunel, and Barlow that the matter continued to be in doubt for so long thereafter.

The discussion on the crystallization of iron had been initiated before the Institute of Civil Engineers in 1842 by Charles Hood,[27] who believed that percussion, heat or magnetism would all cause crystallization at room temperatures. York, the next year, said[28] that hollow railroad axles were better because the process of making them avoids the crystallization of the metal, and the same year the great engineer, Rankine,[29] advocated journals with a large radius in the shoulder so that the fibers of iron may be continuous throughout. With a sharp change of section there "is an abrupt change in the extent of the oscillations of the molecules of the iron, these molecules must necessarily be more easily torn asunder." The Institution of Mechanical Engineers became involved in 1849. In a paper[30] on railway axles, J. E. McConnell remarked that jarring or vibration "tends to deteriorate the quality of the iron by altering its texture from fibrous to crystalline." Robert Stephenson[31] challenged this and a prolonged discussion occurred. At the next meeting, McConnell[32] produced drawings of fractured axles showing the crystalline fracture in the middle of an axle with a fatigue-cracked rim, and the majority of those contributing to the discussion seems to regard it as proved by the appearance of the fracture that shock or vibration crystallized the iron. There was no definition of crystallization, but it is remarked by Cowper that the fiber in wrought iron is composed of the separate particles of iron from the puddling process, elongated and lying next to each other, though not in perfect contact because of cinder. Crystalline iron was that in which the particles assumed any form other than the elongated form. A Mr. Hodge remarked "that to arrive at any true results as to the structure of iron it would be necessary to call in the aid of a microscope to examine the fibrous and crystalline structure"—a prophetic remark indeed.

Stephenson again strongly dissented from the crystallization concept and remarked that he had examined crystalline and fibrous iron under a powerful microscope and no real difference could be perceived between them.

The best specimen he could select of the kind called fibrous exhibited to the naked eye a laminal arrangement of dark and light lines, but the light lines composing the apparent fibre were, in point of fact, as crystalline as the other kind of iron, and therefore, however fibrous it might appear, it was essentially a crystalline mass.

Stephenson was evidently excited about the use of the microscope and suggested:

Perhaps at their next meeting he might have the pleasure of exhibiting to the members, certain microscopic results, but as yet he had not given the subject sufficient consideration to justify him in doing so.

Unfortunately, he did not make good his tentative promise.

Discussion continued for years. An important contribution to it was made by the American engineer J. A. Roebling who in 1860 concluded[33] that:

All fibres are composed of mineral crystals, drawn out and elongated or flattened; and the fracture may be produced so as to exhibit in the same bar, and within the same inch of bar, either more fibre or more crystal. . . . My own view of the matter is, that a molecular change, or so called *granulation* or *crystalization,* in consequence of vibration or tension, or both combined, has in no instance been proved or demonstrated by experiments. I further insist that crystalization in iron or other metal *can never take place in a cold state.* To form crystals at all, the metal must be highly heated or nearly a molten state.

The true explanation for the weakening and embrittlement of iron in service, Roebling thought, was related to the composite structure which resulted from the method of manufacture of wrought iron:

The drawn out fibre is composed of an aggregate of pure iron threads and leaves, enveloped in cinder. [Under vibration the iron must] inevitably become brittle, because the iron threads and laminae become loosened in their cinder envelopes . . . without undergoing any mysterious change in its molecular environment.

At about the same time a somewhat more realistic approach was being taken in Germany. Kohn performed the first controlled tests to find the effect of alternating torsion on wrought iron, the crank-shaped samples being subsequently broken in a hydraulic press, after as many as 128,000,000 reversed torsions. Schrötter, analyzing Kohn's results,[34] concluded that repeated torsion can change the fracture of a bar from fibrous to crystalline and then lamellae. The most important work in this field, of course, was that of the German engineer, Wöhler, whose first paper was published in 1858.[35] He, however, seems to have indulged in no structural speculations.

The best contribution to this mid-century argument was by Kirkaldy in 1862.[36] He reproduced extensive portions of the previous discussions, described many new experiments of his own and then concludes

that it is not to any of those agencies named by others [percussion, vibration, frost, magnetism, etc.], but simply *to the act of breaking,* that are due the different appearances presented to the eye in fractured iron, respectively called fibrous and crystalline. . . .

After describing some experiments on the use of etching, which are referred to in chapter 12, he concludes as follows:

It seems surprising to the writer that such a simple mode as that just described, of examining the texture of iron, had not occurred to any of those individuals who have expressed their opinions for and against the supposed change from a fibrous to a crystalline structure during the time of its being in use, as he believes it would have tended to the settlement of the question long ere now. Perhaps it may be asked, Why such a difference, even in appearance, if no actual change in the structure? The writer replies simply thus,—in the case of the fibrous fracture the threads are drawn out, and are viewed *externally;* in the case of the crystalline fracture the threads in clusters are snapped across, and are viewed *internally* or *sectionally.*

The great metallurgist, Percy,[37] writing in 1864, believed that iron was highly crystalline after being melted and that both fracture and slow etching with acid would develop distinct crystalline markings. Annealing wrought iron at a sufficiently high temperature would give the particles sufficient freedom of motion even below the melting point to arrange themselves in crystals. In a rolled bar, "fibres in the form of elongated crystals pre-exist" and can easily be shown on etching. In the act of rolling or any kind of extension these crystals were elongated but not obliterated and they may always be rendered manifest by sudden fracture.

The introduction of Bessemer steel gave a new material for the engineer to be concerned with. The editor of the *Practical Mechanics Journal* in 1865 concludes[38] that Bessemer steel must be subject to the normal laws of crystallization and attributes the difference in fracture to the slag content. He contends, following Mallet, that the crystals become aligned parallel with the direction of least pressure, the slag acting as a lubricant. He took crystallinity for granted, both in casting and after working. He used the microscope on the fracture and he knew of Sorby's work on rocks. The times were indeed ripe for advance.

Both of the early metallographers Sorby[39] and Martens[40] examined the question. Sorby had a bar of iron attached to a tilt hammer so that it vibrated up and down, breaking after fifteen hours with a crystalline fracture. An examination of a longitudinal section at the fracture showed that the structure was no more crystalline than similar iron in its natural state, although some of the crystals had been slightly separated from each other instead of being in normal close contact. Martens, although noticing a difference of fracture in Wöhler test pieces, found with the microscope no trace of change in the size or arrangement of the crystals, but only in their adhesion and the path of rupture.

Howe sums up the situation in 1890 thus:[41]

As fibre appears to be due to the drawing out of previously equiaxed grains of iron by favorable mode of rupture, we may define $\left\{ \begin{array}{l} \text{fibrous} \\ \text{crystalline} \end{array} \right\}$ iron as that whose grains $\left\{ \begin{array}{l} \text{are} \\ \text{are not} \end{array} \right\}$ readily drawn out into fibres during rupture, or that which $\left\{ \begin{array}{l} \text{can} \\ \text{can not} \end{array} \right\}$ be readily made to yield a fibrous fracture.

Although he quotes evidence to show that crystals of copper and silver can grow at room temperature and that it is unwise to assert dogmatically that there can be no molecular motion, he concludes: "We have, I think, every reason to believe that the granulation and crystallization of iron under vibration in shock is a myth."

It is indeed astonishing after all these clear statements of the case to find the matter of crystallization still being debated actively in 1893.[42] In that year the American Institute of Mining and Metallurgical Engineers had a furious debate precipitated by a chance statement on shaft fractures by the author of a paper on stamps for milling gold ores. The report of the twice-adjourned discussion filled some sixty pages of the *Transactions*. All the ideas, both foolish and correct, of the fifty-year-old London debate were revived but no conclusion was reached. There are few better examples of man's conservative tendency and his willingness to believe his eyes rather than his mind.

11. Idiomorphic Crystals and Surface Solidification Patterns

The conditions of solidification of metals or alloys often indicate their structure even better than does fracture. Because of interdendritic or intercrystalline shrinkage, an alloy will rarely freeze with a smooth surface and even a pure metal will reveal partially formed crystals if the liquid is withdrawn before the crystals have had a chance to impinge upon each other and disguise their geometric shapes.

The formation of a dross in tin and lead which was nothing but impurities in the form of high-melting intermetallic compounds must have been noticed very early. Fuller, for example, in describing the Worthies of Derbyshire says:[1]

The best lead in England (not to say Europe) is found in this county. It is not churlish, but good-natured metal, not curdling into knots and knobs, but all equally fusile, and therefore most useful for pipes and sheets.

Stead reports[2] that for generations tinsmiths have been using surface roughness as an indication of the most fusible solder for their work, finding it easier to adjust the melt to a smooth texture on solidification than to weigh out the two metals. The test, he thought, was of extreme antiquity. The roughness of alloys not of eutectic composition would be due, of course, to the shrinkage of eutectic away from the crystalline pro-eutectic tin or lead and could reveal a departure of 1 per cent or less from the true eutectic composition. Pewterers used a similar test to determine the purity of tin, for the purer it was the cleaner and smoother was the surface of a test ingot. Yet, as with fracture, the crystalline origin of these patterns was very late to be recognized. Even the relationship between fracture and surface patterns was not realized until late in the eighteenth century!

The most noted example of the dendritic crystals that are often clearly visible on the surface of a metal, particularly if solidified under a fluid slag, is the Star of Antimony. This attracted an enormous amount of work by alchemists who regarded it as an almost mystic sign, but not once did it give rise to any useful speculations or experiments on metal structure. The great chemist Boerhaave[3] complained that the regulus had turned the heads of some of the profoundest chemists

and for my own part, when I reflect upon the time and pains I have employed in examining into the nature of this regulus, I cannot help being surprised at my own patience, and can hardly help blushing to think that so great a part of my Life should have been spent in this Inquiry.

The pseudonymous Basil Valentine in his *Triumphal Chariot of Antimony* (1604)[4] gives no crystallographic information, but Nicolas Lémery a century later discusses the Star at some length in his comprehensive treatment on the preparation and medical uses of antimony.[5] The pure metal was made by igniting a mixture of equal parts of crude antimony sulphide powder and saltpeter. On the casting, if the metal were pure,

usually appears the figure of a star, the rays of which are beautiful, large and resplendent, extending from their centre, in the middle, to the extremity of the circumference. . . . The rays form like a sword blade on large cakes of regulus, and in a leaf-like figure approaching that of a lemon tree on *petit pain*. . . . The rays are a little raised in some places and sunken in others, covered with little veins which first appear to be confused but which are nevertheless nearly parallel: they resemble to some extent the veins of leaves on trees.

Lémery believed, however, that the star was only superficial and did not continue into the mass, for the fracture showed the interior to be for the most part a confused mass of little crystals intersecting one another and interlacing, some of the crystals being arranged in order in the manner of rays, but they were not the rays of the surface of the regulus continued. On remelting the metal, a new star would appear on the surface, but the figures inside would be quite different depending on whether the regulus cooled more or less fast.

The star appears only at the surface of the regulus and not anywhere inside because, apparently, only at the surface can the crystals find the opportunity to grow according to their natural determination. Conversely those on the inside, not having the same liberty because they are pressed on all sides, interlace themselves with each other and there results only a confused arrangement.

He spoke truth in these remarks, but the emphasis on external shape instead of internal order prevented him from seeing the identity between the two types of surface characteristics that he studied.

A number of metals exist in nature in the form of idiomorphic crystals, and such crystals have been produced from time to time as a *tour de force* by chemists, initially only by the use of mercury and later directly from the melts themselves. The silver "tree" was discussed in chap. 9. The assayer Samuel Zimmerman in his 1573 *Probierbuch,*[6] describes the growth of crystals of both gold and silver by reacting mercury with solutions of these metals in nitric acid or aqua regia. The famous chemists Macquer[7] and Baumé[8] both comment on the possibility of crystallizing silver and copper from amalgams. Later, Sage describes[9] in detail the growth of polyhedral crystalline forms of silver, gold, tin, zinc, bismuth, and lead, made through the intermediacy of mercury, slowly distilled away. Sage, in fact, discarded dendrites as of no interest and thought the crystals could only form if the mercury had deprived them of part of their phlogiston.

The last quarter of the eighteenth century saw the beginnings of mathematical crystallography and early in the nineteenth century the chemical atom became a definite, quantitatively handleable concept. But neither of these important crystallizations of seventeenth-century corpuscular theories resulted in any immediate clarification of polycrystalline matter or of the metallic state. The eighteenth century is not, however, without considerable interest to our purpose.

As might be expected, mineralogists, most of whose materials are visibly microcrystalline to the unaided eye, describe aggregates of small crystals and early came to realize that crystals could not maintain their normal shape in a compact aggregate. However, the crystallographers exclusively, and the mineralogists almost without exception, limited their interest to perfect polyhedra and rarely regarded the distorted forms in an aggregate as being worthy of the name crystal. An exception is John Hill,* that profuse man of many parts whose exuberant personality seems to have distracted attention from his ideas. His *History of Fossils,* published in 1748,[10] forms Volume I of a large three-volume general natural history. He comments extensively on the microstructure of rocks of all kinds and has very pertinent comments on the formation of growth steps on crystal faces and on nucleation from super-

* I am grateful to Mrs. Joan M. Eyles for calling the works of John Hill to my attention. He has been generally ignored by both scientists and historians. David Garrick's view may be quoted as typical of the scornful attitude of his contemporaries: For physic and farces, his equal there scarce is. His farces are physic, his physic a farce is.

saturated solutions. He realized that fresh crystalline matter formed by addition on the first nuclei that were formed in preference to the formation of new ones, and he thought that material condensed at certain points on the crystal and spread therefrom over the surfaces, thus often giving rise to transverse lines on the facets, which lines

may plainly be discover'd to be no other than the edges of so many strata or thin crusts of crystalline material laid one over the other; and such they really are (tho' not so easily distinguishable) in all crystals even the finest.

There are many growth figures mentioned throughout.

Hill used acid in etching rocks (particularly marble) to distinguish various constituents, and he uses the effect of acid on minerals as one of the identifying characteristics in his extensive classification of them, together with the effect of calcination, the ability to draw sparks from steel, and the appearance under the microscope—the beauty of the latter frequently drawing exclamations of delight from him. He has nothing on metal, but on marble (a rock with a close analogy to a polycrystalline metal, for the grains are the same shape and have resulted from grain growth in the normal fashion for an annealed recrystallized metal) he observed that it was composed of "clusters of angular granulae" of nearly the same size and figure. Though appearing flat to the naked eye, the microscope showed them to be pretty regular parallelepipeds.

There are variations indeed in some places, but these are owing wholly to the hasty concretion of the mass, and accidental jumbling of two or three concretions together.

They somewhat resemble the grit particles in sandstone, but are more closely constipated, and cohere everywhere where they are in contact, which is their whole surface without any interstitial matter. Carrara marble viewed under the microscope is seen to be

compos'd of an infinite multitude of small cubic or something like cubic particles, crouded together into a close union and every where entrenching upon and spoiling one anothers form.

He never uses the term "crystal" for anything but a geometric particle showing nearly planar growth faces. On granite he observed that the different bodies composing it were in intimate and close union and that, "In the general breaking of the stone they are found much oftener to break and tear to pieces than to separate." Granites are "composed of separately concreted Moleculae cohering firmly where ever they are in contact, but never entering into one another's texture." This he believed to have resulted from crystalline particles falling together and

afterward being concreted into a mass, crushing and compacting them
into the irregular directions that they are found in.

It is easy to conceive that the same agent [pressure of consolidation resulting
from the removal of water], acting with the same force on Moleculae less
hardened, would crush them into the very substance of one another, and
could not do otherwise than form exactly such a complex and irregular mass
as we find the stone to be.

This good beginning on the study of polycrystalline rocks and intercrystalline effects was not noticed by metallurgists, neither was it followed up by mineralogists, for the simpler symmetry of idiomorphic
crystals had first to be methodized and catalogued.

In 1775 Pierre C. Grignon published his *"Mémoires de physique,"*[11]
a book on various aspects of iron mineralogy and metallurgy which is
of considerable importance to our subject. Grignon was occupied in
the commercial operation of a blast furnace and forge plant and his
science stemmed from observations on a much larger scale than those
of his laboratory contemporaries. He had the opportunity to see large
crystals in shrinkage heads of large castings and even to take a blast
furnace and, by working the bellows with ever decreasing power, to
allow it to cool extremely slowly and the metal in it to develop very
large crystals. It would be interesting for the progress of physical science,
he remarks,

if all men who employ fire on a large scale as an instrument of the arts would
approach their work with an observant and analytical viewpoint: the bulk
of our knowledge would be greatly increased and the [relative] number of
hydrogenic [i.e., aqueous] systems would diminish.

His recommendation was not followed and even to the present day
chemists remain too much concerned with reactions in aqueous solutions.

The first memoir of structural interest, "Mémoire sur les métamorphoses du fer"[12] had been read before the French Academy in 1761,
but it was not printed until 1775. In it, Grignon remarks that

everything in nature is characterised by an individual essential form on which
chance has no domain. Each has a characteristic predetermined figure which,
in concert with the equally invariable quality of the substance, determines its
properties. . . . All the bodies capable of receiving the consistency of a fluid
by Nature or Art, on condensing take an essential symmetrical form. Every
molecule,* similar to its neighbour, approaches and unites with it, and in

---

* Grignon uses the term "molecule" for a smaller state of aggregation than Réaumur, for
instance. He supposes that they may be of the order of one-thousandth of a ligne, i.e., about
two microns in size.

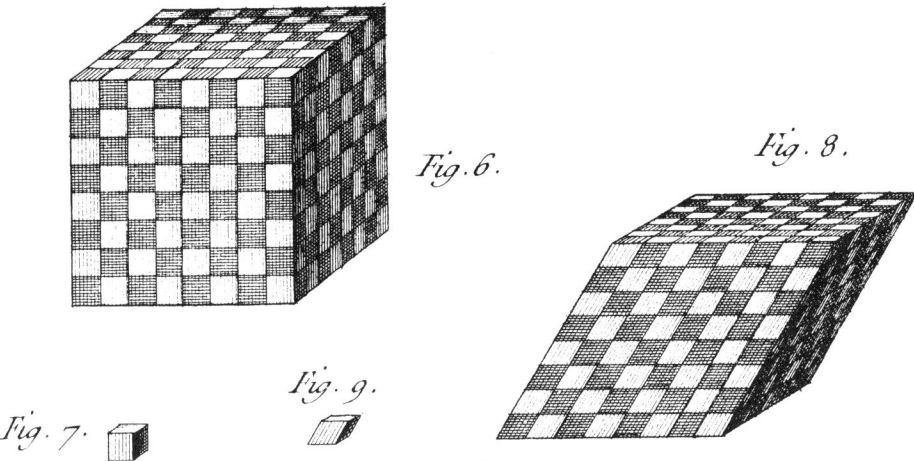

Fig. 68.—_The assembly of polyhedral parts into cubic and rhombohedral crystals. (Grignon, 1775.)_

the measure that the separating fluid is dissipated, they discard the obstacles and draw tighter their bonds with an invariable affinity. They attach themselves to each other in numbers related to their faces and to the opening of the angles of each molecule. This is why, in seeing a body naturally cubic I conceive that each molecule is a cube; a rhombohedral body is composed of rhombs, and similarly others [Fig. 68].

Such polyhedral units (which seem to have been suggested early in the eighteenth century by observations of cleavage fragments) were destined for a time to supplant the seventeenth-century stacked-sphere concepts which are closer to today's views (see chap. 8). A stacked-polyhedron model, essentially the same as that of Grignon, was used to explain the geometry of crystal faces by Romé de l'Isle and Bergman (in 1772 and 1773), and was the basis of the mathematical crystallography of the great Haüy (1784). Haüy was to have much trouble with isomorphism and mixed crystals, but to Grignon, a metallurgist familiar with alloys, the crystallographic model was a natural one:

I am persuaded that having acquired an exact knowledge of the shape of the crystals, or rather of the molecules of each metal, one would be able to discover (from the form which would be taken by the crystals of several metals capable of union when joined together) the kind of metal which would make an alloy, and the proportions of the mixture, by computing the opening and the number of the angles, and the greater or less extension of the faces of the half-breed crystals [_cristaux métis_] of the alloyed metals; because two bodies when united take together a mean configurational form which depends on their proportions.

Remarking that it is well known that all substances like salts **and** acids dissolved and condensed give regular invariable figures, Grignon complains somewhat incorrectly that there had not yet been anything said of metallic crystallizations produced by fire. Cast balls of white cast iron crystallize in concentric radii like all substances combined with sulphur (an analogy with pyrite, antimony and cinnabar, all of which crystallize in a similarly clear-cut fashion). The radial prisms can be varied by the manner of cooling "because the point where the cast iron first cools is where the first condensed molecule attaches itself. . . ." To the degree that the heat retires to the center, the molecules accumulate one on the other following the progress of cooling right up to the center, which is the point toward which they all tend. If for some reason one side is cooled more than the other, the radii all direct themselves at such angles in respect to the positions of the bases of the prisms that those on the coolest side will be larger, and those on the opposite side shorter (Fig. 69).

Gray cast iron also gives a very regular crystallization with each crystal distinct and isolated, but to see it, it is necessary that the cast iron be cooled very slowly during several hours, so that the shrinkage may be considerable and nothing disturbs the order. Dendrites are built up by the accretion of regular rhombs upon each other in successively decreasing size.

The little grottoes that are formed [in gray cast iron] from masses of these crystals present to the lens-aided eye the spectacle of a little metallic forest composed of trees with four sets of branches opposite to each other. . . . These crystals are larger and better formed than those of white iron because they are isolated, they occur in a more homogeneous cast iron, and because the more perfect fusion favors the exact arrangement of the molecules.

Although Grignon certainly realized the importance of shrinkage in uncovering the crystals, he did not specifically state that the difference between the radial crystals of the white iron and the dendrites found in gray iron lay principally in the extent to which the liquid was re moved from intercrystalline spaces before complete solidification.

Wrought iron leaves the refinery as a ferruginous bloom, with the parts separated by slag. This is removed by the pressure of the hammer, bringing the molecules of iron together and making homogeneous matter. The parts free from foreign matter are attenuated to form flexible fibers, each molecule of the metal approaching analogous neighbors. The fracture of iron may be brilliant all over, partially, or not at all. The brilliant reflection comes either from very numerous small facets, or less numerous larger ones—the smaller indicating the greater de-

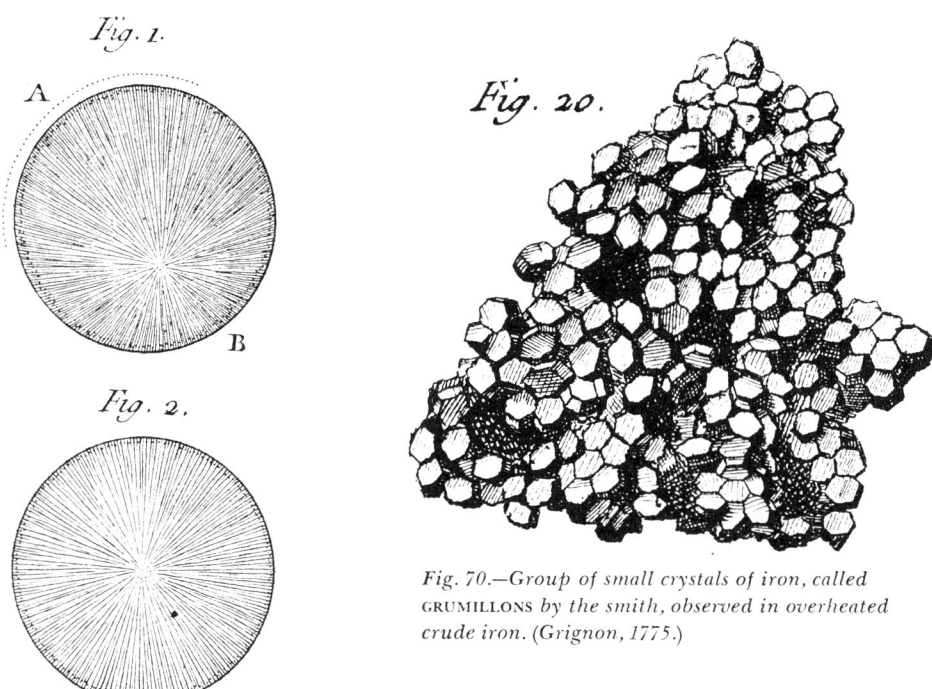

Fig. 70.—Group of small crystals of iron, called
GRUMILLONS by the smith, observed in overheated
crude iron. (Grignon, 1775.)

Fig. 69.—Balls of white cast iron in which the center of crystallization can be displaced by assymetric cooling. (Grignon, 1775.)

parture from a reguline (cast) state and more adherence. If a bar of iron, twice heated and hammered, is then heated to a point where the internal slag melts, the molecules of iron can take their regular form in this slag, which is a solvent and medium for crystallization. If it is then hit with hammer, the iron breaks into a group of crystals which the forge men call "grains" (*grumillons*) (Fig. 70). If the heat is carefully lowered,

the crystals are made irregular by the pressure which they receive in their soft state, and they unite in every direction to form the bar which, on being broken, presents brilliant faces, angles and hollows which are nothing but the surfaces of the crystals seen in different directions.

This is a good description of the hot shortness of overheated steel and its subsequent fracture, after cooling, along intergranular paths. Under the name of "burnt" iron it attracted considerable attention on the part of the older metallurgists and was not well explained until after the growth of metallography. Grignon, however, clearly realizes that the *grumillons* were in some manner crystals, and that their shape depends on the space-filling pressures of their neighbors. A piece of iron that he found in the slag near the taphole of a finery had crystallized

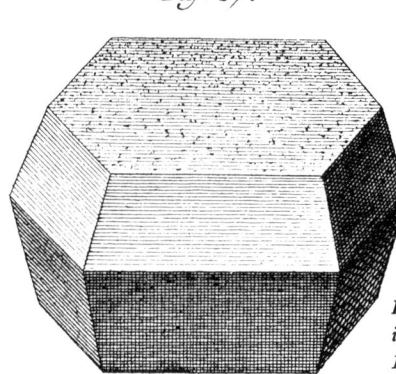

Fig. 17.

*Fig. 71.—Polyhedral crystal of iron like those
in Figure 70, considerably enlarged. (Grignon,
1775.)*

in the slag. "Slag is to iron what water is to crystalline salts." The masses
of these crystals were rarely regular he says,

because the fire which gives birth to them welds them together and mutilates
the angle by the action which it has on their substance. The most regular have
appeared to me to be polygons, hexahedra, formed by several rhomboids
united by their large face [Fig. 71].

It will be seen that Grignon at first believed that iron will crystallize
only from a solvent, and that crystals can be distorted by pressure
against each other. Though he observed and comes very close to describing correctly the typical grain of a polycrystalline material, he
does not think of this as being a stable form in its own right but only
a distorted polyhedral form. He was remarkably perspicacious in pick-
ing a fourteen-sided body as the typical grain, quite wrong in showing
it with such high symmetry and with apexes with more than three faces
meeting at them.

Grignon read a second paper entitled "Mémoire sur des crystallisa-
tions métalliques"[13] . . . to the Academy in 1772. By this time, Grignon
had realized that slag was unnecessary as a solvent and he takes issue
with the crystallographer Romé de l'Isle,[14] who had stated that water
was the principal, or perhaps only, agent in nature for forming crystals,
and that it was useless to try to form metallic crystals in the laboratory
by fire alone in the dry way; the rough-shaped figures produced by art
should not be confused with true crystalline forms, which were the
product of the slow operation of nature with water as an intermediary.
Grignon describes and illustrates a very beautiful crystallization in cast

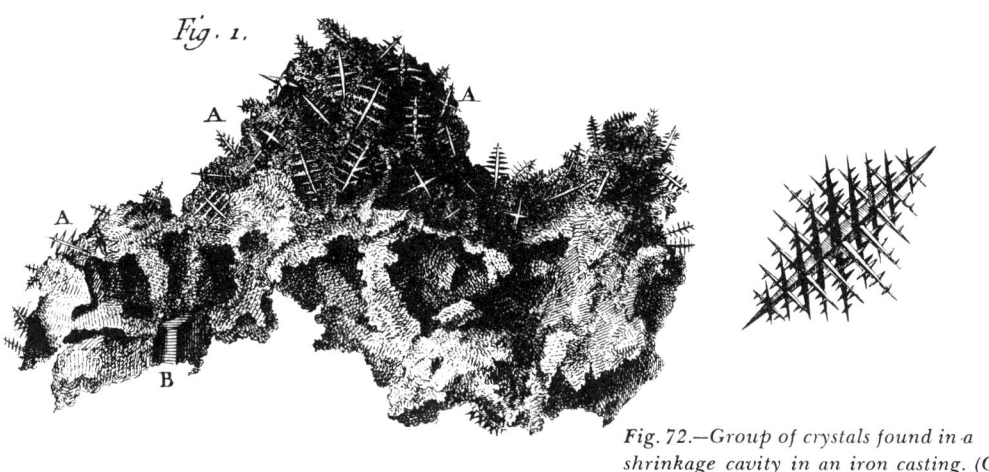

Fig. 72.—Group of crystals found in a shrinkage cavity in an iron casting. (Grignon, 1775.)

iron which had been molten in his blast furnace for several days and cooled over a period of more than fifteen days (Fig. 72). These slowly cooled dendrites had branches on their branches. Water had nothing to do with their formation. "The primary subdivision of the matter of which they are composed had been done by fire, which reduced their metallic substance to a perfect fusion."

This idea of fire as a solvent permeates a paper by Guyton de Morveau, published in 1776,[15] the year following Grignon's book. Fusion *is* solution, De Morveau says, and solidification is a true crystallization which occurs by the evaporation of the superabundant part of the solvent igneous matter. Fire is to metals what water is to salts. Metallic crystallizations have escaped observation for so long because smelting and founding have been relegated to men motivated more by need to play (*jouir*) than by the need to know.

The constancy of form, the regularity based on a secret mechanics (despite the thousand accidents to which the products of fusion are subject) indicate a kind of geometric progress in the composing of the solids which will perhaps eventually serve us as a microscope to perceive the figures of their elements, and which even now allows us no doubt that it is really to this figure that solids owe what we call their properties.

De Morveau found that a piece of blister steel melted in a graphite crucible and frozen under a slag developed a surface which was covered with very regularly intersecting lines, like crosshatching, forming a very perceptible relief which he illustrates, rather poorly, in Figure 73. A little later De Morveau replied[16] to a complaint of his ignorance of

Fig. 73.—*Surface of a regulus of crucible-melted steel as it appears under a magnifying glass. (De Morveau, 1776.)*

Grignon's work that there was nothing in common—Grignon produced superb isolated crystals while De Morveau showed little regulae covered with lines. However, in 1779 he had examined[17] a large plate of cast iron found in the bottom of one of Buffon's furnaces, which left no doubt that the two appearances of crystallization were connected by insensible gradations, so that the difference could only be attributed to accidental circumstances occasioning more or less slow cooling. He illustrates his paper with pictures of the surface of little buttons or ingots of gold, silver, platinum (with no information on how it was melted and crystallized), copper, tin, lead, antimony, and bismuth.

Mongez developed[18] a reliable method of growing metal crystals by melting the metal in a crucible having a hole in the bottom closed with a plug which could be pierced to allow the residual metal to run out when the metal was partly solidified, thus forming a kind of metallic geode lined with crystals. He observed that metals generally formed polygonal pyramids—trihedral or quadrangular—usually joined base to base to give something like an octahedron, while the semimetals were not so regular. Bismuth, for example, formed irregular little hopper-shaped crystals or volute Greek squares adhering to each other. Cobalt crystallized in fasces of needles lying against each other, imitating a mass of basalt columns.

Fifty years later David Mushet published an article (though to do him justice he claims to have written it in 1800)[19] which showed only slight advance on De Morveau's knowledge. He had "a large and beautiful" collection of iron crystals from gun-heads. He saw that the first crystallization effects visible on the surface were linear and the second degree came from the separation of a certain number of lines into quadrangular prisms.

As shrinkage proceeds the quadrangular prisms separate longitudinally and a series of points begins to appear which are the summits of more perfect forms . . . the crystals may now be compared to a spearhead with serrated edges or like some varieties of the fern.

Thereafter, though there are many references to the crystallization of metals, there is little more understanding displayed until the extremely important papers of Tschernoff. Storer[20] made use of Mongez's technique in 1859 to study the crystallization of alloys of copper and zinc. He found that all of the alloys down to 30 per cent zinc were octahedral, composed of octahedra built up with parallel axes to give a striated appearance (i.e., dendrites). By analysis, he concluded that none of the crystals contained more of either copper or zinc than did the liquid and therefore concluded that the alloys were isomorphous mixtures of the two metals. The best crystals came from a 50/50 alloy, separated after solidification by pounding the ingot when hot in the manner that granulated spelter solder has been prepared for the market. At 40 per cent zinc, a peculiar homogeneity occurred with a smooth, compact fracture, while an irregular stringy effect appeared with more copper. This was the composition in wide use as Muntz metal, which can be rolled either hot or cold. Fracture was used to control the composition in the foundry, adjusting the composition of a batch of metal on the basis of trial samples. Storer failed to see the essential difference between the dendrites obtained by draining and the clusters of fibrous grains from a hot short fracture and his observation on solid solutions is his principal contribution.

Largely under the influence of Sir James Hall's Huttonian enthusiasm for artificial rocks and minerals, many chemists and mineralogists attempted to produce synthetic crystals of natural minerals. By 1872, when the observations were compiled into a book by C. W. C. Fuchs,[21] hundreds of minerals had been produced in the laboratory, including the following metals: copper, silver, gold, iron, platinum, "$AgHg_3$," lead, tin, arsenic, antimony, and bismuth. These had been grown from the melt, found in furnaces or produced by various electrochemical tricks. Although it provided additional proof that metals were crystalline and made this a central part of chemical-mineralogical knowledge, metallurgically there was little added to the eighteenth-century observations described above. The general approach of the "experimental geologists" did, however, have considerable influence upon Osmond,

who quotes particularly the works of Vogelsang,[22] Daubrée,[23] Bourgeois,[24] and Fouqué and Lévy[25] on crystal formation (see chap. 14).

Tschernoff,* whose work on solid recrystallization was discussed in
chapter 10, did equally significant work on solidification structures,
which was published in 1878.[26] His interest in the subject had begun
considerably earlier and he transmitted to William Anderson, who
made an English translation of the 1868 paper, some microscopic drawings showing crystallization which were included in Anderson's private-

---

* It should be noted that both aspects of Tschernoff's research arose from difficulties in
handling molten steel in large quantities. The change of scale forced upon the steel metallurgist by the Bessemer and Siemens-Martin processes made inevitable new studies of solidification and of recrystallization. Without these and parallel chemical studies of deoxidation,
the economic potential of the new processes could not have been realized.

† *Note added in proof.*—Since writing the above, the author has seen a copy of the original
pamphlet by William Anderson in the Engineering Societies Library (New York). Its title
page reads: *"Remarks on the Manufacture of Steel and the Mode of Working It.* By D. Cher-
noff, Assistant Manager of the Abouchoff Cast Steel Works near St. Petersburg. Communi-
cated to the Russian Technical Society in April and May, 1868. Translated by W. Anderson,
M.Inst.C.E. London: printed by William Clowes and Sons . . . 1876." This was reprinted
in full in *The Engineer,* 1876, **2,** 1–4, and in extended abstract in J. S. Jeans.[27]
The main text of the translation occupies pages 3 to 21 of the 24-page pamphlet and, except for insignificant editorial changes, is identical with the version that was published by
the Institution of Mechanical Engineers in 1880. There is, however, an important appendix
which was not reprinted (except in Russian retranslation in 1915) and a lithographic plate
containing seven drawings. Only one of these drawings did Tschernoff think enough of to
include in his later paper—that showing the appearance of dendrites—Fig. 74, here copied
from the earlier lithograph. Tschernoff begins the 1876 appendix with the statement:
"The following notes of microscopical observations have never yet been published. The
drawings annexed were made under the microscope, with the assistance of Hartnack's camera
lucida in the years 1868 and 1869. Drawings 1, 2, and 3 are especially remarkable. They represent the efflorescence appearing in the cavities of hard tool steel melted in a crucible, and
allowed to cool extremely slowly in its furnace. The ingot on cooling exhibited cavities
caused by contraction and by bubbles, and these have on their walls or faces arborescent
growths, and on the surfaces of these growths I found microscopic crystals composed of extremely thin transparent plates, having a high refractive power, and extremely hard, so that
they scratch glass."
Tschernoff's drawing of one of these crystals (No. 2) is reproduced in the present book (page
186). Its nature is uncertain: it may be an oxide or a non-metallic carbide. The other kind
of "insoluble substance," shown at a magnification of 155 in Figure 90 (page 206) represents
"the residuum remaining after dissolving steel in weak nitric acid. The substance remaining
is probably silica; it is of a yellowish-brown colour. The particles have the appearance of
little plates, and a general appearance similar to the fracture of the steel dissolved. The study
of such residuum is extremely interesting, and may lead to the elucidation of many problems
concerning the structure of steel."
This additional material shows that Tschernoff was doing serious microscopic investigations before 1870. The non-metallic inclusions and the residues left on dissolving steels in
acid were not, however, of central importance. Though both Martens and Osmond were led
by similar studies to etch polished sections for true metallography, Tschernoff's great contributions to the understanding of the structure of steel were made on the basis of shrinkage
and fracture studies alone.
The author is indebted to Professor George Kurdjumov for a photocopy of Tschernoff's
original papers, and for information about the collected works *Trudi Dmitria Konstantino-*

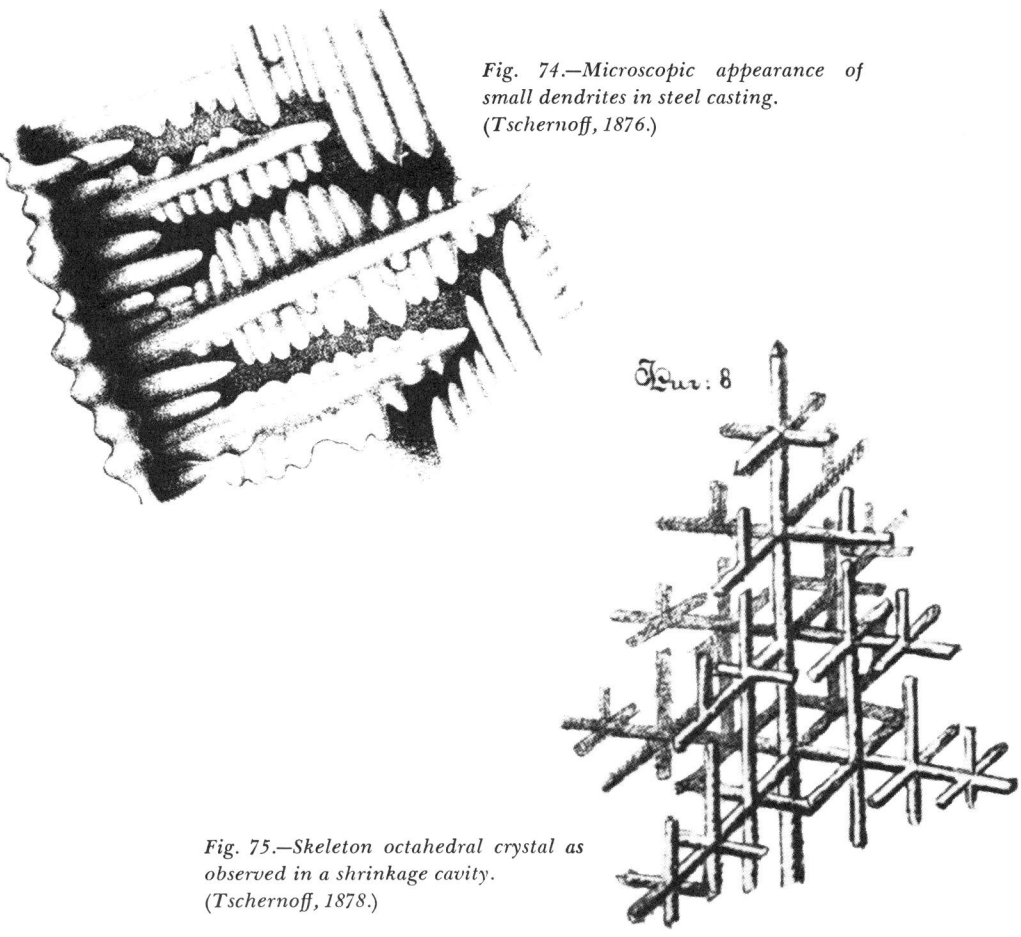

Fig. 74.—Microscopic appearance of small dendrites in steel casting. (Tschernoff, 1876.)

Fig. 75.—Skeleton octahedral crystal as observed in a shrinkage cavity. (Tschernoff, 1878.)

ly circulated translation, but these were not reproduced in the version published by the Institution of Mechanical Engineers in 1880.† Figure 74 shows the appearance of dendrites at a magnification of 165. In 1878 Tschernoff observed the formation of dendritic crystals in shrinkage cavities in steel ingots and inferred from this the manner of growth even when free surfaces were not visible. The crystals in the shrinkage cavity are, he says, of a skeleton or discontinuous type, with larger development in the direction of the octahedral axis (Fig. 75). Branches of second, third, or higher order appeared from the first rudimentary growth, and they become more and more developed as they

vicha Tschernova published by the Russian Metallurgical Society in Petrograd in 1915. This includes a retranslation into Russian of Tschernoff's correspondence with Anderson which formed the appendix to the latter's 1876 pamphlet.

Fig. 76.—Growth of dendritic crystals
from the mold wall (Tschernoff, 1878.)

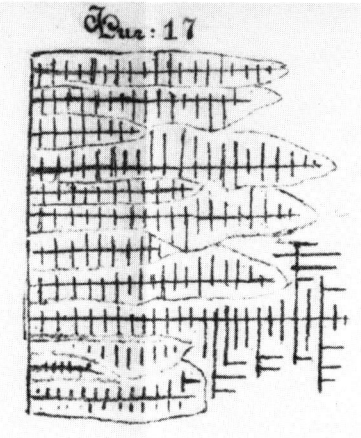

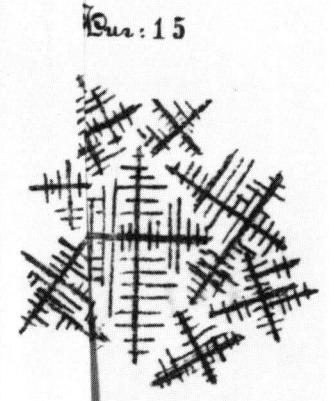

Fig. 77.—Interference of dendritic
crystals. (Tschernoff, 1878.)

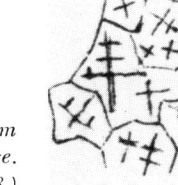

Fig. 78.—Grain shape resulting from
dendritic growth interference.
(Tschernoff, 1878.)

approach the base until they sometimes form the outline of a complete octahedral crystal. The various crystals growing in the center of an ingot intersect each other at every conceivable angle, without order, but elsewhere the principal axes of growth are normal to the cooling surface—radial in small round castings, and meeting along diagonals in a rectangular ingot (Fig. 76). Improperly fed shrinkage leaves local cavities along the planes of contact of the crystals, and thus forms the well-known planes of weakness in such ingots. The confused orientation in the center comes from the slower solidification there, which allows the formation of many centers with perfect freedom of the axes to move around under the disturbances of shrinkage and the settling motion. The lateral axes of the crystal have no relation to each other and the principal axes are of unequal length, showing that the rapidity of growth is not symmetrical. Stages in the growth of the crystals are shown in Fig. 77. The development of the discontinuous crystals was clearer the more carbon in the steel and it was difficult even to find free dendrites in shrinkage cavities below 0.2 per cent carbon.

Tschernoff found that the composition of the separated cast crystals in the shrinkage cavities was the same as the ground mass, however that may vary from melt to melt; consequently the crystals do not represent a chemical compound, although the phenomenon of liquation may suggest this. He realized that a growing dendrite could easily isolate patches of liquid between its branches, which eventually must leave a little local cavity as a result of solidification shrinkage. These cavities would be concentrated at the junction between different crystals, and account for the ease of forming cracks at this location, particularly in the outer acicular parts of ingots.

Tschernoff was inspired by Anossoff, whose work, it will be remembered, was largely concerned with the duplication of the Indo-Persian Damascus steel in the belief that it was the best form of steel for severe service and that the etchable surface patterns represented a practical test of quality. Tschernoff's aim was the purely practical one of building better guns, but he realized as had few before him that the properties of the material depended directly upon structure and that the control of structure, whether during casting, mechanical working, or heat treatment, was the key to a better product. His papers were translated into French and English, the latter by William Anderson, who circulated the first paper privately in 1876. Anderson's translations of both papers were quoted extensively by J. S. Jeans in his influential book.[27] They were read before the Institution of Mechanical Engineers in 1880 and published in their *Proceedings*. It is a tribute to the originality of the ideas that they drew little discussion. Yet structural ideas were definitely coming to the fore, and Tschernoff's papers, joined with Marten's microscope studies that had just been published in Germany, created an atmosphere of encouragement which caused Sorby to publish the results of the work that he had done in 1864–65 and to extend it.

12. Etching for Macroscopic Study

In chapter 1 we have seen how differential chemical attack was used to obtain decorative surface effects on metals, and how in guns and swords an etched structure was regarded as desirable because it indicated either the proper forging of a mixed material or a proper composition and method of fabrication. This undoubtedly led to some control of manufacturing procedures in terms of structure, but not for a long time did it lead to any significant scientific speculations. Toward the end of the eighteenth century there began to accumulate observations by scientists rather than artisans on the relative rates of attack of various materials, and this led first to the study of flow in worked iron and steel and finally to observations of local orientation and composition differences in the manner of the modern metallographer. At the beginning of the twentieth century, etching had largely displaced fracture as a means of revealing structure.

Although the distinction between materials is inherent in the use of etching on the Damascus sword or gun barrel, it was not until Grignon's paper in 1762 (published in 1775)[1] that there was a specific description of the differences between materials. Johann Cramer remarks (1739)[2] that different colors of steel and iron can be seen either on the fracture or on a hardened polished surface.

If then you harden it again by extinguishing it in cold Water, and polish it, the Veins of Iron may be very well distinguished from those of Steel: For, the Iron-ones are more whitish, and almost of a Silver Colour, but the Steel-ones are of a darker Dye and almost of the Colour of Water.

Cramer, who was outstanding for his time in the way he consolidated theoretical and practical knowledge of metals, nevertheless was quite insensitive to structure.

Grignon observed that iron made by ordinary processes was intrinsically heterogeneous (either due to its nature, its treatment, or the effect of fire) and that this heterogeneity could be noticed in the manner in which iron rusted or scaled on heating, for the impure parts are diversely arranged and the first to be attacked by rust or fire. Similarly a piece of iron subject to continuous rubbing, as by the movement of the hand of a workman, eventually developed an uneven surface, with its fibrous structure exposed either by mechanical or chemical action. Grignon investigated the effect of nitric acid on iron, observing that the base of the columns of bubbles came from specific locations on the iron surface, related to those parts of the iron which were preferentially attacked. If the acid were removed and the iron washed after a reasonable length of attack, there would be visible brilliant projecting fibrous parts in straight, contorted, or inclined situations, depending on the different circumstances at the time of the formation of the bloom. In a note to his 1783 French translation of Bergman's *De analysi ferri*[3] Grignon remarked that,

It was a very long time ago that I used nitric acid to observe the constitution of . . . steel. If you place a polished plate of steel, such as a razor, or knife, or other piece, in nitric acid diluted with twice its weight of water, this acid dissolves steel more actively than iron, so that it is formed into veins, cavities and prominences. The field of steel becomes depressed and the iron stands out and is white while the steel is of dark grey colour.

Although Grignon was thus completely aware of the use of acid to reveal chemical heterogeneity of wrought iron and steel, he did not etch the cast materials which he had studied so well by fracture and shrinkage (chap. 11) or notice any effect of crystallinity on chemical attack.

Long after Grignon's work was presented to the French Academy, but before it was published, a paper appeared in Sweden—whence came so many metallurgical advances—which is of great importance for its indirect influence on science, though it seems to be almost unknown. This is the paper[4] by Sven Rinman, "Om etsning på järn och stål" ("On Etching upon Iron and Steel"), read and published in 1774. An English translation of this paper was prepared, probably by Alexander Chisholm for William Lewis, and is preserved with the Wedgwood papers in the British Museum. Because of its importance we give a complete transcription of it in the Appendix to this volume (pp. 249–55). Rinman's work was inspired by a slightly earlier paper by Wäsström[5] describing the newly established Swedish manufacture of Damascus gun barrels to which he (Rinman) had appended some additional remarks on forging—in both senses of the word—Damascus barrels. Rinman also discusses etching in the long section on the corrosion of iron and steel in his book *Försöck till järnets historia*,[6] which was published in Stockholm in 1782 and translated into German both by Georgi in 1785 (an incomplete and poor translation) and by the great Karsten in 1814. Not only is the 1774 paper of great interest in connection with the structure of metals, but Rinman's experiments on the solution of cast iron are obviously a direct antecedent of the famous *De analysi ferri* by Rinman's compatriot Bergman,[7] which led to the identification of material carbon as the cause of the significant difference between wrought iron, cast iron, and steel.

Rinman begins with a purely practical justification, claiming that it was important for many industries (ink-makers, curriers, tinplate manufacturers) to know the differences between various kinds of solutions of iron and the manner of their working.

Were we sure of knowing fully both the principles of the metals and their dissolvents' different habitus in all possible varieties, we might be able to tell beforehand what effect any one ought to produce: But experience convinces us that we cannot thus judge with certainty of the event.

Aqua fortis has a strong attraction for inflammable matter, but the rule that the more inflammable material the more action has many exceptions. Consider gray and white iron, for example. Gray with more inflammable matter dissolves with more difficulty in the acid and

leaves in the solution a black sediment, sometimes of equal bulk and shape with the bit of cast iron itself, consisting of a matter like black lead [*en Blyertslik materia*], which is found to be iron earth overcharged with phlogiston: Take away a part of this grey iron's redundant phlogiston, either by melting it into white iron or by cementation with absorbents in Reaumur's way into malleable iron or steel, and it becomes soluble in aqua fortis without residuum.

Rinman observes that iron and steel dissolve at quite different rates in common parting acid, and that steel, though losing less in weight, is covered with a dark, ash-gray film, while the iron, much corroded, remains white and silvery bright with a few furrows and striae. Five polished pieces of different kinds of iron and steel etched in warm 1:2 nitric acid showed considerable differences in both the rates of attack and darkness of the surfaces. A "burn-steel" (cementation steel) had "small somewhat elevated black points sprinkled evenly all over, and it could be seen how the parting water had dug a little pit round every point." Rinman experimentally etched a composite damasked bar with five kinds of iron and steel, using several different reagents. He noticed that concentrated parting acid gave very sharp edges to the strips but that dilute acid produced a more agreeable effect with better shading. A usual vitriol etch for Damascus blades (an aqueous solution of alum, copper sulphate, and salt) needed a long period to be effective, but adding a quarter of its volume of aqua fortis made it faster acting and gave an agreeable and elegant shading. Similar salts dissolved in vinegar worked slowly and precipitated copper, particularly on the hard steel lines. Commenting that the etched surfaces were more or less black as the metal contained more or less inflammable matter, it was possible, Rinman said, to "judge in some degree of the steel's hardness from the degrees of lighter or darker grey colour which it takes in the etching." Among his conclusions he remarks that "etching affords an easy way of distinguishing the different iron and steel kinds, in regard to their hardness, compactness, and uniform or dissimilar internal properties."

The step from the black material occurring when there is more inflammable matter in the iron and of the identification of the matter like black lead in cast iron "overcharged with phlogiston" is indeed but a small step from Bergman's:

Iron in the state of steel is very near to that of cast iron; it contains, however, many fewer heterogeneous parts than cast iron but more than ductile wrought iron. Steel is richer than cast iron in elementary caloric matter and in the inflammable principle. It appears that plumbago is a principle necessary to iron in these two states but in different but definite proportions for each one.[8]

Though Bergman's quantitative approach using the volume of hydrogen evolved or silver precipitated as an index of available phlogiston is superior to Rinman's, there can be little doubt of its origin in the metallurgist's studies of etching that were precipitated by a practical desire to produce a good copy of a Turkish gun barrel. Here is one more example of an Oriental technique profoundly influencing the development of modern scientific ideas.

The cutler Jean Jacques Perret in 1771 had remarked upon the manufacture of a pseudo-Damascus steel sword by forging a complex faggot, twisting and forging, and recommended the use of nitric acid to bring out the pattern.[9] In 1779 he published a paper[10] (written for a prize competition of the Société des Arts of Geneva in 1777) in which he was more specific, and reports that either under *eau forte* or lemon juice the materials in a forged Damascus steel each take their nuance: "the steel becomes a bluish black, the soft iron finds itself a dark grey, and the black iron a harsh whitish grey."

Guyton de Morveau[11] in his long summary article "Acier" in the *Encyclopédie Méthodique* (1786) includes also some experiments of his own on various matters to do with iron and steel. He uses etching, *inter alia,* for studying differences between quenched and slowly cooled steel and cast iron, on which he notes particularly that it is possible almost immediately to tell the difference between a white and gray cast iron with a drop of nitric acid. He used the lapidary's mill for polishing metals and, following Rinman, he rated eight different kinds of iron in the order of their nuance as precisely parallel to the order of their quality as classed by Buffon and Grignon. De Morveau, much influenced by Bergman, partially frees himself from the Swede's phlogistonism, and is one of the first to insist on a purely material explanation for the difference between iron, steel, and cast iron, which he believed to be the presence of carbonaceous matter.

It is now easy to suggest the immediate cause of the stains which acids always produce on cast iron and steel and which are not observed on touching iron with the same acids: they are due to the precipitation of plumbago.

In 1786 also appeared the remarkable paper by Vandermonde, Berthollet, and Monge.[12] In year II of the Republic (i.e., 1793–94), since "the tyrants of England and Germany" had broken off trade with France, it was necessary for the French to make their own steel, and the Committee on Public Safety published a booklet[13] by Monge, Vandermonde, and Berthollet summarizing their theory for workmen in very clear terms. It concludes with practical hints, among which is a description of a test approved by the Committee, which had ordered its agents to subject all arms submitted for purchase to an acid test. This section reads as follows:

If a drop of nitric acid [*eau forte*] is deposited on a blade of polished iron, and if after it has been left there for two minutes water is thrown upon it, the water will remove the acid and everything it holds in solution, so that nothing but a white spot will remain, a spot of the color of freshly pickled iron.

If the same operation is performed on a blade of polished steel, the acid again cuts slightly into the ferruginous part; but it has no effect on the carbonaceous matter. The latter is therefore deposited during the process of solution and forms a black spot which is not removed when water is thrown on; it even stays there a rather long time because there is some adherence.

In order to be successful with the test, a weak acid, or acid diluted with water, must be employed; for the carbonaceous deposit adheres only if the solution takes place slowly and without too much effervescence. If pure or distilled nitric acid in unavailable, the aqua fortis of commerce may be used, always diluted to a certain extent. One must be careful always to carry the drop of acid on glass or some other material which cannot be attacked by the acid and which contains nothing that could influence the result. A very small drop is enough; it must be spread out rather than be held together in order to mark a larger area; the stopper of a very small flask in which the acid is held is most suitable for the purpose. After having made this test two or three times, both on iron and on steel, for comparison, one has acquired the necessary touch to become sure of one's pronouncements on the basis of the different results obtained. [Trans. A. G. Sisco.]

After this the use of etching became routine and references to it are very common.

Among the many attempts to duplicate Damascus steel which are referred to in chapter 5, the work of Stodart and Faraday[14] on the "alloys of steel" deserves particular mention. They used the action of acids on all of their samples of synthetic wootz, and to show the damask on cast and forged pieces of their various alloy steels. They also used heat-tinting (which they knew to be a result of differences in rates of oxidation of the different parts of the surface); they observed the extrusion and the retention of globules of silver in castings and of fibers in forgings of silver steel (the first alloy steel to become—undeservedly —popular); and they described beautiful dark and white clouds on iron-platinum aggregates made by welding and forging.

Although Biringuccio was aware (1540) of segregation in castings and added tin to the gun-head to correct it, non-ferrous foundrymen seemed to have lagged far behind their ferrous counterparts in studying heterogeneity by chemical attack. Wade[15] in 1854 was apparently the first. He remarks that heterogeneity due to small cavities or to minutely subdivided particles of a white alloy in a bronze gun was hidden by machining, but could be revealed either by exposure to the weather for a few weeks or by the use of concentrated nitric acid followed by either lye or hydrochloric acid for five or six minutes.

In 1801 William Nicholson[16] found that acid worked differently upon the hard and soft parts of a steel of uniform composition, but, in general, all this early etching was essentially chemical in nature, unconcerned with differences in structure except when this was directly related to gross variations in composition. Direct descendants of these methods were used to extract relatively insoluble non-metallic inclusions for separate analysis, first the iron carbides studied in detail by Karsten and others and eventually a large number of intermetallic compounds in other alloy systems.

Investigations of etching in relation to crystalline structure seem to have lagged considerably behind the chemical ones, and it arose from a study of objects in which a pronounced chemical segregation occurs in an extremely coarse crystallographic pattern.

One of the many important interactions with other branches of science that has enlivened metallurgy occurred with the revival of interest in meteorites at the beginning of the nineteenth century. In 1804, William Thomson, an Englishman resident in Italy, examined a sectioned pallasite and remarked on the octahedral crystalline structure that was revealed in the iron etching in acid. This work was published in 1808,[17] but attracted no interest. In the same year, Alois von Widmannstätten, curator of the Royal Mineral Collection in Vienna, etched a section of the Zagreb iron and observed the structure that has since come to be known by his name.* Although both Widmannstätten and Carl von Schreibers, who worked with him, did not keep these matters secret, a complete description was not published until 1820. This was in Schreibers' atlas of meteorites,[18] which contained a number of lithographed illustrations of whole and sectioned meteorites, including some of etched irons, and a superb print (see illustration, p. ii) made directly from the etched surface of the Elbogen iron meteorite. This print, which is on a separate sheet of paper tipped into the bound volume, was apparently one of many made in 1813 by Widmannstätten and Schreibers working together. The clarity of it surpasses even the modern photograph, and it has details that only appear on magnification (Fig. 79). No subsequent prints approach this first one in quality. Schreibers discusses at length the etching of plane sections of the meteorites with nitric acid (which he prefers to sulphuric or hydrochloric),

* Although Sorby referred to a slowly cooled synthetic iron-nickel alloy as having "a structure which may be called Widmannstätten figuring on a very small scale," the term was not commonly applied to artificial structures until after Arnold and McWilliams reported the structure in annealed steels (*Nature*, 1904, **71**, 32). The most important early work was that of N. T. Belaiew, *Kristallisatsia, Struktura i Svoistra Stali* (St. Petersburg, 1909). Belaiew obtained well-developed geometric patterns in a sample of 0.55 per cent carbon steel which he had asked to have left to cool slowly with the crucible furnace in which it had been melted. He was unable to duplicate the structure and eventually he found that his instructions had been disobeyed and that commercial demand for the furnace had resulted in the sample being removed at about 850° C. and air cooled, thus preserving the pro-eutectoid ferrite in clear form (*Mining and Metallurgy*, January, 1930, pp. 69–70). One of the first applications of X-rays to orientation measurements was the elucidation of the Widmannstätten structure in terms of the mutual orientation of the crystals of the two constituents (Young, 1926, followed immediately by the broader study of Mehl and his collaborators) and even to the present day is an important objective of metallurgical research, although the scale of interest has shifted nearer to the atomic. See R. F. Mehl's "On the Widmannstätten Structure," *The Sorby Centennial Symposium on the History of Metallurgy*, ed. C. S. Smith (New York, 1965), pp. 245–70.

Fig. 79.—The Elbogen Iron Meteorite. The original is a direct typographical imprint from the etched surface. (Schreibers, 1813.) ×2. See also illustration, p. ii.

cautions against the use of an improperly prepared or damaged surface, and mentions the possibility of heat-tinting to distinguish between the constituents.

In the days before photography the invention of the method of copying metal structures by printing was itself a great advance. Schreibers says that after the etching has progressed to a certain point (about 1/144 inch at the most depressed parts, which should not be exceeded if the intermediate parts are to print)

the whole design can be directly printed from the surface with printer's ink. The most elevated parts will print strongly, the less elevated ones more weakly and the depressed ones do not print at all. Since they all alternate in an

151

orderly way and are connected with each other, we obtain in this way not only a quite perfect and accurate reproduction of the etched surface but also a true picture of the natural structure of the mass as brought out by etching.

Schreibers concludes that the structure is

to be considered as characteristic of meteoric iron, and as occurring exclusively in such iron. It seems to point to a qualitative and quantitative relationship, to a combining and breaking apart of chemical and structural constituents according to definite laws of affinity and crystallization, and to a process that cannot be explained by analogy with anything quite similar that occurs on our planet.

Schreibers has a footnote which seems to be the only contemporary description of Widmannstätten's discovery and we therefore reproduce it *in extenso:*

The discovery of this peculiarity of native iron of truly meteoric origin has been reasonably well known for several years. It was made by Director Von Widmannstätten as early as 1808 during his physical and technical investigations on the Agram iron. We had no intention of keeping it a secret. On the contrary, we spread the word to all our scientific friends as the occasion arose; and the mass, a sizeable plane surface of which had been etched (as mentioned above) in order to bring out the structure, became in 1809, as it had been before, part . . . of a public exhibit of minerals in the Royal Mineral Cabinet. The same year, Herr Von Widmannstätten had an opportunity to confirm his interesting discovery on an exceptionally beautiful piece of Siberian iron from Von der Null's collection . . . in 1810, on a piece of Mexican iron which Klaproth had just given to the Emperor's collection; then, in 1812, on the large native-iron mass given by the City of Elbogen, in Bohemia . . . and finally, in 1815, on the piece of Carpathian iron. . . . Since, upon etching of flat and polished surfaces this structure appears as palpable configurations (in *bas relief*) which, depending on the duration of the treatment, are more or less raised or depressed, Herr Von Widmannstätten had at once, when working with the Agram mass, the happy idea to obtain a completely true and easily multiplied picture by making prints of such plane surfaces with printer's ink, using the mass itself as a natural printer's form or stereotype. The good results of this procedure persuaded us in 1813 to make a sufficient number of these direct prints of the large etched plane surface of the Elbogen mass, on which the structure is especially beautiful and clear, in order to use them as illustrations for a treatise which we were then thinking of writing and publishing. But events and circumstances of the time* impeded our work, which required a series of painstaking and uninterrupted experiments and investigations, and finally made us give it up completely (which is just what had happened in 1809 to my earlier, similar undertaking), so that this autograph was buried unused in the files until this day when

---

* Perhaps the international confusion culminating in the Congress of Vienna. The 1809 interruption would have been attributable to Napoleon's arrival.

we finally had another incentive (albeit too unexpected and peremptory) for getting it out.

In the meantime, the news was spread by word of mouth, by strangers and travelers. In the end it was Chladni (who had witnessed our experiments during his stay in Vienna in the spring of 1812), who not only personally discussed the subject but stimulated public discussion. Thus, opinions and experiments were expressed by Neumann of Prague. . . . (1812, *Hesperus*, No. 9), Schweigger (1813, *Journal für Chemie und Physik*, vol. 7), and Chladni himself (1815, in *Gilbert's Ann.*, vol. 50), and our Herr Von Hammer mentioned the subject in connection with conjectures on the oriental damascened blades (1815, *Fundgruben des Orients*, vol. 4; excerpts, *Hesperus*, No. 9). Later on, I myself completed the spreading of information in Germany and France by conversation with scientific friends during my journey to Paris in 1815; and further details became known when I occasionally distributed a page of autographic prints among my correspondents there and in England, which stimulated the remarks of Gillet de Laumont* (*Journal des Mines*, vol. 38, Sept. 1815), Sömmerring (lecture at the Bavarian academy of sciences in February 1816; reprinted in *Bibl. univers.*, vol. 7, and in Schweigger's *Journal für Chemie und Physik*, vol. 20), Schweigger (his *Journal*, vol. 19), and Leonhard (his *"Taschenbuch für Mineralogie,"* vol. 12). [Trans. A. G. Sisco.]

Little less important than the discovery of the structure was the development of the method of reproducing it for record and distribution. Up to this point, the only reproductions of structures had been through the artist's eye and the etcher's needle, the engraver's burin, or the lithographer's crayon. Not only in cheapness but in accuracy, these direct prints were far preferable to anything that was available before the invention of good photoengraving. Schreibers' 1813 print of the Elbogen iron meteorite is the first—and the best—metallurgical application of a rather old technique.†

It is not unlikely that decorative impressions of textured objects can be found in very early ceramics. In China ink rubbings on damp paper were made from stone surfaces at least as early as the second century A.D. and the technique has long been a common means of copying any surface with relief or intaglio details. Even today some Oriental scholars

---

* De Laumont's illustrations are reproduced and his work is discussed by C. S. Smith, "Note on the History of the Widmannstätten Structure," *Geochimica et Cosmochimica Acta*, 1962, **26**, 971–72.

† It would throw interesting light upon the technique of the early European typefounders if some metallurgical detail could be found in the impressions of early type. Though modern type alloys do not shrink, the first alloys, probably tin-rich ones based on pewter, could have developed dendritic or granular surface patterns if carelessly cast in large sizes. No such features have been reported however, and they are not visible in the samples of fifteenth-century printing that the writer has seen.

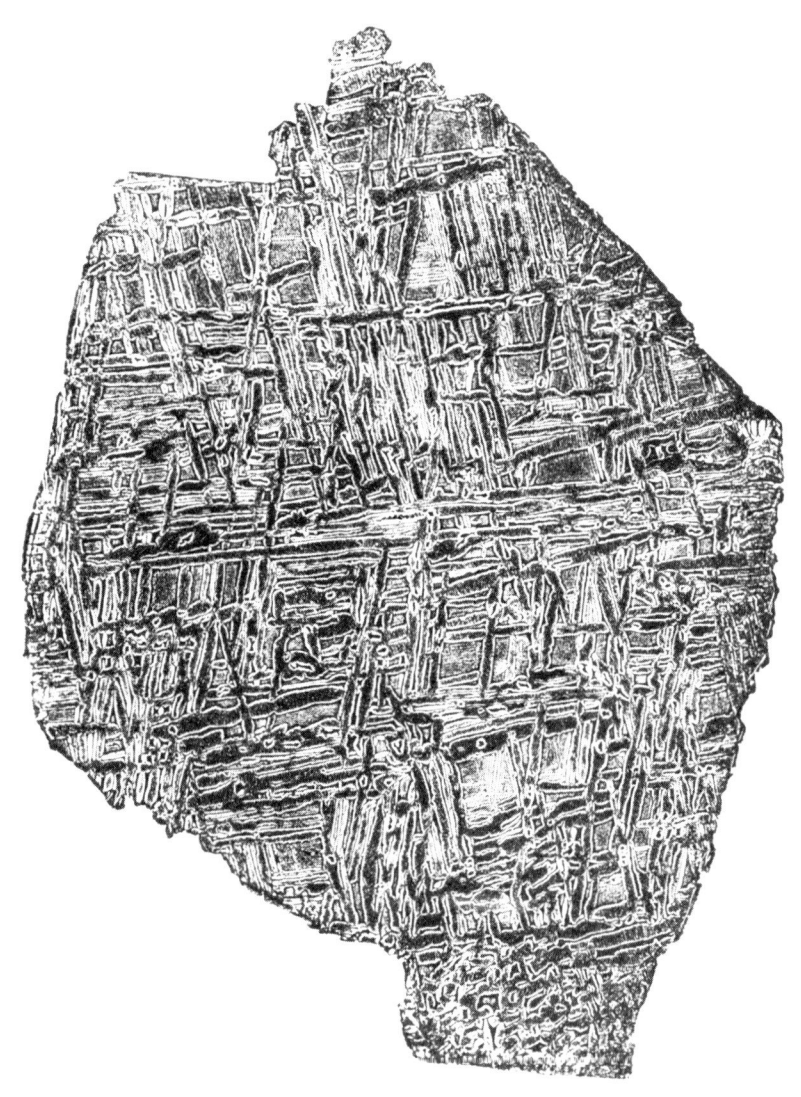

Fig. 80.—The Lenarto Iron Meteorite. Printed from metal stereotype, probably cast directly against the etched metal surface. (Partsch, 1843.)

prefer rubbings to photographic copies of coins, tiles, and similar objects. It is likely that rubbings of Japanese swords exist, far antedating any European copies of metal surface detail, showing metallurgical features of the *yakiba* as well as engraved designs on the blade and the filed surface of the tang. In Europe as early as 1555 impressions of leaves were made on paper by inking them with printer's ink and pressing a piece of paper thereon, whereupon "you will find the natural form of the leaves down to the smallest vein, so that they will seem beautiful and with all their natural marks." A green ink was used to make a verdant frieze around the room in winter.[19] In the eighteenth century a number of botanical books were published with illustrations printed in this way—the first being the *Botanica in originali* of J. H. Kniphof in 1733.

For twenty years after Schreibers there were no more prints of meteorites, then in 1843 Partsch gave a clear stereotype print of one[20] (Fig. 80). Shortly thereafter M. W. Haidinger[21] illustrated his book on mineralogy with some illustrations of meteorites that had been printed from lead plates into which the etched meteorites had been impressed. He followed this with a long series of papers in the *Sitzungsberichte* of the Viennese Academy, containing prints of meteorites, most of which were made by stereotyping, although a few were made directly from the meteorites and, in 1864, from copper electrotypes of lead impressions, an English process invented in 1847.

In an 1855 paper Haidinger[22] included prints from an etched iron bar, taken from the firebox of a steamship, which had become brittle after many years service and had a coarse crystalline fracture—the crystals being as much as one-third of an inch across in some zones. The metallurgist, Prestel, who had sent the sample to Haidinger had found parallel lines which he thought might be related to meteorite crystallinity. Actually, of course, they were stringers of slag, unrelated to the crystalline fracture. The rather poor print shows no crystalline features whatever, but it has the distinction of being the earliest published structural imprint from a man-made metal surface.

By this time, however, the printers themselves were becoming increasingly alert to the need for new methods of graphic reproduction generally. Shortly after Partsch and Haidinger's first work on meteorites, Leydolt[23] experimented with the printing of etched stones, particularly agates, the structure of which, he showed, could easily be recorded on paper rubbed against its inked surface. If the stone were strong he recommended the use of either a typographic or a copper-plate printer's press, for positive or negative printing, respectively. Either electrotyping, stereotyping, or transfer to a lithographic stone could be used if the stone would not withstand the force of the press. All this work was done in Vienna. Leydolt worked in close co-operation with Von Auer, who was manager of the Austrian Imperial Printing Works and very active in studying all possible methods of printing. In 1852, when shown some impressive English lace catalogues printed by lithography, Von Auer surprised his customer by producing direct copies on paper within twenty-four hours. This he did by a newly invented process called *Naturselbstdruck* ("Nature self-printing"), which was obviously a direct outgrowth of his work on meteorites and mineralogical samples. It involved passing the object to be copied through a roll press in contact with a lead plate, which took a detailed impression of the surface. The lead plate could be stereotyped or electrotyped, while if the objects were very fragile they themselves could be coated with gutta percha and electrotyped. A display of earlier prints (including fossil, fishes, and animals, but no minerals or metals) had been shown in the 1851 London exhibition and gained a prize medal. In 1853 when Auer published his *Polygraphische Apparat*,[24] he included fine examples of plates printed by virtually every method known at the time. He shows, as examples of nature printing, several botanical specimens, two superb agates, a fossil fish, and the meteorite reproduced in Figure 81, which he had evidently obtained from Partsch, though he gives no details of its preparation. Auer was inordinately proud of his invention and regarded it as one of the three important moments in all history—the invention of writing, that of printing itself, and his nature printing!

John Bradbury, an Englishman who had worked in Auer's plant, claimed in 1852 to have invented the process himself, and set up shop in London where he published several fine books. Auer was rightly enraged at the theft and to protect his claims he published a fine polyglot book in 1854[25] which contained some large, particularly fine prints of rocks of various kinds but no meteorite.* As we will see in the next chapter, Sorby knew of Partsch's work and made use of the process himself. Being unaware of Haidinger's work he believed his prints to be the first of non-meteoric metal. The only earlier print in Britain was the extremely poor one of a meteorite given by Smith in 1862.[26]

* The history of nature printing is discussed by Hardie[26] and by Fischer.[27] The latter gives an extensive bibliography of books or journals illustrated by the method.

Fig. 81.—An iron meteorite. Printed from a lead plate bearing impression of etched surface. (Auer, 1853.)

However, to leave typography and return to metallurgy, we revert to the period immediately following Widmannstätten's discovery. In 1816, Wollaston,[29] probably after hearing a rumor of the Viennese experiments, used acid in a clumsy way to detect crystallinity in what he presumed to be a mass of native iron. He found that under the hammer it showed a tendency to break into regular octahedra and tetrahedra and "the crystalline surfaces appear to have been the result of a process of oxydation which has penetrated the mass to a considerable depth in the direction of its laminae."

In the following year, J. F. Daniell[30] discussed at length the manner of solution of crystals, and pointed out how the geometric forms are just as determinate as in the case of growth. He found that a lump of bismuth acted upon by nitric acid for a few days was left with its surface covered with small cubic figures which presented the same curious linear arrangement which is observable in the artificial crystallization of that metal.

Antimony was dissolved in the same way; and the parts which most resisted the power of the acid, presented a series of rhomboidal plates.

Nickel was covered with perfectly defined regular tetrahedra one-twentieth of an inch long, but no regularity was observed in the relative position on the mass. Other metals did not give satisfactory results. Daniell proceeds to explain the phenomena on the basis of Wollaston's spherical packing hypothesis, which had been presented just a few years earlier. Daniell is practically the first man to state clearly the lack of the necessity for a relationship between regularity of internal structure and that of external shape.

We have seen that crystallisation does not necessarily include symmetry of outward shape, but that a mass of metal which has been melted in a crucible, and in cooling has adapted itself to its cavity, has its constituent particles as regularly arranged as the most mathematical figures of a crystallised salt.

He speculates on the nature of a nucleus and comments on the similarity of structure between liquid and solid metals. Of greatest interest is his discussion of grain boundaries and the geometry of intercrystalline junctions, a subject of which twentieth-century metallographers, including the writer, have made much. Discussing growth from several nuclei, he points out that each will continue to attract particles from the medium until two come in contact.

A greater number [of particles] will be collected at the point of junction, than at any other, and they will therefore arrange themselves in the least possible space. Accordingly we find, that whenever a crystal is attached to another, in such a manner, that their axes run in contrary directions, if we pull the two asunder, we shall invariably be presented with a regular hexagonal arrangement at the point of junction, whatever be the form of the crystal, the nature of the substance, or the direction in which any part of it would be disposed to separate by mechanical force. This observation has been repeatedly verified upon carbonate of lime, selenite, fluorspar, quartz, topaz and other mineral bodies.

Daniell was wrong in packing more particles at the point of junction than in the crystal, but he was absolutely right in regarding this as a configuration that tended to the least possible space and in seeing that the geometry of intergranular junctions was independent of the structure of the crystals themselves. It was almost a century before this concern with grain-boundary equilibrium angles reappeared in scientific literature.

An even more important paper followed from Daniell's pen in the following year.[31] This, entitled "The Mechanical Structure of Iron Developed by Solution . . . ," clearly states the importance of the subject and the value of the method of macroetching.

In prosecuting my inquiries into the resistance which mechanical structure offers to chemical action, I have been led to bestow considerable attention upon the difference of the molecular arrangement of various kinds of iron. No subject stands more in need of illustration, nor is there any, perhaps, which is more likely to lead to useful practical results, than one which concerns a substance of such primary importance to the arts.

Daniell was disappointed not to see well-defined polyhedra, but at least he discerned the importance of what he saw.

Although mathematical solids were not discovered by the solution of iron, yet a difference of structure was plainly discernible in the different varieties submitted to the experiment, which is well worthy of attention.

He found gray cast iron, after it had been immersed in dilute hydrochloric acid and the layer of spongy graphite removed,

to be covered with small irregular ridges, which, when examined with a magnifier, presented the appearance of bundles of minute needles.

A bar of wrought iron

presented the appearance of a bundle of fasces, the fibres of which it was composed running in a parallel and unbroken course throughout its length. At its two ends, the points were perfectly detached from one another, and the rods were altogether so distinct, as to appear to the eye, to be but loosely compacted.

White cast iron of a radiated fracture

appeared to be composed of a congeries of plates, aggregated in various positions, sometimes producing stars upon the surface, from the intersection of their edges. It exhibited altogether a very crystalline appearance, but no regular patterns were discoverable.

He comments for the first time on the lack of connection between fracture and other measures of structure. A bar of cold short iron with a fracture much resembling antimony, on testing with acid

proved to be fibrous but not so perfectly so as the first specimen of bar iron. The course of the fibres was very much broken, the acid having dissolved out small cavities which cut them short. It was a square bar, and the alternate sides were more acted upon than the others, so that the fibres would moreover appear to have been flattened.

A rod of *hot short iron* presented at the end of the operation, a closely compacted mass of very small fibres, perfectly continuous. The congeries was twisted, but the threads preserved their parallelism. A portion of a gun-barrel was submitted to the experiment. The metal was remarkably free from particles of an extraneous nature. The texture proved to be fibrous, but the threads were not regular or straight. They were generally disposed in waved lines, and the whole together was very compact.

A mass of steel just taken from the crucible in which it had been fused was subjected to the action of muriatic acid. It was of a radiated texture, the upper surface being marked with rays which proceeded from the centre to the circumference. It was readily acted upon by the solvent, and when withdrawn, presented a highly crystalline arrangement. It appeared to be entirely composed of very bright and minute plates which reflected the light in every direction. The laminae were very thin, and there was no order discoverable in their mutual positions.

A specimen of cast steel which had been subjected to the action of the tilting hammer of a very fine white granular fracture was next examined. It was not easily acted upon even by strong muriatic acid, and it required the addition of a small quantity of nitric acid to effect its decomposition. When the acid was saturated, the metal still presented a compact appearance; nothing of a fibrous structure was visible; but in one or two places where the acid had acted with most energy it had detected the edges of laminae, which appeared to form plates of the extent of the whole surface.

The blade of a razor composed of Wootz steel presented the same appearance, differing in nothing except three large notches in the back at right angles to the edge.

The blade of a razor of inferior description presented a fibrous texture of waving lines. Deep notches in the back similarly placed were likewise visible in this. It was sufficiently evident that the fibrous texture of this razor was owing to the admixture of iron and to the imperfection of the process for converting it into steel.

A bar of steel of an even granular fracture was broken into two. The two pieces were heated in a furnace to a cherry red. In this state one of them was plunged into cold water, and the other allowed very gradually to cool by the slow extinction of the fire. They were then both placed in muriatic acid, to which a few drops of nitric acid had been added. The last was readily attacked, but it required five fold as much time to effect the saturation of the acid as the first. When the solvents had ceased to act they were both examined. The tempered steel was exceedingly brittle, its surface was covered with small cavities like worm-eaten wood, but its texture was very compact and not at all striated. The untempered steel was easily bent and not elastic, and it presented a fibrous and wavy texture.

Speculating on the fibrous nature of many materials, Daniell believed that good iron, like good rope, should have the particles well interlaced and compacted. This he thinks is the advantage of the method of gun-barrel manufacturing, and perhaps Damascus swords, whose structure seems to answer to small rods of iron and steel welded and twisted together, could actually be made that way—a somewhat tardy suggestion. Finally:

The good qualities of steel seem to depend for different purposes upon a varying mechanical arrangement of its particles. This difference of structure is conferred by certain regulations of temperature. We find that the same bar of metal suddenly cooled from a high temperature is possessed of a quite different texture, and different mechanical properties from those which characterise it, when gradually lowered. May not the qualities of cast iron vary also with the rate of cooling? And might not a proper regulation of heat improve the fibrous texture, or even confer a certain degree of malleability?

Daniell's descriptions are clear, the subject obviously important, and the medium of publication authoritative and widely distributed, yet the seed fell on barren soil and was slow to germinate. He discussed the polyhedral and spheroidal theories of crystallization in a paper in the 1831 *J. Royal Institution*, but a decade later in the introduction to his own *Study of Chemical Philosophy* (1843), Daniell summarizes[32] his experiments in two sentences and he no longer implies they may be of use. He had even retrogressed to the view that the markings on etched iron "cannot exactly be called crystalline because they do not present exact geometric figures."

During the next four decades, which saw a great systematization of chemical knowledge, a vast increase in the engineering testing and utilization of materials, and rather wide use of etching to reveal chemical heterogeneity, there was no further development of etching in relation to crystallinity or fiber. Finally, however, the increased interest of engineers in materials made some advance inevitable. In 1860 Lüders[33] described the appearance of the now-famous bands associated with his name, which could be seen in stretched steely iron covered with a uniform scale, or by etching with nitric acid, when they became visible as a light color on a dark ground.* They could not be seen on all specimens, and needed a very careful etch. Lüders realized that it was not a result of previous heterogeneities and attributed it to localized deformation on the shear planes and as a result of molecular commotion which left the metal somewhat harder and differently etchable.

At about the same time the Scotch engineer David Kirkaldy[34] became interested in the reports on the crystallization of iron by vibration. Kirkaldy had heard from "Mr. Tomlinson" (a source that has not been traced) that "the fibres [in commercial iron] may be made evident by the fracture of the bar under tension, or by acting upon it with dilute muriatic acid." Studying the reports of the English and French discussions on fatigue (see chap. 10), he concluded with Robert Stephenson

---

* Somewhat earlier Piobert had described the pattern of lines surrounding the point of impact of a projectile on an iron plate. (*Mém. de l'Artillerie*, 1842, **5**, 505.)

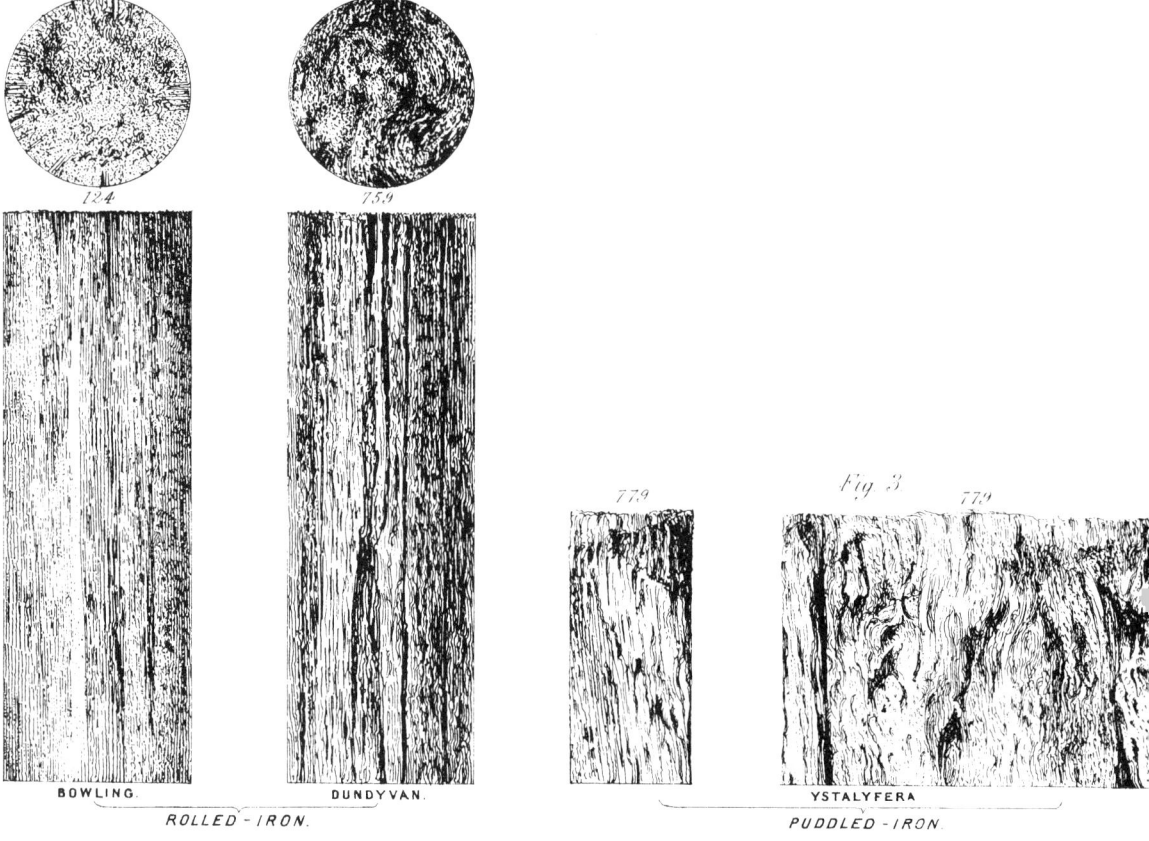

Fig. 82.—Texture of wrought iron developed by etching in acid. (Kirkaldy, 1862.)

that there was no change in structure, but only in the manner of fracture. Proceeding, he remarks:

It occurred to the writer that much additional information would be gained by immersing a few pieces of the various classes in dilute hydrochloric acid which, acting upon the surrounding impurities, would expose to view the metallic portions alone for examination. To the beautiful specimens thus obtained he now solicits attention. . . .

Some of his illustrations are shown in Figure 82. These experiments were probably done between 1859, when an early report which does not include etching was read to the Institution of Mechanical Engineers in Scotland,[35] and 1862, when the experiments were published. His book was an important contribution to the experimental knowledge of engineering materials and had wide influence. Many people followed

Kirkaldy's lead and macroetching became a fairly standard means of studying the quality of metal, but Kirkaldy himself did not use etching in his later comparison[36] between the mechanical serviceability of cast and wrought iron in 1876 or in the material that he contributed to the 1867 revision of Peter Barlow's *Treatise on the Strength of Materials*.[37] Yet he certainly did appreciate its value in 1862:

The effect produced by repeated rolling is at once apparent when we compare a highly worked or superior quality with a slightly worked or inferior. In the former the appearance presented by the bar is that of very fine and straight long hairs or threads lying close together as in [Fig. 82 *1*]. In the latter, to some extent, there is the same thread-like appearance but indistinct and wavy, very irregular in size, in some parts lying close, in others far apart, as in [Fig. 82 *2*]. Puddled iron rolled, or wrought iron in its lowest state, present more a wooly than a thread-like appearance, as in [Fig. 82 *3*]; the latter development (threads) seems as if commencing to be formed by the first rolling, the former, or wooly resemblance, to be due to the puddling process. . . .

Kirkaldy's important contributions to fatigue were discussed in chapter 9.

Shortly after Kirkaldy's study of semifabricated material, Henri Tresca published his work on the flow of material during various kinds of deformation.[38] He took the viewpoint that under sufficiently high pressure (really pressure gradient) solids will flow like liquids, and he showed dramatically the manner of flow of stacked plates of metals extruded in various geometries. Then he remarks that, in the case of iron, it is not necessary to use separate plates to reveal the flow, for chemical action will show it. Iron is not homogeneous and foreign substances participate in the molecular movements produced by flow in all parts of the mass. Any differences in chemical nature of the metals allow the diversity and hence the results of flow to be rendered manifest by the action of chemical reagents. Tresca preferred to etch in a dilute solution of bichloride of mercury, which was used on a machined flat surface, dry-polished with emery powder. When the first traces of action appeared, the sample was transferred to water to develop the structure, then dried, varnished, and photographed. Figure 83 is typical of Tresca's illustrations. Iron which had not been forged too extensively would show a succession of separate and well-defined lines, but in rolled rods the details of the coloring could rarely be distinguished without the aid of a microscope and could not be photographed. In a discussion on this paper Mallet[39] expounds the view that fibrous and crystalline states are

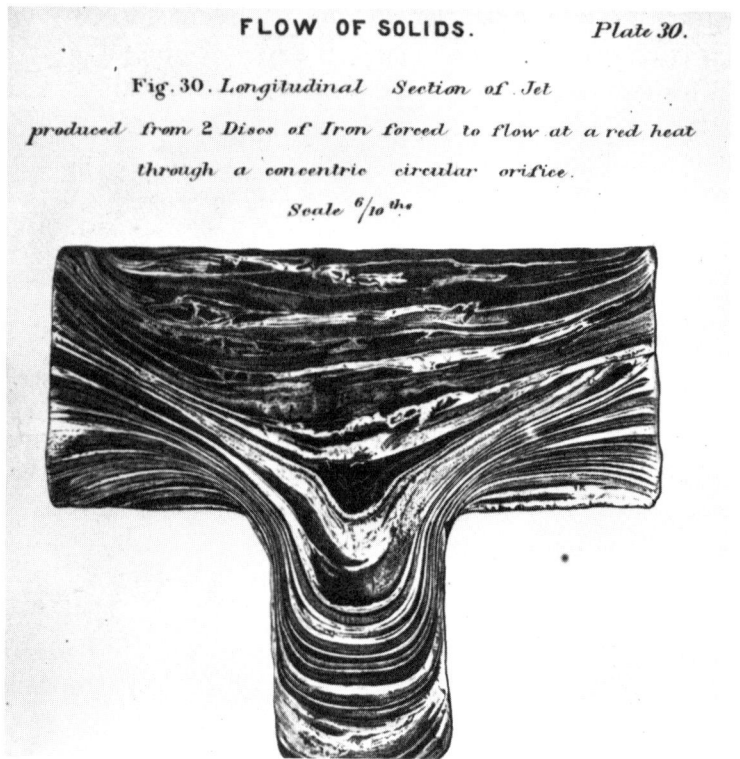

Fig. 83.—*Flow patterns in iron revealed by etching in mercury chloride solution. (Tresca, 1867.)*

both crystalline and the difference is only one of malleability. The fibers in wrought iron were simply the original crystals elongated. A gold crystal, he remarks, even though it elongates, still maintains its crystalline condition, as indicated by the directions of the striae produced upon its surface with acid. Previously Mallet had stated[40] that

there appears to be no more good reason to doubt the truly crystalline arrangement of the molecules of iron than there would be to doubt the isolated octohedral crystal of native gold, was truly a crystal, because, by the blow of a hammer, we can flatten it into a spangle. . . . Metallic crystals are . . . therefore susceptible of distortion (to almost any extent, in the more malleable metals), and of re-formation, without external change, except as to form, in the mass itself.

In a later paper[41] Tresca describes the study of a wide variety of metal deformation methods—punching, forging, machining, etc.—in all of which he shows the flow of the fiber of iron developed by etching well-polished sections, the structure being shown either by a deposit of copper from solution, by the action of an acid, or better still by the action of bichloride of mercury, which clearly shows the arrangement of the fibers, enabling one to trace through all the deformations of a piece the molecular displacements which would have remained undetermined without this demonstration. He remarked that the hydrochloric acid etch was so effective that the surface could be inked and proof prints taken from it in which the directions of the fibers are perfectly distinguishable. The mercury chloride etch more clearly and delicately defined the fibers.

Up to this point—the beginning of the last quarter of the nineteenth century—most investigations of structure had been macroscopic and superficial. Though there had been many astute observations, the science of the structure of metals was almost nonexistent. Yet much was known in a disconnected kind of way and the situation was ripe for the detailed attention by individuals whose background was more scientific than practical, and a period of rapid development followed. The individual whose work precipitated this was Henry Clifton Sorby, to whom the next section of this book is devoted.

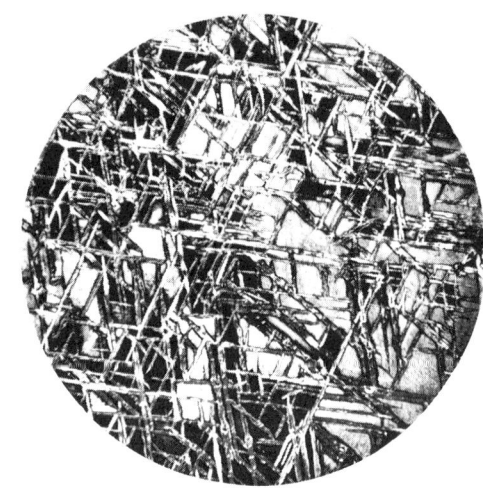

White cast iron and meteoric iron, ×9. Photomicrographs made by Sorby in 1864.

SECTION IV

THE WORK OF
HENRY CLIFTON SORBY

Photo courtesy of Professor A. G. Quarrell

I N PREVIOUS chapters we have seen how by the middle of the nineteenth century there had accumulated an extensive background of fracture tests for control purposes and of the use of etching for decoration and for identification. Much useful speculation on the structure of metals had been indulged in, but there had been little attempt to associate observations and theory in a manner which placed the former in proper perspective and gave the latter an air of reality. Against this background the work of Henry Clifton Sorby (1826–1908) is truly outstanding.

13. Metallography in Sheffield, 1863-87

Henry Clifton Sorby was one of those English amateurs of science who have contributed so much to its advance. He was a man of independent means and devoted his life wholly to science. Being free from professional responsibilities, he was able to follow a particular subject only as far as his interest took him, whereon he abandoned it for some other field. He was a pioneer, not a consolidator, and his highly diverse interests gave a unique color to all of his achievements.

He was a superb technician, though he was interested only in the knowledge to which the techniques led. He is most famous for his work in petrography and metallography, but he made important contributions to biological microscopy (particularly the technique of differential staining), microspectrographic analysis (including the identification of human blood), the archeology of masonry, flow and sedimentation in rivers, and ancient natural history.

He was born on May 10, 1826, in a country house on the outskirts of Sheffield, and he lived in Sheffield all his life.* He was descended from a long line of master cutlers, and his income arose chiefly from the family business. Sorby's formal education was at the Collegiate School of Sheffield (now the Sheffield Grammar School), which was followed by a period of tutoring in mathematics by the Rev. Walter Mitchell, who instructed the young Sorby in chemistry and anatomy as well as mathematics and was the coauthor of a popular treatise on crystallograpny.[3] Sorby was inspired to devote himself entirely to a scientific career, and he studied many subjects, not to pass an examination, but to qualify himself for a career of original investigation. His first paper was on agricultural chemistry, but this was soon dis-

---

* Some of this biographical information is taken from the unpublished biographical notes prepared by Dr. D. W. Humphries for the British Association meeting in Sheffield, August, 1956. Desch in the Second Sorby Lecture (1921)[1] discusses Sorby's metallurgical contributions.

Since the first printing of this book, there has been published a full biography of Sorby by Norman Higham, *A Very Scientific Gentleman: The Major Achievements of Henry Clifton Sorby* (London, 1963), and three important biographical articles.[2]

placed in his interest by geology. A chance contact on a journey from Scarborough to York brought him in contact with W. C. Williamson (later professor at Owens College) who was adept in preparing thin sections of wood, teeth and bones, and other biological material for microscopic study. Williamson taught this technique to Sorby and in 1849 Sorby was applying the method to rocks. He used polarized light —both parallel and convergent—and found how to distinguish between calcite, quartz, and calcedony by methods which he described in a Geological Society paper in 1850. He continued to work on geological subjects, though his papers aroused no enthusiasm. In 1852, on a journey to Germany with his mother, he encountered the German mineralogist Ferdinand Zirkel, who, fresh from graduate school, was so inspired by Sorby's conversations that he consecrated his entire life to the study of microscopic geology, reducing it to a formal discipline and gathering around him many workers. Shortly after his return from Germany, Sorby revived an old interest in meteorites, and from there he passed naturally to artificial metals. It is unfortunate that there was no Zirkel in metallurgy to follow up Sorby's wonderful beginnings, for his work in this field remained relatively without effect for almost twenty years.

Sorby spent a quiet and uneventful life in his native Sheffield, but he was on good terms with many of the leading scientists of the day and frequently traveled to London and elsewhere to attend scientific meetings. He was President of the Geological Society, the Royal Microscopical Society, the Mineralogical Society, and was a member of numerous other national and foreign learned societies. Rather surprisingly for a man who in his earlier years was of simple, mildly eccentric habits, he later become active in various civic activities pertaining to science and engineering, his most important office being that of President of Firth College, which later became the University of Sheffield, during the critical years 1882 to 1897.

One of his most important connections, which in one form or an other he maintained throughout his entire life, was with the Sheffield Literary and Philosophical Society. This society was of a kind very common in nineteenth-century industrial England and provided a local focus for many intellectual activities. Sorby became a member in 1847, served it frequently as secretary, and was president for at least seven annual terms. In its heyday this small society brought to Sheffield the leading professional scientific lecturers of the day, had papers of high caliber by its own members on all aspects of the arts and sciences, and started collections which later formed the nucleus of the Sheffield City

Museum and the university library. There were talks on local geology, astronomy, archeology, chemistry, and many other topics. As was natural in the city of Sheffield, metallurgy occupies a significant place on the program. For example, on June 5, 1860 (when Sorby was secretary of the Society), there was a paper "On Ornamental Etching in Sheffield," by John Holland.[4] This discussed the three forms of etching then in use—the ornamentation or trade-marking of manufactured articles, the etching of plates for printing "trade pattern books" (i.e., manufacturers' illustrated catalogues) and the production of pictorial works. The art of ornamenting polished-steel articles by etching had been practiced in Sheffield not more than about seventy years—the earliest etcher remembered by the oldest living razor manufacturer being a drawing master named Edward Goodwin. Razors and other blades were decorated with transfer portraits or pictures of the Battle of Trafalgar, for instance, as well as with the maker's name or other written matter. Sheffield was supplying the best and the largest quantities of steel plates used by engravers in London and in the United States, and Holland drew a nice contrast between two products of the action of aqua fortis on Sheffield steel which were displayed at the meeting—an engraving of "Bolton Abbey in the Olden Time" and a saw made by the firm of Messrs. Sorby. Though it appears that Henry Sorby did not pay much attention to the activities of his ancestral cutlery firm, it seems high likely that the quiet evenings spent at meetings of the Literary and Philosophical Society planted in his mind some ideas on the action of acids on metals which were ready to be developed when the intellectual need arose.

On August 4, 1863, a paper was read before the Literary and Philosophical Society by William Baker,* entitled "On the Processes Employed in Refining Iron and Steel." To judge from the minutes,[5] this was a very good review of the subject. It included a paragraph on the constitution of iron and steel which could easily have led Sorby to apply his new petrographic techniques to metals. However, Sorby had recorded his first observation exactly eight days earlier, and Baker had seen the etched specimens on the very day of the discovery. Sorby was a friend of Baker, and the conclusion is almost inescapable that Baker's preparation for his Society paper led to discussions with Sorby on iron and steel which started the latter's microscopic studies.

---

* Baker was at first employed by a firm of lead pigment manufacturers and later practiced on his own as an analytical chemist. He met a tragic end in 1878 by a fall when sliding the banisters at his club.

Sorby's diary[6] for the period 1859 to 1908 is still preserved at the University of Sheffield.* The entry at the end of July, 1863 (most exciting for a metallographer to come across nearly a century later!), reads:

"Discover the Widmannstättischm fig. in Ⓛ iron. To Chadburns, showing them to Stuart. Baker to tea and show them to him and walk to Little Mattock." Although Sorby evidently mistook streaks of slag in wrought iron for a true Widmannstätten pattern, these lines, written on July 28th, mark the very day on which modern metallography was born. For eighteen months thereafter references to metallographic studies are frequent, but usually in the briefest possible form, conveying no more information than "etch iron," "etch and print," "expt. with irons," "draw cast iron," and similar phrases. Immediately before July 28, Sorby had been concerned mostly with experiments on the effect of pressure on the equilibrium between solids and solutions, which he was exploring for its probable geological interest. Pressures up to 100 atmospheres were obtained by sealing filled glass tubes when supercooled and allowing them to warm up to room temperature. There were occasional references to meteorites, and it was to meteorites that he turned his interest at the conclusion of his work on artificial irons.

Sorby had many visitors and evidently talked freely about his work on iron. On September 10, 1863, he noted, "Make out change of structure in hardened steel." On September 18, "Kirkaldy stays with me several hours looking at irons." Kirkaldy had published his studies of etching of wrought iron the previous year (chap. 12). He had seen

---

* The writer is indebted to Dr. A. W. Chapman, Registrar and to Professor A. G. Quarrell, Chairman of the Department of Metallurgy for providing access to this interesting document, and to the latter for making the photograph of the page reproduced here.

† The symbol Ⓛ, "hoop L," is the mark of a Swedish wrought iron made of Dannemora magnetite at the Löfsta Iron Works which was particularly valued by the Sheffield steelmakers for cementation. Percy[7] gives an analysis of it. Stuart, who appears often in Sorby's diaries, was undoubtedly Graham Stuart, F.G.S., F.C.S., a well-known geologist and one of the proprietors of the Literary and Philosophical Society.

Sorby's specimens at the British Association the previous month and undoubtedly had there arranged for a further discussion. On October 22 and 23, Sorby visited Bessemer. On November 13, "Baker comes in morn. to look at irons. Contrive a new improved illuminator for micros."

The failure of others to take up and extend Sorby's metallographic work has often been interpreted as being due to an original failure to appreciate its value. Yet a number of enthusiastic contemporary comments leave little doubt that it was briefly appreciated, even though it was not flattered by imitation. Two newspaper clippings, bound with Sorby's own collection of his writings show that he exhibited some of his etched samples at a soirée of the British Association held at Newcastle-on-Tyne in August, 1863. Sorby was one of the three secretaries of the Geological Committee of the Association for that meeting and his diary records that he was preparing "mic. ob." (i.e., microscopic objects) of iron during the week immediately preceding his departure for Newcastle. The London *Times* which carries a quite full report of the meeting does not mention Sorby or give any details of the soirée. The *Sheffield Telegraph* on September 4, 1863, reports[8] that Sorby's sections of iron and steel attracted unusual attention:

These specimens exhibited a most interesting structure. Chemists, engineers, geologists and physicists all obtained some hints or [*sic*] the molecular constitution of iron which opened out a new field for speculation. . . . The line of research opened out by Messrs. Sorby and Stuart seems likely to throw some light on the cause of change from fibrous to granular fracture in axles and other masses submitted to vibration.

The *Sheffield and Rotheram Independent,* of the same date, has more detail. It reported that Sorby's samples

engaged the attention not only of the physicist, as giving some insight into the molecular structure of various kinds of iron, but the engineer and chemist found these remarkable objects equally deserving their notice. With great skill and ingenuity Mr. Sorby has prepared thin slices of wrought iron, Bessemer steel, blister steel, cast iron—the same decarbonized, and Spiegel eisen. These are carefully etched, and each displays a structure peculiar to itself, in some cases resembling the crystalline surface hitherto supposed to be only shown by meteoric irons. . . . Messrs. Fairbairn, [John] Scott Russell, and others, considered the steps which had been taken to throw light upon the internal structure of iron likely to open a new field of investigation, which would be a most important one for scientific as well as practical men.[9]

In January, 1864, Sorby received two letters (now preserved by the Sheffield Public Library among a fascinating collection of letters to

Sorby, but unfortunately containing none from him) making it obvious that his microscopic work on steel was well known and much was expected of it. Thus, John Percy writes[10] on January 20, 1864: "I am glad to hear about your iron researches. You are the very man for such an investigation. I know, even without the aid of the microscope, numerous varieties of appearances on fracture." After some comments on the distribution of carbon in cast iron and the effect of sulphur on it, Percy recommends Sorby's examination of cast steel before and after hammering to check Caron's chemical observations (see chap. 14, p. 205), and after his signature he returns to add, "There are 1000 questions for you relating to the structure of metals & their mixtures & compounds."

Walter White, then assistant secretary and librarian of the Royal Society, wrote to Sorby January 23, 1864, thanking him for sending a ditty and suggesting that he write a ballad about his own scientific discoveries:

> Hard crystals I anatomize
> And Iron and Steel too
> And Stubborn rocks I shave and grind
> And look them through and through.[11]

In view of these comments it can hardly have been in an atmosphere of indifference that Sorby presented his paper at the British Association meeting in Bath in September, 1864, but the record is not impressive. The *Times* merely reports that "Mr. H. C. Sorby F.R.S. exhibited and described a number of microscopical photographs of various kinds of steel." The official report of the meeting[12] carries only the following short abstract.

The author first briefly explained how sections of iron and steel may be prepared for the microscope so as to exhibit the structure to a perfection that leaves little or nothing to be desired. He then exhibited a series of microscopical photographs, taken under his direction by Mr. Charles Hoole, illustrating the various stages in the manufacture of iron and steel, and described the structures which they present. They show various mixtures of iron, of two or three well-defined compounds of iron and carbon, of graphite and of slag; and these, being present in different proportions, and arranged in various manners, give rise to a large number of varieties of iron and steel, differing by well-marked, and very striking, peculiarities of structure.

Further practical interest was shown by the chemist George Gore[13] who, in January, 1865, requested the loan of some of Sorby's photographs showing the structure of Bessemer steel, which he wanted in connection with his consulting practice on a job concerning phosphorus in steel for a firm employing "upwards of 3000 workpeople." Sorby also had an inquiry[14] from W. Vivian of the Mwyndy Iron Ore Company in Glamorganshire, who wanted to know how to "print off" the structure from a steel specimen. Vivian's interest, however, may have been general rather than metallurgical, for he remarks that he has already printed from foliage and grass and wanted to add iron to his collection.

It seems, then, that many people appreciated the value of Sorby's studies at the time they were done, but none of them had sufficient interest to develop the technique independently, and metallography lay dormant. Even the lucid description of polishing and etching techniques written by Sorby for the fourth edition of Beale's *How To Work with a Microscope* (London, 1868)[15] failed to arouse interest, probably because the microscopists, though numerous as both amateurs and professionals, had by that time become almost exclusively biological in their interests.*

* Because this contains the first description of Sorby's methods we reproduce it in full:

"*Of the Microscopical Structure of Iron and Steel.*—The microscopical structure of iron and steel is best shown by polished surfaces slightly acted on by very dilute nitric acid. The section should be cut in the required direction by means of a saw, and ground or filed down to a convenient thickness, and fixed to a piece of glass. The upper surface should then be filed level, and dressed with coarser and finer emery paper, and afterwards ground smooth with a bit of a soft Water-of-Ayr stone about ¼-inch square. Every trace of roughness should then be removed by means of rouge and water on cloth; for unless the surface be extremely well polished the structure of some kinds of iron cannot be seen. It must not be that sort of polish which merely gives a bright reflection, but one which may show all the irregularities of the material, and is as far removed as possible from a burnished surface. All trace of the rouge should be washed off, and care used not to touch the surface with the fingers, which is then acted on with extremely dilute nitric acid. If the action be allowed to proceed too far, the most important points in the structure may be entirely obliterated; and therefore it is well to take the section out of the solution and examine it under water in a glass trough, and again act on it with acid, time after time, until the structure is seen to the greatest advantage. The section must then be well washed and quickly dried by wiping the surface with a handkerchief; and after slightly rubbing it on soft wash-leather, a thin glass cover must be mounted over it with Canada balsam.

"Of course such sections must be examined by reflected light. For this purpose no illuminator is better than the parabolic reflectors supplied by Messrs. Beck, which were in fact first made for me, for that special purpose. I afterwards added another small reflector, inclined at an angle of 45°, attached to a moveable arm, so that we may see an object by direct reflection. The general construction will be seen from fig. [84a], copied from Mr. Richard Beck's paper in the Transactions of the Microscopical Society [actually *Quarterly Jour. Microscopical Soc.*, 1865, pp. 117–18].

"The small reflector is seen at *m*, with a semi-cylindrical tube *x*, to shut off the light reflected by the parabola. When the latter is used the small reflector is turned away by means of the milled head *w*, into the position indicated by the dotted lines. The difference between the two illuminations will be seen from figs. [84b and 84c]. When the parabola is used, light

Sorby's first representations of his microstructures were sketched by hand, but he soon hit upon the idea of printing, without magnification, directly from the etched surface. On February 2, 1864, he wrote in his diary, "Etch iron for printing and have some excellent prints which I show and explain at Phil. Soc. in even." A report of this in a weekly journal, *The Reader,* on March 5, 1864, refers to Sorby's display at the British Association meeting at Newcastle the previous autumn, and describes the new method of obtaining prints from etched surfaces. Sorby had already identified some constituents, for the report says:

When iron is converted into steel by cementation, three distinct crystallized compounds are formed, two of which are readily dissolved by diluted nitric acid, whereas one is scarcely affected by it at all.

The next issue of *The Reader* (March 12, 1864) contains, imprinted in its columns, actual impressions from Sorby's etched steel samples.[17] These prints,* to the historian of metallography most interesting, are reproduced in Figure 85. Later, Sorby writes in answer to a carping correspondent in the same journal that "prints had been taken from meteoric iron in Vienna as early as 1843, and published in Partsch's work on meteorites. He only claims novelty as far as iron and steel are concerned."

Evidently he had not heard of Schreiber's beautiful 1813 print of a meteorite or of Haidinger's print (albeit a very poor one) from etched wrought iron published a few years earlier, both of which are discussed in chapter 12.

This method of reproducing structures was inadequate for it allowed no magnification. In January, 1864, Sorby was experimenting with the

---

passing from *d* is reflected from *f*, fig. [84*b*], and if the object *b* has a polished surface, is again reflected to *e*, quite outside the object-glass *a*, so that a polished surface appears black, whilst at the same time a rough surface appears white or coloured by diffused reflection. When, however, the small reflector *g*, fig. [84*c*], is half over the object-glass, the light is reflected through the other half of the lens *c*, in such a manner that a polished surface appears bright, and a rough surface comparatively dark. We can thus distinguish at once the difference between black slag and that very hard constituent of some kinds of steel, which remains so bright and polished after having been acted on by cold diluted acids, as to look quite as black as the slag by ordinary illumination. In fact, I may say that in studying irons and steel such an illuminator is indispensable."

Sorby had noted in his diary on September 23, 1863, "expt. with mirror & write to R. Beck," and he refers to improvements on November 13 which are undoubtedly the origin of the illuminator.

* Another print from one of these samples appears (photolithographically reproduced) in Sorby's 1887 paper before the Iron and Steel Institute; the other sample is reproduced in negative and slightly enlarged, from an unstated source, in Desch's Sorby Lecture.[18]

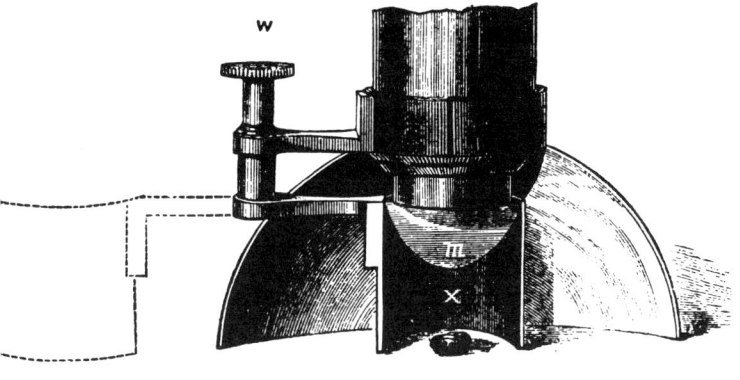

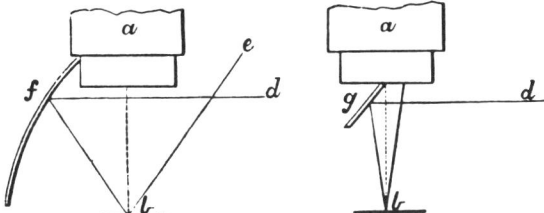

Fig. 84.—Beck's parabolic reflector with Sorby's flat mirror m. The diagrams show the paths of light with normal and oblique illumination. (From QUARTERLY JOURNAL MICROSCOPICAL SOC., *1865.)*

Fig. 85.—Nature prints, made from steel specimens etched by Sorby, imprinted directly into the pages of the weekly Reader, *March 12, 1864. Enlarged 3/2 in reproduction.*

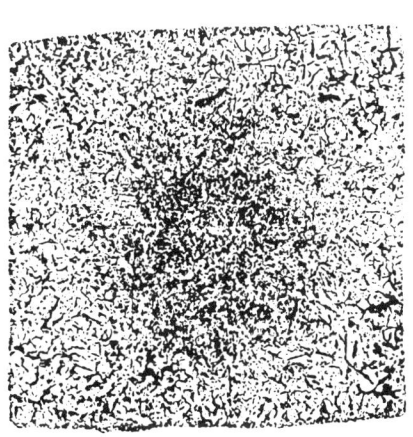

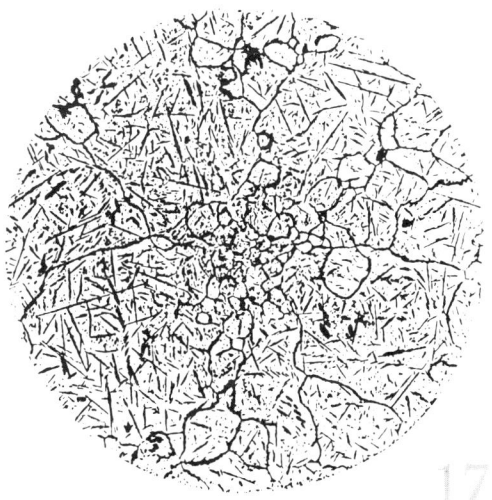

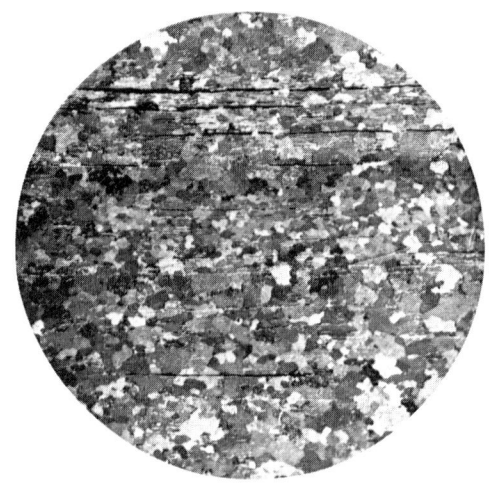

Fig. 86.—*Photomicrograph of Bowling bar iron, longitudinal section, ×9. (Sorby, 1864. Reproduced from woodburytype accompanying 1887 paper.)*

projection of sections on a screen, but between August 9 and 22, 1864, he made almost daily visits to his friend Charles Hoole, who was listed in Sheffield directories (1864–87) as a commercial photographer. It was no easy matter then to make photomicrographs, for neither adequate artificial illumination nor fast plates were available. There is no record of Hoole's techniques, but it is probable that he used wet collodion plates. Sorby experimented with an oxyhydrogen light for projection at this time, and perhaps used it for photography, though his detailed 1887 paper mentions only the optical train, not the source of light. It should be noted that these photographs were obtained by oblique illumination, although Sorby used a vertical illuminator in studying samples visually. On August 18 and 22, 1864, he reports several good photographs. For two weeks thereafter there are frequent references to photographs, and on September 8 he wrote the paper which he presented at the British Association meeting the following week. These photographs were destined not to be published until 1883, when they were first reproduced (as halftones) in the United States.[19] The same illustrations accompanied Sorby's 1885 paper to the Iron and Steel Institute, but the preparation of a fine set of woodburytypes delayed their appearance until 1887. Three of these are reproduced in Figures 86, 87, and 88. Prints of the photographs—identical with Figures 1–4, 7, 8, and 14–17 of the 1887 publication—were sent on loan to the

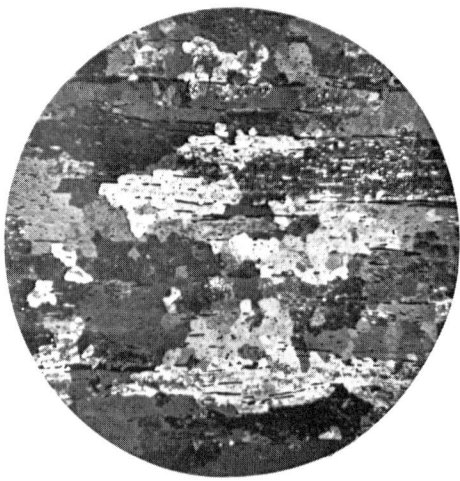

Fig. 87.—Photomicrograph of wrought-iron armor plate, longitudinal section, ×9. (Sorby, 1864 [Cf. Fig. 86].)

Fig. 88.—Photomicrograph of blister steel, longitudinal section, ×9. (Sorby, 1864 [Cf. Fig. 86].)

Science Museum in South Kensington in 1876, where they are still preserved. They are mounted on cards with printed titles, evidently having been prepared for exhibition and perhaps dating from the British Association meeting.

Sorby's interest, never remaining long on any one subject, seems to have reverted to meteorites by the middle of 1865 and his diary thereafter contains little further reference to iron and steel until 1884. He had, however, given a lecture[20] on October 20, 1882, at Firth College, Sheffield (shortly after he became its president), undoubtedly under the influence of reports of Martens' work, which had been published in Germany in 1878.

The diary records that Sorby gave a lecture on iron on April 23, 1884, at an unnamed soirée and in October of the same year he wrote[21] to John Percy, who had been trying to revive Sorby's interest in metallurgical problems. On April 10, 1885, he remarks "Greenwood* came to tea to see iron etc., and said he could not have believed it possible to see what is shown by my sections and that no one would have supposed they were metals." Between March 30 and April 20, 1885, the diary reports almost constant work on iron and steel and the preparation of his paper, which is undoubtedly the one that Sorby pre-

* Supposedly William Henry Greenwood, Sheffield metallurgist and author of *Steel and Iron* (London, 1884).

sented to the Iron and Steel Institute at its spring meeting in 1885. This was issued as an eight-page preprint, but was destined never to be published in the *Journal*. A copy[22] of it is preserved at the University of Sheffield Library, from which it is reprinted in full in Appendix B of the present work, pages 256–60. Sorby took advantage of the delay occasioned by the reproduction of the photographs to entirely rewrite this paper in February and April, 1887, and it finally appears in the second volume of the *Journal of the Iron and Steel Institute* to be issued that year.[23] Prior to this, however, Sorby had made (in the autumn of 1885) a series of observations at high power which were presented at the spring 1886 meeting of the Institute. These studies, showing that the previously observed lamellar structure of the pearly constituent was due to actual alternate layers of iron and of iron carbide in ratio of about 2 to 1, were more promptly published.[24] They appear in the same volume as the "Presidential Address"[25] by John Percy, which contains some very complimentary references to Sorby's work. Even this paper elicited only some unpertinent remarks from Bessemer by way of discussion. In the following year the full paper (incorporating all of Sorby's results to date and including many illustrations) was available. It was placed upon the program, but the new president merely announced that the paper would be taken as read and printed in the *Journal!*

Except for slightly more detail on the "pearly constituent" and some pregnant comments on the hardening of steel by quenching and the effect of alloys thereon (which are included in the 1886 note), all of Sorby's work is contained in this 1887 paper. It is difficult to tell when he did the work described therein, but it seems that most of the observations, if not the interpretation, dates from the work of 1863 and 1864: the photomicrographs are the ones made with Hoole's assistance in 1864 and the samples were certainly the ones he collected then. The letter[26] from W. P. Beale of the Parkgate Iron Works, Rotherham, to Sorby, dated November 10, 1863, lists samples of wrought-iron bloom bar, and armor plate, with their markings, many of which can be identified in the paper.

The report of the lecture[27] at Firth College in October, 1882 (delivered before Sorby had recommenced active work for the Iron and Steel Institute papers), refers to seven well-marked constituents and to the recrystallization of iron on annealing. The lecture probably contained all of the essential observations incorporated in the complete 1887 paper except for the detailed structure of the "pearly constituent" revealed by high magnification and first described in the 1886 paper.

What were Sorby's discoveries? Not as a result of philosophical speculation but principally as a result of accurate observation possible for the first time because of his superior technique, he was able to relate the well-known properties of iron and steel in their various forms to specific visible structural changes. He effectively laid the bogey of crystallization under shock or vibration, he showed that metals were undoubtedly crystalline and that distorted crystals, being in a state of unstable equilibrium, would recover their normal crystalline polarity whenever the circumstances were such as to permit this rearrangement. He clearly realized that a sequence of constituents appeared as carbon was added in increasing amount to iron, and that the "pearly constituent" was itself composed of lamellae of iron containing virtually no carbon alternating with a compound of iron and carbon, and that this structure resulted from the decomposition of a constituent that was homogeneous at higher temperatures, while both iron and iron carbide were themselves stable at all temperatures. The hardness of quenched steel was due to the suppression of this decomposition, and tempered steel became softer because of the separation of the two constituents to a greater or lesser extent. Although Sorby's ideas on the hardening of steel are clearly stated in the 1885 paper, his wording is most succinct in the brief report of his lecture to the Yorkshire Geological and Polytechnic Society in 1886 and we quote from their *Proceedings*:[28]

The laminae [of the pearly constituent] are often of extreme thinness, those of the soft iron being often about 1/40000th inch in thickness, and those of the hard substance only 1/80000th, so that we have alternating ridges and grooves about 1/60000th inch apart. The only satisfactory explanation for this remarkable structure appears to be that at a high temperature a stable compound of iron and carbon exists, which is not stable at a lower temperature, but breaks up with the two substances named above, which certainly are stable at both a high and low temperature. . . .

Independent of the interest that may be attached to such a remarkable structure, it seems to me that the facts throw much light on the hardening and tempering of steel. It is possible and indeed probable that when red hot steel is suddenly cooled by plunging it into cold water, the compound of iron and carbon which is stable at a high temperature is suddenly fixed, before it has, so to speak, time to break up, and retains properties intermediate between those of the soft iron and the intensely hard but brittle compound, that is to say, combines great hardness with strength. On again raising the temperature somewhat, so as to temper the hardened steel, we can easily understand that the two constituents separate out to a greater or less extent, so as to give rise to a structure like that met with when the steel is slowly cooled. As far as I can learn, this view of the subject agrees perfectly well with all the facts seen by studying different kinds of iron and steel with high magnifying powers.

Sorby saw that iron containing a high nickel content did not recrystallize as did pure iron on annealing and cooling. Aware of the observations of Forbes and Tait on the discontinuity in the change of electrical properties of iron with temperature change, he saw the connection with his structural changes and believed that recrystallization occurred as a result of this allotropic change which he compared to that in yellow mercuric iodide.

Taking, then, all these facts into consideration, it seems to me nearly certain that the separate grains seen in fig. [86] are separate, though imperfectly developed, crystals, and that this crystallisation may, and often does, take place after the iron has been hammered or rolled, but is still hot. This crystallisation may in great measure be due to a strong tendency in distorted crystals to recover their normal crystalline polarity. When distorted the particles must be in a state of unstable equilibrium, and we can therefore readily understand why recrystallisation so easily takes place whenever the circumstances are such as to permit the particles to rearrange themselves in a state of stable equilibrium. It appears to me that this is a general principle of great importance in connection with the mechanical properties of worked iron, probably often overlooked.

It is, however, quite possible that the recrystallisation of iron on cooling from a high temperature may also depend on special properties of the metal. According to Mr. George Forbes, when a hot iron wire has cooled to a dull red heat it suddenly brightens for an instant, and at the same time its conducting power for electricity suddenly increases. Professor Tait has also shown that the thermo-electric properties likewise suddenly change at about the same temperature. These facts make it extremely probable that a great molecular disturbance takes place in iron at a dull red heat similar to the change which occurs when yellow mercuric iodide formed at a high temperature turns red on cooling. Such a change, combined with the tendency to restore the equilibrium of distorted crystals, would, I think, explain why the structure of the iron itself in cold bars, worked when hot, is very little, if at all, related to the previous great changes in the shape of the mass, and so very similar to that seen when complete recrystallisation has undoubtedly occurred. This, of course, would easily explain why the structure of a transverse section differs so very little from that of a longitudinal section. Preparations made from specimens cut transversely show the relative amounts of iron and slag much the best, because the thin fibres of slag are so easily detached in polishing when they lie lengthwise, whilst they are well supported on all sides when they lie end on. The relative amount of this slag in particular layers is then seen to be sometimes so great that it really forms a large part of the entire bulk, and one almost feels inclined to look upon such layers, not as iron, but as impure slag. [Extract from 1887 paper, pp. 263–64.]

Sorby not only identified for the first time seven microscopically distinct constituents whose names have since become commonplace, but he used the structures observed to deduce directly the sequence of physical events that had given rise to the structures. His papers not only record observations, but they put in order the facts of the appearance of new phases in specific ranges of temperature and composition, the effect of alloying elements on transformation rates, the effect of cold deformation in making the grains "broken up, distorted and elongated," and of annealing in causing recrystallization to an essentially equiaxed structure. He saw that on hot working the grains were themselves ductile and did not slide over each other.

He uses the term "grains" repeatedly for microcrystals, and regarded them as being separate though imperfectly developed crystals which, though irregular in form, fit closely to one another in every direction. He believed that plates in the "pearly compound" had a definite orientation relationship to their parent crystal—a realization of the most important phenomenon of epitaxy, though misplaced in this particular instance.

His photographs showing the extreme lack of uniformity in wrought iron and in unmelted steels must have had a disturbing effect upon iron manufacturers and it is hardly surprising that his work was endorsed by Bessemer, whose steel was at least relatively clean.

An important part of Sorby's contribution was his polishing technique, in developing which he was undoubtedly aided by the existence in Sheffield of many cutlery and engraving-plate manufacturers. After preliminary sawing and filing he rubbed the surface on a sequence of emery papers, finishing with the smoothest employed in preparing steel plates for engraving. Then he removed the modified surface by using a fine-grained Water-of-Ayr stone—a stage on which he insisted, and the lack of which produced some very curious results in Wedding's photographs, for example—followed by polishing with crocus and rouge on wood, or sometimes by a relief polishing by rouge on parchment. Etching was done usually with dilute nitric acid.

It is interesting that Sorby's petrographic background led him to cut relatively thin slices of his steel samples and to cement them on glass plates for polishing and finally to cover the finished surface with a cover-glass cemented with Canada balsam. Thanks to this technique,

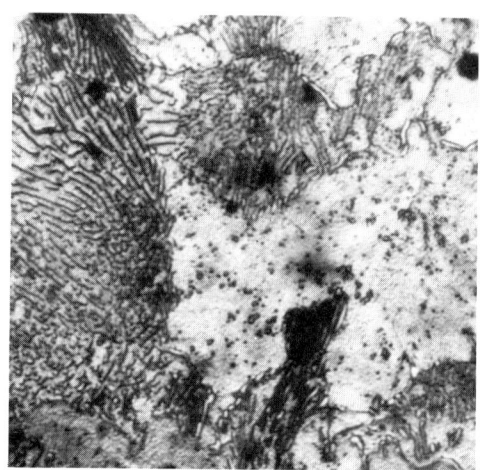

*Fig. 89.—Specimen of heat treated cast iron, polished and etched by Sorby; photographed in
1953 through the cemented-on cover-glass with no further treatment, ×500.
(Courtesy of Professor A. G. Quarrell.)*

some of his samples (which are preserved at the University of Sheffield)
can still be seen in the condition in which they left the master's hands
nearly a century ago. One of these has been recently photographed
(Fig. 89) and shows that the quality of finish was little inferior to that
given by the best methods today.

Why did Sorby's work presented in 1864 have so little influence?
The 1885 reprint seems to represent about the same stage of development as in 1882 and there is nothing in the brief record of this to show
that Sorby was doing any more than consolidating and reporting the
ideas and observations that he had made in the sixties. The paper was
completely rewritten and enlarged before the final publication in 1887.
Behind the change were the observations at higher power, made possible by the improved vertical illuminator, and above all the fact that
Percy and others had persuaded Sorby that the subject was important
and should be presented fully. Though most of the observations date
from 1863–65, it seems likely that it was not until the eighties that
their basic significance was fully appreciated even by Sorby himself.
It is interesting to speculate on whether the British Association Lecture
of 1864 would have aroused more interest had it been given a more
theoretical slant.

It is apparent that the work did not pass unnoticed in 1864, but,
indeed did create a certain amount of excitement among several of

the leading engineers and metallurgists of the day. The formal publication as a brief unillustrated abstract was not calculated to achieve wide notice,* but many people must have tried to see structures in steel after hearing of Sorby's experiments. The 1864 abstract does not describe his polishing technique or emphasize the importance of avoiding a distorted or burnished surface. Without definite knowledge of the fine results that could be achieved, there was little incentive to persevere. Even Sorby saw no reason to take time from his other studies, and it was not until after the work of Tschernoff and Martens had appeared that those Englishmen who knew of Sorby's work prevailed upon him to publish it in a suitable form. The intervening twenty years, however, had resulted in a kind of supersaturation, and once man's attention was focused on the structure of metals, the aggregation of knowledge was extremely rapid. By 1900 the situation was probably little different from what it would have been had there been a continuous, but of necessity slower, growth following Sorby's first announcement.

* The impossibility of reproducing structures except by engraving or lithography would have made their reproduction expensive and subject to misinterpretation by the engraver. Halftone photoengraving had then barely started, and even the 1885 paper was delayed for two years while the Iron and Steel Institute investigated the then new woodburytype process. This involved the preparation of an intaglio impression on a lead plate of a light-hardened gelatine relief into which pigmented gelatine was poured and paper applied to pick up locally varying amounts of pigment. Though expensive, it was a very satisfactory process, as the fine condition of Sorby's prints in the Institute *Journal* testify to this day.

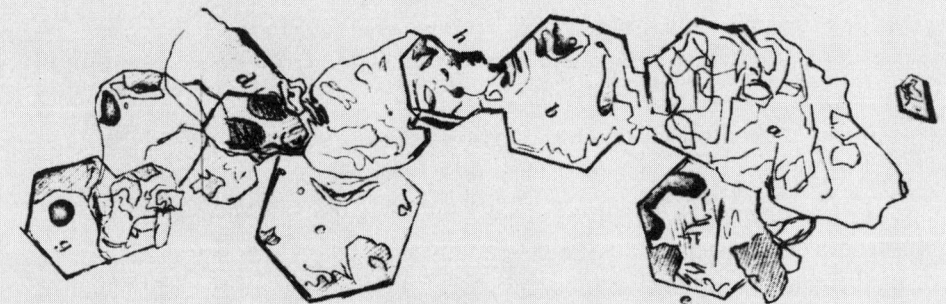

Crystals from shrinkage cavity in steel ingot, ×500. Tschernoff 1876 (cf. pp. 140 and 206).

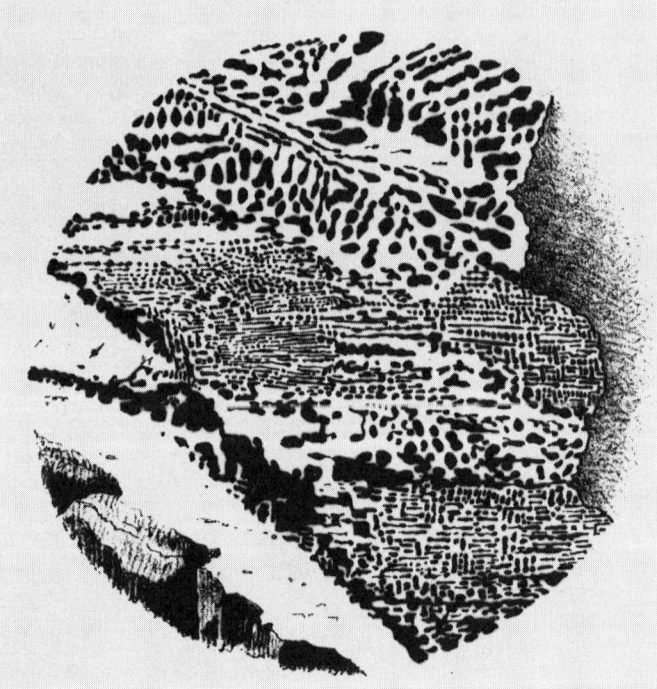

Microstructure of spiegeleisen, ×100. Martens (cf. p. 212).

THE PERIOD
OF ACTIVE
OBSERVATION

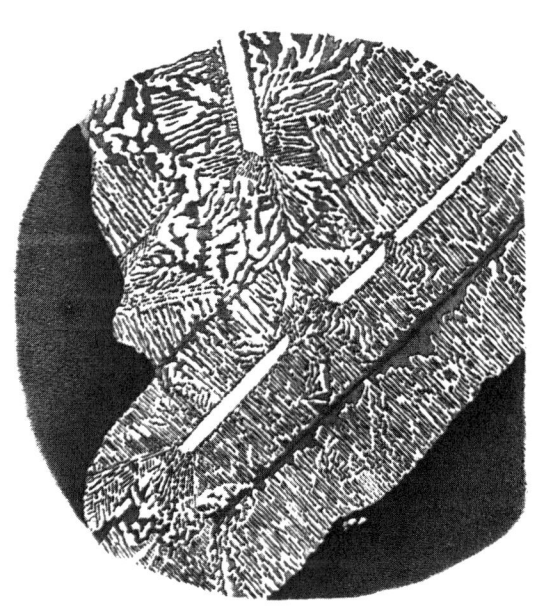

ALTHOUGH this history is concerned in detail only with the structure of metals, it is necessary to pay some attention to general advances in chemistry and physics, for these had a profound influence on metallurgical thinking. In chemistry the development of the atomic theory to a quantitatively useful concept was of utmost importance, as was, on the practical side, the growth of analytical chemistry and its application to the raw materials and products of the metallurgist. Not only did it permit the smelter to know accurately what he should obtain from a given batch of ore, but impurities could be accurately controlled and in many cases the peculiar behavior of various kinds of iron and steel that for centuries had merely been given a name such as hot short or cold short could be specifically related to a chemical composition. The fact that identical composition did not always give similar properties gave incentive for structural studies. Most large metallurgical works had an analytical laboratory by the mid-nineteenth century, and it was through the chemist that science first was introduced into production. It was mainly the engineer who first attempted to relate structure to the mechanical properties of metals. The physicist, as we will see, made important advances in understanding the properties of metals, but his structural contributions for a time were very limited.

As the last third of the nineteenth century began, conditions were ripe for an active disciplined development. Prior to this, individual men had done what interested them and structural observations were mostly a sideline of other activities, for structure did not in itself seem important. In the development of any branch of science (just as in the development of a new phase in a transforming alloy), there comes a change from random fluctuations depending on highly localized concatenations of conditions and energy to a point where the structure of knowledge, like that of matter, begins to develop an ordering force of its own and to advance almost independently of the wishes of individuals. Men like Hooke and Réaumur, for instance, made observations on structure that had an amazing individuality and would not have happened without them: neither, in fact, did their observations have much direct influence on others. Sorby also was unique in his viewpoint, his technique, and his association with the steelmakers of Sheffield. No one else could have seen, thought, and done what he did. The growth of metallurgical industry toward the end of the nineteenth century, the advances in knowledge of physical properties, and the widespread emphasis on chemical analysis in both laboratory and plant all made it inevitable that the relationship between structure and properties would be realized and exploited. As the structure of metallographic knowledge itself began to crystallize, one observation leading obviously to another, there would have been relatively minor changes of progress even if the protagonists had been different men. Once a fundamental idea has been grasped and clearly stated, it requires little more than time for the details to be filled in. After Sorby the advance of metallography becomes in a way less interesting, partly because it is more familiar to a writer in the mid–twentieth century, partly because the literature is too voluminous to be studied in detail. This, the concluding section of the book, will therefore deal only with some of the more influential workers.

14. Chemistry and Physics in the Nineteenth Century

Once the heritage of misleading ideas on the nature of the chemical elements had been swept away by Lavoisier and a reasonable table of chemical elements had been produced, the quantitative study of their combinations became possible. The old tables of affinity were recast by quantitative methods. Berthollet, whose background was strongly metallurgical and who had played a major role in unraveling the mystery of steel, knew that many compounds could have variable composition and he saw that a reaction depends upon relative amounts as well as affinities. Though true, these concepts did not lend themselves to easy theoretical approach, and the fixed simple combining proportions of Richter, Proust, and particularly Dalton served to exclude other chemical thinking.[1] It became possible to represent compounds with symbols and to show the sequence of compounds with various simple proportions. The molecule became supreme. Not only were variable compositions excluded from chemical interest but also the crystal itself was forgotten, for the interesting aspect of any material was the simple molecule and it mattered little how the molecules were themselves arranged.

Dalton's atomic theory was really more a theory of the molecule than of the unit atoms that compose it. It was the molecule rather than the atom that dominated nineteenth-century chemistry and physics. It was a great advance to be able to put in quantitative terms the vague idea of corpuscular aggregates that had first been seriously proposed in the seventeenth century. The physical molecule of Avogadro provided all that was necessary for the development of kinetic theory of matter, and when it was realized that the same material could exist in two different allotropic modifications it was easy, and even sometimes correct, to attribute this to a different arrangement of atoms in the molecules. As long as there was some state of aggregation beyond the atom, all changes could qualitatively be ascribed to it. The frequency with which solids could have non-stoichiometric compositions, a point which was strongly urged by Berthollet in opposition to the simplifications of Dalton and Berzelius, was either ignored or partially explained as due to a solution of molecules in a matrix. A great deal of the argument on the nature of steel revolves around whether steel (i.e., the composition capable of maximum change of properties on heat treatment, a little less than 1 per cent carbon) constitutes a compound and around the question as to the molecular form in which carbon exists in steel in its various hardnesses.

The idea that atoms aggregate into molecules which can vary in the number and arrangement of the few atoms they contain is adequate to explain many physical properties. While geologists and metallurgists were struggling to develop the idea of crystals, the physicist seems to have been satisfied with molecules, and completely ignored the spatial relationship of these molecules. Crystalline properties were simply vectorial, and there was nothing which forced the acceptance of a space lattice even to account for the properties of single crystals, much less for the nearly isotropic properties of a polycrystalline body. Far from accepting crystalline order, to quote Von Laue,

among the few physicists who were at all interested in crystallography [there were some who] adopted the opposite view, that in crystals, as elsewhere in matter, the molecular centers of gravity were distributed irregularly and that only the parallel placing of preferred orientation in the molecules produced anisotropy.[2]

So great was the influence of chemical theory that even when crystalline order was well accepted it was supposed to be of molecules rather than atoms. Osmond, for example, seems not to have been disturbed at all when he found that his alpha and beta iron were crystallographically identical: The molecular arrangement could be different, without the crystalline form being changed.* No one seems to have bothered about the difficulty of building a continous lattice from mixed molecules of different sizes and shapes, and the geometric significance of the existence of crystals of variable composition seems to have attracted no atten-

* There is a good discussion of allotropy, polymorphism, and isomerism, as indeed of many other aspects of our subject in *Molekular Physik*[3] by O. Lehmann (1889). Lehmann was professor of *electrotechnik* in the Polytechnic Institute at Dresden and his book shows an appreciation of both theoretical and practical knowledge that is most unusual for his time. He even includes a chapter on biological shapes, somewhat foreshadowing D'Arcy Thompson.

tion. The constancy of composition of ionic compounds in both solution and crystalline form supported the molecular view. Groth,[4] who contributed so much to the understanding of the interaction of the physical and chemical aspects of crystallography, remarked that it was not altogether impossible that the molecules of sodium chloride that occurred in a crystal contained a large number of atoms of each kind combined together, so spatially arranged that they acted like a sphere in the structure as a whole.

It was not until the first results of X-ray diffraction became known, with the concept of the whole crystal being the molecule, that the framework of physical thinking could tolerate the solid solution in anything approaching the modern viewpoint. The less precise thinking of the metallurgist accepted solid solutions by analogy with liquids without considering the geometric difficulties of the molecular hypothesis. The achievements of molecular theory made belief in the molecule inescapable and it is a small wonder that some of the slight discrepancies resulting from its simple application to solids were unnoticed for so long. There is no doubt also that the molecule was partially an inheritance of Haüy's primitive polyhedra, his geometrically shaped integrant molecules which stacked up regularly to form the crystal. Haüy was influenced by observations of cleavage geometry and growth faces in building up his crystallography. But Weiss, who translated Haüy into German (1815), said that integrant molecules were not necessary, and decried mechanistic atomistic views. Omitting primitive forms entirely, Weiss related everything to formal relationships between axes and coordinates. The relationship between optical axes and crystal form (observed by Brewster) was also important, for it removed the emphasis on shape, and crystallography without polyhedra became possible. Like all advances it was obtained at a price, in this case a loss of atomistic thinking on physical problems. Von Laue has remarked in his *History of Physics* that no physical phenomena of the nineteenth century required the concept of a space lattice. This seems rather a narrow definition of physics. The phenomena existed everywhere: it was interest that was lacking. The success of the molecular theory (particularly the kinetic theory wherein physics and chemistry so triumphantly and beautifully joined) had induced an almost complete blindness on the part of people who called themselves physicists to the crystalline properties that were well known to engineers and chemists.

The association of the hypothetical *molecules integrant* of Haüy and the definite chemical identities, atoms, had been postulated by Seeber[5] in 1824, following Wollaston's (1812) geometrically ordered assemblies of spherical or spheroidal Daltonian atoms. The idea lay dormant until it was taken up again by Barlow[6] in 1883, who gave the essentially modern picture of the cubic metals and simple ionic crystals. This relationship between lattice points and atoms as distinct from molecules was further developed by Sohncke (1888) and by Groth[7] in 1904. Groth realized that the idea of a molecule was purely arbitrary in the case of crystals of many simple compounds, but the idea did not become widespread until after X-ray diffraction had given a new reality to space lattices. Even as late as 1890 the great scientist Lord Kelvin (William Thomson)[8] could discuss polycrystallinity as an essential theoretical concept, without any reference to the by then well-established structure of real matter.

The scientific world is practically unanimous in believing that all tangible or palpable matter, molar matter as we may call it, consists of groups of mutually interacting atoms or molecules. . . . It is, indeed, very difficult to imagine equilibrium, static or kinetic, in any irregular random crowd of molecules. Such a crowd might be a liquid—I can scarcely see how it could be a solid. It seems, therefore, that a homogeneous isotropic solid is but an isotropically macled crystal; that is to say, a solid composed of crystalline portions having their crystalline axes or lines of symmetry distributed with random equality in all directions.

Glass, he believed, may have homogeneous crystalline portions very small in comparison with the wave length of light but containing very large numbers of molecules.

In the development of metallurgical science a very important part was played by the determination of various physical properties of alloys as a function of their composition. Matthiessen,[9] who in 1859–63 did the first systematic electrical conductivity measurements on series of alloys, divided physical properties into those that did and those that did not indicate the chemical nature of alloys. He believed that the various kinds of conductivity curve for alloy series represented chemical compounds, mechanical mixtures, or "solidified solutions" of one metal in another. By solidified solution he means "a most intimate mixture, such as would occur in the sudden conversion of a liquid into a solid . . . in fact, a perfectly homogeneous diffusion of one body into another." Later (1867) he added that "even under the most powerful microscope it would be impossible to distinguish the components of a solid solution."[10] He gives glass as an example of such a body and in alloys only those whose electrical conductivities are linear functions of the volume composition such as lead-tin, which he certainly did not examine microscopically for it would not meet his definition. His solid mechanical mixtures are really solidified two-phase liquids. Matthiessen discovered the rapid decrease of conductivity of a metal on adding small amounts of another metal to form what we today know as a true solid solution range. Arguing that this could not be due to a compound because of the small amounts of addition necessary, he believed the effect to be due to an allotropic modification (for it was well known how small amounts of an element could modify another) and that further additions formed solid solutions with linear conductivity variation. His conductivity curves led him to suggest that all combinations of these effects were possible in different systems, and several in a single system in different composition ranges. It was some decades before metallographers were able to show that Matthiessen's general approach was correct even though he misinterpreted most systems in detail.

Equally important as a background to the understanding of the structure of alloys were the chemical studies of segregation. Levol's work[11] in 1852 created much interest, for he found that only one alloy would freeze without partial separation into constituents of different melting points. This (actually the eutectic) he thought to be a compound, Ag_3Cu_2. Guthrie's thermal studies later showed that eutectics did not correspond to simple atomic proportions:

It is curious that chemists paid so little attention to the liquation methods for desilverising copper—a sixteenth-century development, still in use in the nineteenth century—for the behavior of the copper-lead alloys used would have made it perfectly obvious that a homogeneous liquid alloy could separate into chemically and physically distinct phases on partial or complete solidification. Karsten, in 1825, under the impact of the new atomic theory, postulated[12] the existence of the compound PbCu which split up into $Pb_{12}Cu$ and $PbCu_{12}$, which conform approximately to the compositions observed in practice. His fellow metallurgist Percy[13] later took issue with this, and proved that the solid alloy was a mechanical mixture by careful polishing experiments which, depending on the amount of smearing, could confer the color of either copper or lead to the surface of an alloy with 21.5 per cent copper. Although he used a microscope on the fracture of this alloy (which was leaden in color, showing no trace of copper) Percy seems not to have used one on the surfaces that he carefully polished with charcoal powder. The equally interesting Pattinson process for desilverising lead by partial crystallization inspired even less attention, and, of course, the host of metallurgical processes depending on immiscible liquids, whether metal, matte, speisse, or slag, never presented any special structural challenge. Some microscopical studies of the appearance of metallic iron during the reduction of its ores were described by Jules François in his *Récherches sur . . . le traitement direct des minerais de fer dans les Pyrénées* (Paris, 1843).

It was natural in the chemical climate of the time to attribute any metallic anomaly to molecular rearrangement, and even simple metals like copper, lead, and zinc were at one time or another supposed to suffer an allotropic change. It was superior observation, not intellectual creation, that distinguished Osmond's rightly famous work from that of his predecessors.

After the work of Fuchs on steel, discussed below, Bolley[14] attempted to account for the change in properties and fracture of zinc on annealing in terms of allotropy.

In 1881 and 1882 Kalisher[15] described some interesting experiments on the "molecular structure" of metals, which were destined to be overshadowed by the contemporary beginnings of microscopic metallography. Zinc, he showed, when annealed at a high temperature after rolling developed a number of property changes which are due to a molecular change, and it becomes crystalline. Etching with copper sulphate solution left the crystallization clearly visible on the surface. Crystallization began at about 150° C. and increased up to the melting point and was accompanied by a change in sonorousness. There was no trace of crystallization below 150° C. At temperatures below 300° the crystals were like frost flowers, becoming sharper, almost individual, above 300° In his second paper Kalisher concluded that all metals (except cadmium and tin) were not crystalline in the rolled state but that they became so on annealing. He worked with iron which showed some crystalline structure, though steel had none either annealed or unannealed. Copper sheet also showed a structure (developed electrolytically in copper nitrate or without electrolysis in iron chloride or ammonia). Copper and brass wire were not crystalline on annealing, but the striated structure of the drawn wire did change to an aggregate of little parts which by their separate appearance and the regularity of their packing seemed to approach a definite state. Although he found traces of a crystalline structure in a copper alloy with 40 per cent zinc but none with 43 per cent, though the latter showed etch figures, Kalisher states "that investigation under the microscope, even in cases where the crystal structure is unmistakably distinguishable to the naked eye, is extraordinarily difficult." He concludes that the crystalline state is the natural state for most metals and that they become more so under the influence of heat. He believes that many of the changes of properties by working and annealing are due to this, and suggests that the methods of etching which had been frequently used by mineralogists should be useful to physical and chemical studies of the nature and constitution of other materials.

Kalisher's papers just failed to be a major contribution to the science of metals: His observations were soon completely overshadowed by those of the English and French metallographers.

A prior application of allotropic theory to steel, particularly interesting because it associates molecular and crystallographic change is that of J. N. von Fuchs.[16] He was convinced that iron was dimorphous, existing in both isometric and rhombohedral states of symmetry. Wrought iron was the former, while cast iron the latter. The difference was attributed to the presence of carbon, perhaps because rhombohedral graphite gives iron the disposition to crystallize in the same way. There are many other changes of solid bodies known below the melting point and it is not surprising that iron would change. He assumed that iron was amorphous immediately at the welding heat. Steel is an alloy, Fuchs thought, of both forms, and the changes in properties on heat treatment come from a change in their relative amounts, the cubic being in excess in the soft and the rhombohedral in the hard state. The reason why steel can retain permanent magnetization while iron loses it is perhaps because the two kinds coexist in steel in a state of stable opposed stress. The change of iron on fatigue might also be due to such crystallization.

In 1865 a paper on the constitution of iron, cast iron, and steel was read by C. E. Jullien[17] to the Société des Ingenieures Civils in Paris. The paper was regarded as important enough to be allotted the major part of five sessions of the society. It stimulated a great deal of interest in the nature of the hardening of steel and directly influenced both Tschernoff and Osmond.

Jullien believed that metals never combine with each other and that molten alloys (including carburized iron in all its forms) are simple solutions. He contends that (1) solution is not necessarily a liquid composition: it is a chemical state differing from combination; (2) when a liquid solution solidifies, if it is converted into a mixture (of pure solid crystals) it invariably happens that the solubility is less than in the same solvent when liquid; (3) all solid bodies have two structures—one crystalline, the other amorphous: the amorphous is obtained either when a liquid is cooled (sometimes fast, sometimes slowly, depending on the material) or when the liquid is dissolved in a solid body; (4) the size of crystals in a crystallized solid body is proportional to the time of their formation. A compound to Jullien was characterized by a definite proportion of the constituents, and by properties different from them, whereas a solution (*never* crystalline) can be of any composition, sometimes within a maximum solubility, sometimes extending completely to the other component, and has properties intermediate between those of its components. With this preliminary he goes on to establish an essentially structural theory of hardening of steel.

Thus if, of two bodies which form between them a liquid solution, one crystallizes on rapid cooling and the other crystallizes on slow cooling, it is evident that (depending on the limiting speed with which the solidification of the composite body takes place) it will possess, when cold, the properties of the component which has crystallized. Further, if the cooling is not fast enough to make one component crystallize, and is too slow for the other to crystallize, then the solid mixture will share the properties of the two amorphous components.

A liquid solution can thus give rise to three solid mixtures that are physically different, depending on the speed of solidification. Metals crystallize on slow cooling and become amorphous if quenched; if they are amorphous, they become crystalline on a sufficiently long subsequent annealing. Conversely, carbon and the neutral silicates crystallize if rapidly cooled, and are amorphous if slowly cooled or annealed (a statement based on the change of diamond to graphite on annealing).

Since copper crystallizes with difficulty on annealing but tin does so easily, and since tin is liquid in bronze at a red heat, if it is cooled slowly, tin will crystallize and give its structure to the alloy; conversely, rapid cooling leaves the tin amorphous and confers a fibrous texture. Jullien denies the existence of iron carbide since only graphite or carbon could be extracted from slowly cooled iron. He observed that at a cherry-red heat carbon would penetrate iron with a gaseous or liquid mobility, and hence concluded that the carbon was liquid, and dissolves as such in the solid metal. Fibrous steel is "a solution of amorphous carbon and amorphous iron." Slowly cooled steel is "a solution of amorphous carbon in crystallized iron, while steel hardened by quenching is a solution of crystallized carbon in amorphous iron." The crystallization of iron is visible because it occurs on annealing, that of carbon invisible because it happens on rapid cooling. Jullien also applied his theory to the formation of glass and the vitrification of porcelain.

Jullien had approached this theory in his book on the metallurgy of iron published in 1861.[18] At that time he thought that hard steel was a solution of crystallized carbon in *mou* iron (by which he apparently meant a true soft state of matter opposed to the crystalline and perhaps amorphous), whereas the hardness of quenched steel depended on the presence of crystallized carbon in solution. Diffusion was somehow related the thermal emissivity.

Jullien's theory is wrong in almost every detail, yet it was important for attacking the intrenched carbide theory, for emphasizing the high mobility of carbon even in solid iron, and particularly for its insistence that the state of aggregation changes during the heat treatment of solids. Jullien's book on the metallurgy of iron was noted by Tschernoff (chap. 10, n. 21), who emphasized the fact that the change in the crystalline condition of the carbon only occurred when the steel was heated above the critical point which he denoted as *a*. Tschernoff refers to another 1865 paper by Jullien (the original of which has not been traced) entitled "Les Affinités capillaires et les phénomènes de la trempe mis en présence," in which, in a manner somewhat reminiscent of Joseph Black, Jullien discusses thermal effects accompanying hardening:

. . . he deduces the very probable conclusion that steel, in cooling from a red heat, appropriates a certain amount of latent heat, the quantity of which is directly dependent on the rate of cooling; so that the quicker the steel is cooled, the greater quantity of latent heat it will contain; but if the rate of cooling diminishes below a certain limit, then the latent heat all escapes, and no hardening can take place. The actual hardening Jullien explains by the supposition that the carbon assumes an abnormal crystalline condition.

The first observation of the critical point in steel was made by J. F. Angerstein in 1777, who remarked that the right moment for quenching a piece of steel is visible to a trained eye "as a kind of fluttering, as if a cloud of dust or a faint shadow were rapidly flicking over" the steel as it was being heated. (See C. Benedicks, *Metallographic Researches* [New York, 1926], pp. 205–17.)

To return to physics, after the initial discovery by Seebeck in 1822, physicists studied thermoelectric power in alloys, both solid and liquid. Nairn[19] in 1780 had observed permanent dimensional changes of iron wires after electrically heating; in 1869 Gore reports:[20]

While making some experiments on heating a strained iron wire to redness by means of a current of voltaic electricity, I observed that, on disconnecting the battery and allowing the wire to cool, during the process of cooling the wire *suddenly elongated* and then gradually shortened until it became quite cold. [Gore's italics.]

This length change was confirmed by Barrett (1873)[21] and Norris (1876),[22] (the latter looking for the effects of heat treatment and attributing the discontinuities to thermal effects accompanying the "decrystallization and crystallization which occurs during heating and cooling"), but it seems to have been the thermoelectric measurements of Tait[23] that convinced physicists generally that there was an interesting phenomenon here. The development thereafter was mostly metallurgical, however.

Elastic properties were extensively measured in the nineteenth century and departures from elasticity, particularly the elastic aftereffect, attracted the attention of many leading physicists. Coulomb, Biot, and Poisson assumed a particulate structure of matter and realized that there could be either an ordered or disordered arrangement, but they seem not to have considered the intermediate stage of order corresponding to a microcrystalline body. Biot[24] has a very clear distinction between elasticity and cohesion, and his whole viewpoint was one which augured well for solid state physics in the nineteenth century. He concludes, however, that

there are so many states of equilibrium possible between all of the forces by which the particles are animated, but these forces are too unknown and too numerous to enable the results of their combination to be calculated in advance from their posture and position.

This remark of Biot's was prompted by the observation that the old alloy used for cymbals (78 copper, 22 tin) behaves in an inverse fashion from steel, namely it becomes softer on quenching than if it is slowly cooled, and the appearance of its fracture changes.

Savart[25] observed in 1829 the anisotropy of elastic behavior in plates showing Chladni figures and was led therefrom to some very acute comments on the structure of metals. This led him to the first definite description of the structure of most metals.

Metals have hitherto been regarded as the solid substances which most closely approached conditions of homogeneity; they have been regarded as assemblages of an infinite number of little crystals united together without order and at random, and there was no suspicion that in any mass of metal whatever there could be differences of elasticity or cohesion as great as, or perhaps greater than, those observed in a fibrous body such as wood.

He tested plates cut from various positions in a large cylinder of lead and, showing that these plates were not isotropic but had elastic axes which changed with direction of excitation, he noted that

these facts and many others of the same kind show clearly that metals do not possess a homogeneous structure, but only that they are not crystallized regularly. There remains thus only one supposition to be made, namely that they possess a semi-regular structure, as if, at the moment of solidification, there were formed many distinct crystals, of a substantial volume, but whose homologous faces were not turned to the same points in space. According to this idea, metals would be like certain grouped crystals, in which each one, considered individually, presents a regular structure while the mass as a whole appears quite confused.

A lead ingot, the crust of which was pierced and drained when partly solid, shows

numerous octahedral crystals arranged in parallel rows, crossing rectangularly, which form a great number of distinct systems, corresponding in position to the little furrows that had appeared on the opposite surface of the solid crust.

Seen with the magnifying glass, the little crystals which compose each of these systems seem to be grouped around three directions at right angles to each other, and they are arranged so that their axes are parallel to each other, hence they seem to touch only at their solid angles. If we suppose that the three directions of each system can have a direction which is indeterminant in relation with those of the neighboring systems, one will have a sufficiently clear idea of the semi-regular crystallization of a mass of lead.

He studied the effect of various casting conditions, including vibration while the metal was solidifying, which he found to disturb the crystalline system and to produce a high uniformity of elasticity—an observation which he thought might be important in the arts. He showed that a slight amount of hammering of a cast lead disk left the elasticity relatively unchanged while rolling gave a regular structure. A rolled plate of zinc had a uniform preferred orientation so that the sheet of metal, although highly polycrystalline, behaved as if it had crystallized regularly in its whole extent. He then studied the influence of annealing, which he found was small or nonexistent with metals that had not been cold worked, but plates first cold worked and then annealed had both the frequency and the orientation of the elastic nodes changed. He found the effect to be different in different metals and to be sensitive to impurities, being very small in the case of most alloys. He showed that the elastic behavior was not peculiar to metals, and concluded that there was a heterogeneity in nearly all solid bodies except deposits of pulverulent matter like chalk or sealing wax which was isotropic because the particles of cinnabar prevented the resin from taking a regular arrangement. Finally, Savart shows that metals resonate with less facility immediately after casting than after some hours or days.

This seems to result from the fact that during the act of solidification many particles are, as it were, caught in positions which they tend to leave later, and that they do not attain a state of equilibrium for a long time.

It will be seen that Savart touches on almost all of the questions of concern to the present-day metal physicists. He mentioned preferred orientation in castings and worked materials, and its change on annealing; the difference between static and dynamic modulus; structural relaxation on aging; and other aspects of internal friction.

Another important, though uninfluential, book published in this decade was the *Lehre von der Cohäsion* of M. L. Frankenheim.[26] Frankenheim was well aware of the nature of polycrystalline aggregates in metals, and discusses the effect of mechanical and physical properties, although in a somewhat hypothetical way. He realized that the last parts of a metal or alloy to solidify would be in intercrystalline spaces, forming a mortar (perhaps of different composition) between the crystals. But conditions were not ripe for the physicist to be concerned with such complex things, and elasticity had first to develop as an ideal science before the interesting anomalies discovered by Savart could be exploited.

Michael Faraday studied the magnetic properties of single crystals of several metals, and reported the magnetic anisotropy of bismuth (*Philosophical Transactions*, 1849, **139**, 1–18, 19–36). The lack of directionality in bismuth that had been quickly solidified was due, he believed, to the fact that it was either amorphous or finely crystalline. Thinking it possible "that thin wires, which by the action of acids exhibited fibrous arrangements, might have their particles in a state approaching to the crystalline," he submitted bundles of platinum, copper, and tin wires to the action of the magnet but could find no anisotropy.

We cannot here do more than mention the development of the formal mathematical theory of the elasticity of crystals, a subject that attracted some of the greatest mathematicians of the nineteenth century* and which is central to many present-day problems in the physics of solids. Its roots were in Boscovich's giving, in 1758, mathematical form to Newton's concept of the forces between elementary particles.[29]

* For details see Todhunter and Pearson[27] and Love.[28]

Boscovich's theory involved periodic interactions between several points simultaneously and was a truly great intellectual achievement, which is only now beginning to take its proper place in the history of science.

Crystal elasticity theory, like the theory of space groups mentioned earlier, remained somewhat outside the mainstream of nineteenth-century physics, although simple elasticity and elastic anomalies were frequent objects of experimental study. Coulomb had observed elastic hysteresis in his first experiments on torsion in the eighteenth century. The elastic aftereffect in quenched steel was first studied by Weber in 1835 and in more detail by Kohlrauch (1863) and Lord Kelvin (then William Thomson) in 1865. This prompted the theory of Boltzmann,[30] who was proud of the fact that his theory had no connection with any physical hypothesis. Clark Maxwell (1878) developed a rather similar theory[31] but this was based on a physical model: in fluids there are groups of molecules continuously accumulating and breaking up, while in a solid (even a homogeneous one) there are different groups of molecules which differ in respect to their amplitudes of oscillation and the amount of strain in the original configuration. In some groups the ordinary (thermal) agitation of the molecules is liable to accumulate so much that every now and then the configuration of the group breaks up, and this whether it is in a state of strain or not. However,

we may suppose that there are other groups, the configuration of which is so stable that they will not break up under any ordinary agitation of molecules, unless the average strain exceeds a certain limit, and this limit may be different for different systems of these groups.

If these groups are disseminated through a substance in such abundance as to form a solid framework, the substance is solid, and can only be permanently deformed by a stress greater than a critical one. Mixtures of stable and unstable groups will show the elastic aftereffects; after permanent deformation, some stable groups may retain original configuration while the unstable ones will be broken and re-formed. Vibration, temperature increase, or moisture will break up the unstable groups, allowing the stable ones to again assert their sway and produce recovery. Although Maxwell's molecular "groups of greater stability" are formally microcrystals, and those unstable ones are formally dislocations in today's terminology, the theory was without any real structural basis. It was a fine qualitative explanation in principle; but it could not be used, for it lacked definable, measurable, units. By the twentieth century this viewpoint had decayed to the imaginary assembly of complicated networks of dashpots, springs, ratchets, and flip-flops used by some mathematical rheologists to get equations that would match the measured stress-strain-time-elongation behavior of any given polymeric plastic.

In 1887 Barus,[32] a most indefatigable measurer of physical constants of metals and alloys, verified Maxwell's theory by experiments on torsional relaxation of steel in various heat treatments. He also showed the formal similarity between this and Clausius' theory of electrolysis.

Distributing unstable molecular configurations uniformly through the substance of a rigid metal like steel is analogous to that of dissolving molecules of acid or salt in a non-conductor like water. These added molecules are the unstable groups with which Clausius' theory deals.

The interest of physicists in real matter continued to decline until the introduction of X-ray diffraction in 1912. The standard texts have no structure in them. Poynting and Thompson's important *Properties of Matter*,[33] even in the ninth edition of 1922 devotes only half a page to Ewing and Rosenhain's results on slip, while that most popular text by Searle[34] nowhere used the word *crystal*. However, the apathy was soon to disappear and the excellent book by Ewald *et al.*[35] in 1929 provided a fine introduction of structural realities to physicists.

This is not to say, of course, that physicists were not doing useful work on the structure of matter or crystals. Indeed, the whole development of crystal elasticity, the wonderful mathematics of space groups, and the studies of crystal optics were all essential to subsequent advance. But physics was concerned with perfection, and most of the properties which affect the use of metals, and hence are the concern of metallurgical engineers, are due to imperfections. At long last, however, these have entered the realm of respectable physics.

CHEMICAL ANALYSIS OF EXTRACTED COMPOUNDS

In the meantime, advancing chemical knowledge was laying the ground for later important contributions to metallography. The early work of Rinman, Bergman, and others on the action of acids on steel was the beginning of work by a number of chemists who extracted and analyzed carbides and sometimes attempted to explain the properties of steel in terms of definite carbide molecules. Despite the importance of this technique in connection with steel, it was not used on other alloys. In retrospect it is amazing that so little attention was paid to the rather obvious changes in shape and size of the extracted carbides under various conditions of heat treatment. The emphasis, moreover, seems to have been entirely on the particles that could be extracted and none whatever on their matrix. Nevertheless, before carbides were actually observed under the microscope, there was considerable knowledge on the form in which carbon existed in steel. The chemical work on the extraction of carbide begins with the work of Rinman in 1774 (see chap. 12), who showed the difference between the action of acids on irons and steels, and Daniell, who first noticed the difference between the residues left by acid solution of quench-hardened and annealed steel. After the initial identification of carbon as the supreme factor in the composition of steel, the argument soon developed as to whether the hardness was a result of the formation of a chemical compound or due in some way to the mechanical dispersion of the carbon. For example, Karsten,[36] whose early interest in this subject was shown by the notes he appended to his German translation of Rinman,[37] showed that chilled cast iron contained chemically combined carbon only, while the same iron slowly cooled was much softer and contained graphite. Since the same sequence of treatments on a steel produce similar changes of hardness, it was easy to think that carbon became transferred from the uncombined to the combined state by quenching, and the variation of density as well as electrical conductivity when it was first measured seemed to support this. Karsten, perhaps the leading metallurgist of the time, was firmly convinced of the existence of iron polycarbides of definite (though unknown) composition which tended to dissociate more and more to graphite at higher temperatures. He performed numerous experiments to saturate iron with carbon by diffusion, and to analyze the residues of various sierugical products dissolved in acid. At one point he thought he had isolated a definite compound Fe_3C, but later retracted it. Berthier,[38] Gurlt,[39] and many others extracted and analyzed carbides. Chemical theory was such that the existence of compound molecules was presumed and the existence of a solid solution in a rather narrow temperature range and based on a different crystal structure modification from the one stable at room temperatures was hard to conceive. Bréant's work (see chap. 4) in which structure, chemical theory, and practical observations were combined, seems to have been forgotten.

The chemists who extracted carbides from steel and analyzed them were misled by the partial reaction of iron carbide with their solvents and it was actually not until the work of Abel[40] and Müller[41] that sufficient precautions were taken to give a residue accurately reflecting the separate carbide phases present in the steel. In the ordinary solutions the carbide was likely to be partially decomposed, emitting hydrocarbons, and partially retained, while some finely divided elemental carbon could also be formed. Carbon in hard steel (i.e., in the form that we now call martensite) was almost entirely liberated as gaseous hydrocarbon.

An important incentive to research was given by the French chemist H. Caron,[42] who knew of the work of Berthier and Karsten with their residual carbides. In his experiments he found no definite polycarbides, but a residue varying in amount with the quality of steel, and with the shape and dimension of the sample, and with the solvent used, which led him to believe that carbon existed merely as mechanical mixture in the metal. The amount of carbon in the residue, (which contained also iron and silicon) varied from 0.825 per cent of the original cemented steel bar to 0.560 per cent for the same material forged, while only traces were found after hardening. From this he concluded that the qualities which distinguish steel increase proportionately with the amount of carbon combined more intimately with the iron. On tempering, the amount of insoluble carbon increased with the duration and the intensity of the heating. He concludes:

The affinity of carbon and iron is thus so feeble that heat alone (when it is not carried to the point of melting the metal) suffices to disunite them more or less completely and change the qualities of the steel.

At nearly the same time, and apparently independently, L. Rinman[43] (1865) also observed the effect and he called the two types of carbon "hardening carbon" and "cement carbon," (*härdningskol* and *cementkol*, respectively) the latter because it occurred in cemented blister steel. Caron[44] later, observing the change of volume on hardening steel and the carburization of a bar of iron hammered on a layer of charcoal dust in an anvil, concluded that the shock of the quench, like hammering, was responsible for producing the intimate combination of iron and carbon. This theory was championed in 1879 by Ackerman.[45] In the discussion of this paper J. W. Spencer[46] accounts for all of the carbon in all forms, and shows that the "missing" carbon, i.e., that not shown by the color test that was then common in industrial practice, is a characteristic of hardened steel alone, for the color test shows only carbon in compound form.*

* The colorimetric method for determining carbon in the form of carbide in steel was devised by the Swedish metallurgist Victor Eggertz (*Jern-Kontorets Annaler,* 1862, p. 54). He arrived at it by combining the earlier acid extraction tests with Liebig's color test for organic substances.

*Fig. 90.—Insoluble substance
remaining after dissolving steel
in nitric acid, ×155.
(Tschernoff, 1876.)*

Similar measurements were made by Osmond a few years later[47] and they led him directly to the use of the microscope for observing the actual distribution of residual carbides in steel and so to even more important problems. In 1875 Tschernoff, in the notes submitted to William Anderson at the time of the latter's translation, published a drawing (Fig. 90) showing the microscopic appearance at a magnification of 155 of the insoluble substance remaining after dissolving steel in nitric acid, with another showing some hexagonal plates and some comments on their chemical nature. (See p. 140.)

In 1890 even Howe was of the opinion that the examination of residual components as a skeleton after solution of the matrix was a more promising method than the examination of a polished surface. In this he was wrong. Though extraction methods had played an important role in preparing the ground for metallography, they were soon supplanted and were little used except for studies of non-metallic inclusions. It is indeed interesting to see that at the present time they are being reintroduced as an important technique in combined electron microscopic and electron diffraction studies of fine precipitate particles in many types of alloy.

Before proceeding to discuss the German and French engineers under whom microscopic metallography became an established discipline, we must mention some other workers whose observations and ideas are of interest although they lack both definiteness of theory and adequacy of experimental technique.

Perhaps under the influence of earlier work on the extraction of carbides, Lampadius in 1827[1] makes a rather clear structural observation: "the best cast steel has so fine a fracture that the platelets of iron carbide that are mixed in it can be clearly disclosed only with a powerful microscope." It is, however, far from certain that Lampadius was actually able to distinguish carbide in the fracture from the other constituents.

An anonymous popular book, *Microscopic Objects, Animal, Vegetable and Mineral* (London, 1847), advocates the examination of fractured surfaces of cast iron and states that "So important has a microscopic examination of [cast iron] been deemed by the Government, that Her Majesty's Honourable Board of Ordnance has directed Mr. Pritchard to construct a microscope for the Arsenal at Woolich."

In 1867 F. Kohn reported[2] the use of a pocket microscope at the Paris Exhibition of 1867 by Eduard Schott, manager of Count Stolberg's iron foundry at Ilsenberg in the Harz. Considering the widespread use of the naked-eye appearance of fracture as an indication of quality and treatment it is, says Kohn, "astonishing that metallurgists should have neglected the use of the microscope for such a time." Schott had been able to deduce the quality (i.e., carbon content) and methods of manufacture of many hundreds of steel samples, purely on the basis of studies with a small pocket microscope (actually a hand lens). He clearly misinterpreted the structures and related what he saw to the practical problems of steelmaking.

Only a month later, Kohn reports[3]:

The little pocket microscopes for examining the crystalline structure of iron and steel recently mentioned in our pages, are multiplying in this country [England] with astonishing rapidity.

After mentioning some manufacturers he goes on:

The price of such lenses of suitable size and power is from 14*s*. to 18*s*.; and their services, after very short practice, are so important that no iron or steel master, once accustomed to their use, likes to miss his pocket microscope for an instant.

Somewhat earlier the same year Kohn had reported[4] enthusiastically on "the Berlin castings" displayed by Schott's foundry at the Exhibition. These castings were of a kind that became famous in the nineteenth century for exquisitely finished surfaces and intricate designs almost akin to jewelry.* Schott.maintained that the excellent results came about partly from a careful control of the sand (made by incorporating ground natural and artificial stones, selected by quantitative porosity measurements) but principally from the control of the metal. Schott controlled the temperature of casting by observing the increase in temperature of a fixed weight of water, and selected mixtures of pig iron to have the desired melting point. But melting point alone was not enough, and Schott found that by careful examination of the surface of a test casting for shrinkage and crystallization he could obtain a uniform iron. Schott believed that the patterns that he saw on the surface were the result of a splitting of the surface, "a tearing asunder of the skin with fissures in various directions." The figures, he says, may often be traced even after the setting of the iron, when they form projections on the surface. Two of his drawings, representing a charcoal and a coke iron of No. 2 quality are shown in Figure 92.

After discussing the various surface features of iron during and after solidification and the appearance of its fracture, Kohn proceeds:

M. Schott maintains that all crystals of iron are of the form of a double pyramid, the axis of which is variable as compared with the size of the base. The coarser kinds of iron have crystals of about twice the height of pyramids as compared with the finest qualities of crystalline iron. The more uniform the grain, the smaller the crystals, and the flatter the pyramids are, which form each single element, the better the quality, the greater the cohesive force and the finer the surface of the iron. The character of steel has been found in no way different as far as the nature of its crystalline structure is concerned. The pyramids of steel vary in size of axis and in shape. They

---

* The American Cast Iron Pipe Co. (Birmingham, Ala.) has a collection of such "cast iron art" assembled by Gustav Lamprecht. They issued an informative catalogue in 1941. Figure 91 shows one of the objects in their collection that was cast at Ilsenberg somewhat before the time Schott was making his observations.

Fig. 91.—Cast-iron medallion made in the Ilsenberg foundry. 7 x 10 cm. (Lamprecht Collection, Birmingham Cast Iron Pipe Co.)

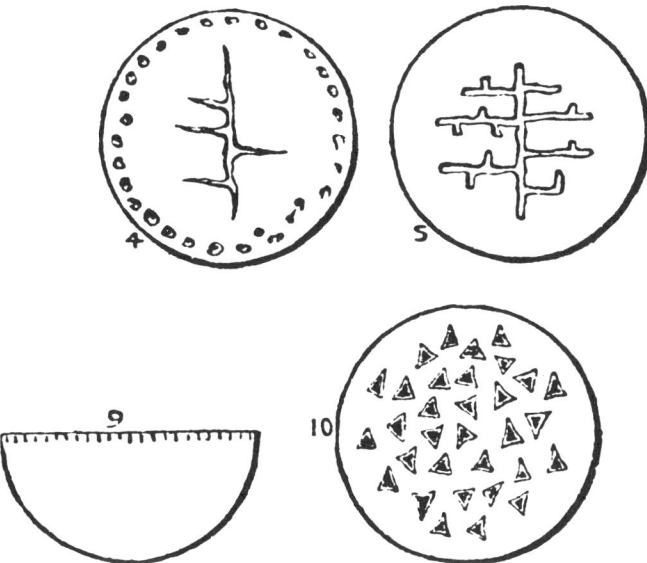

Fig. 92.—Appearance of surfaces of test castings of iron. (Schott, 1867.)

are more or less pointed, according to the nature and quality of the steel; they are of more or less uniformity in size and relative position, according to the mode of manufacture and treatment. For steel, the parallelism and regularity of the relative position of the crystals is one of the most important elements of its quality, and this can be very well observed under a microscope of sufficient power.

Kohn remarks that Schott's castings were admired by every visitor to the Exhibition and had been awarded a gold medal. The order books of the establishment were crowded at a time when other foundries were slack— a tribute to the advantage of scientific control.

To collect sound information about these details, to study the reason for the different phenomena which present themselves in the foundry, and to arrive at a scientific mode of commanding success under all circumstances, will make our future foundry practice in [*sic*] a very different position to that which it occupies at present. We shall have a science instead of a craft, and knowledge instead of habit and mere mechanical skill.

At the Vienna Exhibition of 1873, a particularly fine exhibition of art castings was made, and *Engineering* printed a series of five articles[5] which, though unsigned, were written by Eduard Schott, probably translated by Kohn. A German edition of the same material was published in Braunschweig as a little pamphlet bearing the title *Die Kunstgiesserei in Eisen* which carries Schott's name on the title page[6] and reference to the earlier English publication. The article includes descriptions and drawings of the crystalline patterns on the surface of solid and solidifying cast irons of seven different kinds. Four of his drawings are reproduced in Figure 93. Schott discusses the formation in shrinkage cavities of fir-tree–like crystals formed of octahedral needles at certain distances, arranged around a central axis and forming also an octahedral-like space. He states that "the assertion might almost be made that cast iron is nothing else but a compound of bar-iron crystals and graphite." From the standpoint of control of the foundry he observes that it would be most instructive if composition and microscopic investigation of crystalline formation could be related to each other.

Schott's work is not in itself an important contribution to the knowledge of the structure of metals, for most of his observations had been made before and in some cases better understood. It is interesting to see that Schott's desire for careful control of his iron was motivated by his desire for a finely finished art product. His aim was not engineering, but aesthetic qualitative characteristics. In the case of the Swedish work on Damascus gun barrels, a desire for a surface texture led to important chemical advances, and in a parallel way Schott's "art castings in iron" led to important structural studies. As Bergman followed Rinman and Wässtrom, so did the German metallographer Martens follow Schott. It was Schott's rather vague discussion of structure that provided the incentive for Martens' more scientific approach, and with Martens metallography finally came into its own.

Adolf Martens' initial article[7] (published in the *Zeitschrift des Vereines deutscher Ingenieure* in January, 1878) is the first paper on microstructure of metals to be widely noted by both scientists and industrial metallurgists alike, and it marks the transition from sporadic observations of structure to a planned attempt to study it in relation to properties. For reasons that are more related to the general state of knowledge at the time than to the excellence of Martens' own papers,

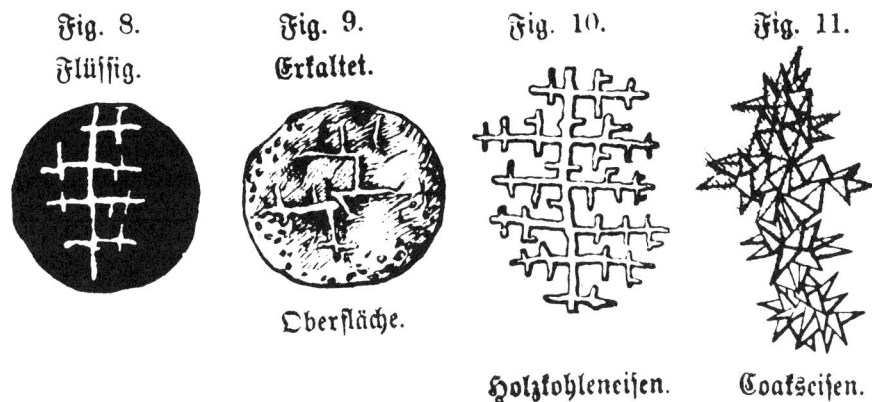

Fig. 8. | Fig. 9. | Fig. 10. | Fig. 11.
Flüffig. | Erfaltet.

Cberfläche.

Holzfohleneifen. | Coakfeifen.

Fig. 93.—Surface textures on iron castings. (Schott, 1873.)

they were received with much interest and they precipitated a wave of studies in other countries. Martens* was working at the time as a bridge engineer on the Prussian state railways, but he did all his metallographic work in his spare time at his own expense. Even his appointment in 1884 as director of the Königlich Technische Versuchanstalt did not greatly facilitate his research and it was not until E. Heyn, a student of Ledebur, came to work in Martens' institute that metallography really flourished in Germany.

Martens was led to study the microstructure of iron by reading Schott's work which we have just discussed. His first words are, in fact, thus:

The first impulse to these investigations, which are not to be considered conclusive in any way but merely as introductory, was given by Mr. E. Schott's remarks on the value of the microscopic investigation of iron for practical ironfounding in his valuable and interesting work on "Die Kunstgiesserei in Eisen." The object of this article is principally to show the practicability of the method, to induce further investigation, and to indicate the manner in which satisfactory results can be obtained.

Martens studied acid-extracted residues, fractures, and ground and polished surfaces. He gives sixteen drawings of structures at magnifications up to 300. He did his grinding and polishing on glass plates with emery, finishing off with putty powder or other polishing material, also on glass. Like Sorby, Martens cut a thin slice and cemented it to a glass plate. On fractures, he observed the shape of graphite, noticing that the particles were not flat but bent and wrinkled in all directions, and com-

* A biography of Martens, with a full bibliography of his metallographic papers, is given in *The Metallographist*,[8] 1900, **3**, 177–81. It is probably by Albert Sauveur.

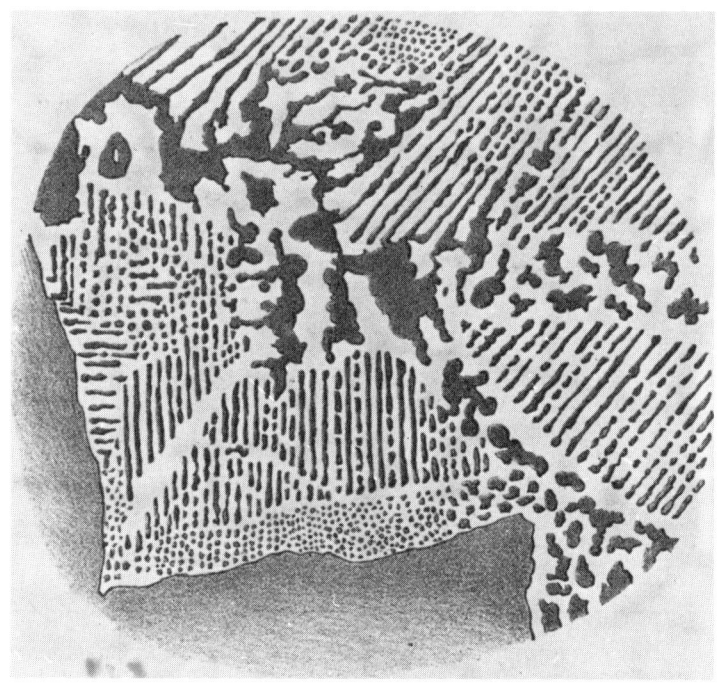

Fig. 94.—*Microstructure of spiegeleisen, etched with magnesium sulphate,
×200. (Martens, 1878.)*

posed of small crystalline scales, mostly in the form of equilateral triangles. Spiegeleisen showed on its fracture regularly arranged little globules (presumably interstices in a dendritic eutectic) as well as well-formed platelike crystals. Gray cast iron showed fir-tree–like crystals like those described by Schott, but Martens found by examining etched sections that they had marked composition gradients within them. After discussing fracture, he proceeded to that class of microscopical investigations which he considered the most useful for practical purposes, the investigation of ground and polished surfaces. Etching was done with hydrochloric or salicyclic acids of utmost dilution—1 part of acid in 9 parts of alcohol and then 10,000 or 15,000 parts of water, acting for as long as three and a half days. He observed the eutectic structure in spiegeleisen (Fig. 94), the distribution of graphite in gray and mottled cast iron (Fig. 95), and a "knitted" structure that was probably fine graphite.

He was fully aware of the value of microscopic studies.

A careful observer of all these results cannot but come to the conclusion that in pig iron the various combinations of iron are only mechanically mixed, that during the process of cooling or crystallisation they arranged

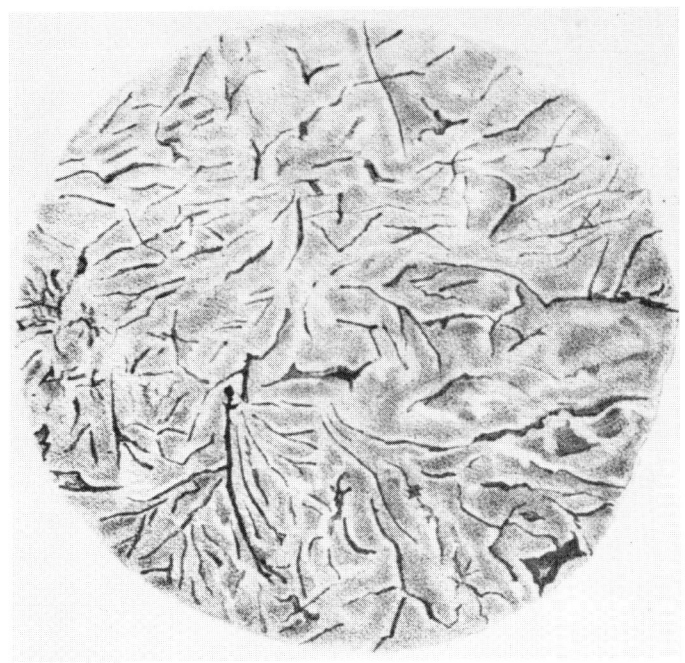

Fig. 95.—Microstructure of gray cast iron, showing graphite flakes, ×25. (Martens, 1878.)

themselves with most astonishing regularity, and that the microscopical investigation of iron has a very great chance of becoming one of the most useful methods of practical analysis.

He emphasizes the importance of comparative studies of the kind pointed out by Schott and concludes with a statement that combining microscopical with chemical analysis had for some time been his principal objective, for he believed that practical results may be obtained in this way.

In a paper published shortly thereafter, Martens discusses[9] at some length the microstructure of spiegeleisen, and reproduces numerous excellent drawings. Again he depended upon the appearance of the fracture as well as sections and he correctly concluded that spiegel was a mechanical mixture of a chemical compound of iron and carbon with iron containing no combined carbon. The constituents of the mixture dispose themselves regularly in a definite relationship with each other into different structures, the carbide being rhombic and the iron in dendritic forms tending to octahedral and belonging to the cubic system.

Martens' work attracted considerable attention and he published several more papers. In 1880 he described both the appearance of metal broken by repeated impact[10] (which he eventually concluded on the basis of fracture studies alone was without doubt due to a molecular transformation) and further studies of cast iron.[11] By this time, too, he had devised a microscope specifically for studies of metals[12] and for recording the results by photography. Later, in collaboration with the Carl Zeiss establishment he designed what was to be the best (though perhaps not the most convenient) equipment for metallographic photomicrography for several decades.

Martens wrote a large number of papers describing the structure of various ferrous materials, only a few of which need mention here.[13] Most of his papers were illustrated by drawings, or later, photographs. He usually dealt with both sections and fractures, for he continued to believe that much could be learned from fractures, and indeed some of his observations on crack propagation will repay study today.

Martens deserves much credit for conceiving early and publicizing the importance of metallography, as well as for establishing it on a firm basis in Germany. Nevertheless, despite the observations that he made, he seems to have made remarkably little contribution to the advance of metallurgical theory. His approach was too practical. Lacking in theoretical motivation, he is in a quite different class from Osmond, for example, whose experiments led directly to important advances in the understanding of steel and metals generally. According to the biographical note writen in 1900 for the *Metallographist:*

He [Martens] has repeatedly contended that metallographists should devote their energy to a thorough investigation of the best methods of developing the structure of the polished metallic surfaces, and that they should for a while content themselves with a description of what they see, avoiding hasty inferences.

In his writings Prof. Martens has adhered to these precepts, for his articles have been mostly of a descriptive character, often perhaps too minutely so, and he seldom draws any conclusions, or advances any theory.

If we may be permitted some criticism, we would venture to say that the fear of hasty theories, while generally a very wholesome one, may yet be carried too far, and such seems to us to have been the case with Prof. Martens. When experimental facts have been recorded, sufficiently numerous and concordant to warrant the putting forward of a theory, it is right and desirable to do so, for whether eventually proven false or true, it will suggest inquiries which cannot but help in advancing our knowledge.

If on the contrary, the experimental evidences be left scattered and no effort made to connect them under some theory, they will be productive of little good.

Not until Heyn began to publish his work at the turn of the century[14] does one find really perceptive microscopic studies in Germany. In the meantime developments in England and especially France had raised important basic questions and in partially answering them had laid the foundations for the modern science of metallurgy.

Martens' work soon became known in England. On the invitation of Percy, Wedding[15] in 1885 gave a report, based largely on Martens' work

Fig. 96.—Section of broken iron connecting rod. Etched, reduced one-third. (Hill, 1882.)

on the properties of malleable iron deduced from its microscopic structure. He discusses the formation of crystals and grains, recrystallization, and the nature of fiber (which he realized did not occur without slag). He confused iron with iron carbide, and believed pearlite to be homogeneous (i.e., non-crystalline) iron. The paper is illustrated by some color drawings, with misplaced captions. In the discussion, Sorby pointed out that the specimens were improperly prepared, did not show the true structure, and that Wedding should have used a vertical illuminator. Five years later, Wedding published thirty-eight good photomicrographs in his *Das Kleingefüge des Eisens* (Berlin, 1890).

In the United States the first use of etching seems to have been by A. F. Hill[16] in 1880–82. In his first paper he published some nice engravings of etched sections of punched, drilled, and riveted joints to show differences in distortion, much as Tresca had done earlier. In the second he describes a postmortem examination of an iron beam involved in a catastrophic steamboat accident. He attacks a writer who had advanced the old crystallization theory and gives engravings which show the etched structure of iron after a series of good and bad heat treatments. Figure 96 shows his macrograph of a fractured connecting rod. His samples were carefully prepared by planing and polishing on a lens grinder and they show the crystalline structure better than any other early macroscopic studies. In the explanation of them Hill talks of segregation of chemical compounds during crystallization obviously following Tschernoff[17] and reproducing his pictures of extracted carbides (see pp. 186 and 206) but not mentioning his name.

Martens' results were first reported in the United States in 1883 by Bayles,[18] who also summarized Sorby's work (publishing his photo-micrographs for the first time anywhere) and added a few useful comments on technique of his own. Bayles notes that his paper should be received with the understanding

that as yet very little has been accomplished in the way of practical micro-scopic analysis. . . . Nevertheless, there is no longer any question as to the important place the microscope must hold henceforth in metallurgical in-quiries, nor as to the magnificent field it has opened for investigations of an entirely novel character, the results of which cannot but prove of great value to the practical metallurgist.

The next American paper, by F. L. Garrison,[19] adds nothing but some photographs under poor oblique lighting of overetched irons and steels.

The year 1885 was an important one for our subject. In England, Sorby presented a detailed account of his work before a national meet-ing of the Iron and Steel Institute; in Sweden, Brinell reported his studies of the critical point and grain size changes on the basis of frac-ture; and in France, Osmond and Werth published their *Théorie cellu-laire des propriétés de l'acier,* which appeared in the July–August issue of the *Annales des Mines.*[20] Osmond had deposited a sealed document with the Académie des Sciences on July 9, 1883, in which, he claims, he recorded the essentials of this paper. It is interesting to note that Floris Osmond,* trained as an engineer, was working at the Creusot plant and credits J. Barba, the chief engineer (a man who had himself made im-portant chemical studies on the effect of heat treatment on the amount of carbon in solution and the effect of cooling upon it),[22] with creating a favorable environment. Somewhat later,[23] he remarks that

in France, M. Barba introduced in 1880 the use of the microscope in the works of Creusot, and gave the first impulse to the labours of Messrs. Osmond

---

* At the centenary celebration in 1949 Albert Portevin[21] gave an excellent outline of Osmond's scientific philosophy and work, and made an interesting comparison with Henry Le Chatelier. This report contains a bibliography of Osmond's 191 papers. There is also a brief biographical note by Sauveur in the *Metallographist* (1900, **3**, 255) and a fine appreci-ation of Osmond by his friend Roberts-Austen in the latter's presidential address to the Iron and Steel Institute at its Paris meeting in 1900.

and Werth, which have been pursued since that time by methods perfected under the influence of Dr. Sorby.

It is also significant that this, like most other early metallography, was concerned with the most complicated commercial material, steel. This seems always to have incited more scientific investigation than the simpler materials which could have been more easily studied. Indeed, however illogical it may be, the scientific study of simple materials seems almost to have marked the beginning of declining interest, for it is only when much is known that a clearly simple approach can be planned.

Osmond was a prolific writer but the essence of his developing ideas can be seen from four papers—the *Théorie cellulaire* of 1885; *Transformations du fer et du carbone dans les fers, les aciers, et les fontes blanches,* 1887;[24] *Méthode générale pour l'analyse micrographique des aciers au carbone* (1895 and 1901);[25] and *Sur la cristallographie de fer,* 1900.[26] Osmond started by combining the chemical and structural viewpoints, then developed thermal analysis and combined all approaches to give a true understanding of steel. Very little of his work was on nonferrous metals or alloys. In his first paper Osmond was profoundly influenced by Tschernoff (whose paper appeared in French translation in 1880) and by the studies of carbide extraction referred to in chapter 14. There is nothing to indicate that he knew of the works of Martens or Sorby when he began, although he later refers to Sorby as the founder of microscopic metallography.

Osmond begins his first paper (written jointly with Werth) by discussing the various means of revealing the cellular structure of steel and proceeds to discuss the changes of state resulting from cold work and heat treatment in the light of the cellular theory. He rejects fracture as a useful means of study, discusses at considerable length the Weyl method of electrolytic extraction of constituents which leaves behind the carbide, the shape of which can be observed under the microscope. It was this that led him to the realization that in an annealed steel—but not in a quenched one—the carbide was in a flaky form, which he believed formed the envelopes of the granules of iron.

Our elementary polyhedra are thus formed from a granulation of soft iron generally sheathed with a surrounding of carbide of iron, the thickness of which varies with the hardness of the steel and with the physical conditions which determine the molecular arrangement.

He then proceeds to a biological analogy which, like most analogies, was both useful and misleading:

By analogy (which is perhaps somewhat forced but which has the advantage

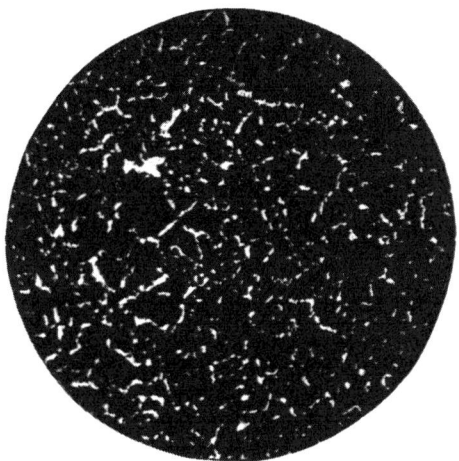

Fig. 97.—Network of iron carbide left in situ after dissolving a steel slice cemented to a glass slide, ×45. (Osmond, 1893.)

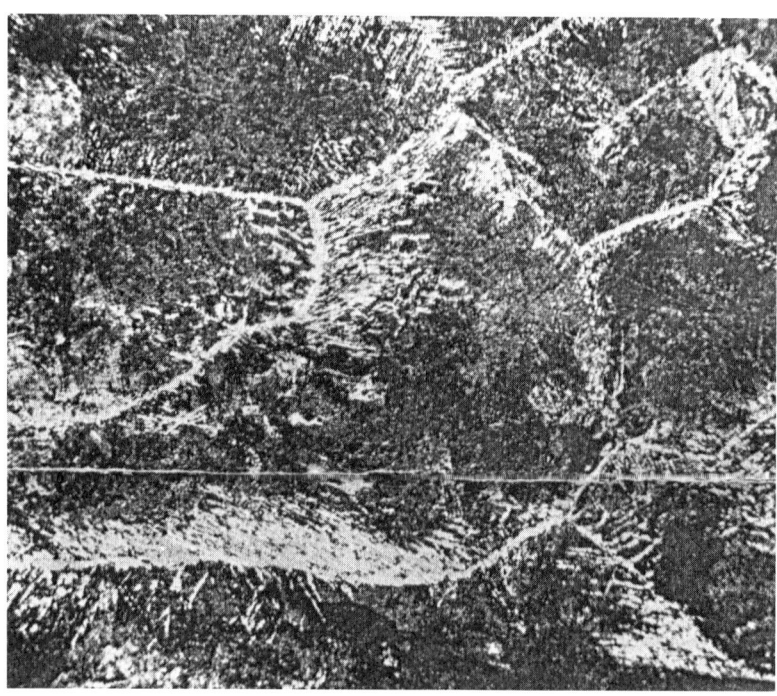

Fig. 98.—The "cellular" structure of steel. Etched with nitric acid, 36° Baumé, ×18. (Osmond, 1885.)

of applying a striking expression to observed facts) we will compare the iron granulation to the nucleus [*noyau*] of organic cells and the carbide to the cell wall [*envelope*], and we call the whole a single cell [*cellule simple*].

By cementing a thin slice of steel (note the petrographic influence again) to a glass slide with Canada balsam, and dissolving away the metal with nitric acid, a network of iron carbide was left *in situ* adhering to the balsam* (Fig. 97). He then proceeds to the etching of polished sections which he notes was an old technique, referring particularly to the extensive collection of mercuric chloride–etched samples (supposedly Tresca's) displayed at the Paris Exhibition of 1878. More often, however, deep-etching had been used in order to show the defects in steel by exaggerating them, while he (Osmond) was concerned with the normal structure. His preferred etching reagent was concentrated nitric acid, which quickly rendered the surface passive, repeated two or three times if necessary, washing in a lot of water after each immersion and drying with alcohol and ether.

A cast steel with 0.5 per cent carbon showed the elongated equiaxed polyhedra of Tschernoff. The microscope (Fig. 98) showed that there was no carbide on these boundaries and these grains therefore represent composite cells, with weak joints between them. (The unit cell in this description was actually the intradendritic spacing, the whole dendrite the composite cell, while the weak joints correspond to the grain boundary network.) A forged steel also had polygonal cells but they were very much smaller than in the casting, and annealing made them still smaller. On quenching, the cells disappeared, the surface being smoky, covered with shallow furrows, irregularly chevroned—the first hint of a martensitic structure. Tempering at a red heat gave a structure like annealed steel but a bit finer. Cold working produced diagonal flow-lines somewhat schistose in appearance, looking like Tresca's flow markings, but obviously having nothing in common with quench-hardening as Tresca had supposed.

Osmond concludes that his study of polished samples etched with nitric acid has permitted the identification, in every case, of the composite cells with the "grain" of steel and so allowed a precise definition of this age-old but elusive quality. Osmond at this time regarded the carbide as forming an actual cement which joined the iron grains to-

---

* Osmond soon abandoned the method as difficult and uncertain, but it has recently been revived as the useful extraction replica method of the modern electron microscopist.

gether.* He recognized that the homogeneity of steel was not what it is in liquids—steel is homogeneous when its cells, each with their complete envelope, form a connected tissue without geometric order.

Osmond and Werth made extensive studies of the variation of the amount of extracted carbide with heat treatment, and the amount of combined carbon as determined by the Eggertz solution-color method, which demonstrated conclusively that there was no loss of carbon but only a change of state on quenching. They conclude that carbon can be in two distinct forms: one really combined with the iron as carbide, which they called *carbone de recuit* because it predominates in annealed steel where it forms the cement, and another, *carbone de trempe* (English, "hardening carbon"), which was probably dissolved in the iron, perhaps related to occluded hydrogen, but not combined with the iron. This latter was dominant in the outer regions of quenched steel and disseminated throughout the bodies of the cells. To distinguish between the theories of Caron (who supposed changes in the state of carbon to be responsible for property changes on both cold work and quenching) and Tresca,[27] who postulated a molecular transformation (following Jullien and Füchs, see chap. 14), the authors used solution calorimetry, which was being used in many chemical researches at the Sorbonne by Troost and his collaborators, following the lead of Berthelot. Osmond and Werth found that the amount of heat evolved on dissolving steel filings in copper ammonium chloride was 4 to 6 per cent greater if the filings were dissolved in the cold worked condition, and 5 to 15 per cent more (depending on the carbon content) if quenched, than if annealed. This they believed to indicate two different allotropic states of iron, a view supported by the changes in the density and in the thermoelectric, magnetic, and chemical properties. These phases were dubbed alpha and beta and an argument was started thereby that was to rage for four decades and in many countries.

After discussing the effect of silicon, sulphur, phosphorus, manganese, and their relations with carbon and iron and each other, the authors remark that iron sometimes, but rarely, has a cement analogous to steel, for the carbon content is too low, and the easily oxidizable elements are almost never found free in a condition to form the cement. However the slag, particularly in puddled iron, does form a cement, producing a network of low resistance, easily disclosed by fracture, deformation, or

---

* The term used is *"ciment."* This is not to be confused with cement carbon, which refers to carbon introduced in the cementation process and eventually gave rise to the word "cementite" for iron carbide, Fe_3C. This term was introduced by Rinman (see chap. 14) .

TSCHERNOFF'S DIAGRAM.	When the temperature does not exceed point a, the steel does not harden on quenching. When the temperature rises from 0 to b the texture of the steel is unchanged.		As the temperature descends from c to b, the steel becomes crystalline. As the temperature rises from b to c the texture of the steel becomes amorphous.
	0 ———— a	**a ———— b**	**b ———— c**
State of cement........	........Solid and rigid..........	........Solid, then pasty........	........Liquid................
State of core..........	Not modified if starting with annealed steel; if starting with a quenched steel, the heating produces a new state of equilibrium between α and β iron. _[Perhaps undergoes an allotropic transformation.]_	Becomes proportionally more and more plastic as the temperature increases.	Melts in the cement.
Transformation of the structure by cooling {Slowly..	..........None..........	The grain becomes more regular, the composite cells tend to disappear. Hardens more and more easily on quenching.	The composite cells reconstitute themselves. Burnt steel. Quenched metal has no body.
{Rapidly.	..........None..........		
Mechanical working.....	Difficult with work-hardening and the appearance of schistosity.	Becomes more and more easy. (Normal working range.)	Difficult, but possible with precautions ......Impossible.
Welding.............	........Impossible........	........Detectable........	Welding zone. Welding easy, but the metal can no longer practically be "regenerated."

Vertical side labels: CELLULAR THEORY · Blue Heat. · The carbide of iron commences to dissociate. · Fusion of the cement. · Fusion of the steel.

Table I. Osmond's cellular theory of steel compared with Tschernoff's diagram.

attack with acids. This insistence on a cement or some adhesive material to fill in the spaces between the crystals seems remarkably naïve, for they should have realized that pure metals were polycrystalline but cohesive, and it is perhaps a residue of earlier thinking where crystals had to be polyhedral and could not possibly stick together except in unusually regular geometries.

The authors summarize their theory and relate it to that of Tschernoff in a table, a translation of which is given in Table 1. Osmond and Werth believed that simple cells formed during solidification just as they did in rocks—indeed the mineralogical analogies are frequent throughout all of Osmond's writings—the iron formed first, and the liquid was left in capillary spaces in thin layers that eventually solidified. The unit cells or globulites are not independent but are related like magarites, crystallites, or microlites in rocks. These three terms were borrowed from the mineralogist Vogelsang (via Michel Lévy[28]) for complex crystallites in which unit crystals are arranged in parallel clusters, in radial stellate configurations, or in regular successive accretions passing into a crystal properly called. As Tschernoff had shown, complex

*Fig. 99.—Fracture of burnt steel.
Magnification not stated, but apparently
about natural size. (Osmond, 1885.)*

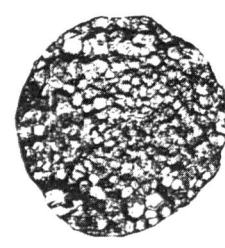

*Fig. 100.—Lead balls pressed together to simulate grains.
Magnification not stated, but apparently about
natural size. (Osmond, 1885.)*

aggregate structures (dendritic grains) could either be in the form of treelike agglomerations following three axes, or form floating centers which can grow in all directions to give polyhedra.

Tschernoff's point *a* marks, according to Osmond and Werth, the temperature at which the dissociation of iron carbide "acquires a sufficient tension to manifest itself in physical effects" while the point *b* corresponds to the melting of the cement. If steel is heated much above this it becomes burned, and its fracture shows polyhedra of a shape (Fig. 99) which can be exactly imitated by pressing balls of lead together in a mold (Fig. 100)—an experiment which had been done with other materials by Kepler in 1611, by Stephen Hales (who was interested in the shape of living cells) in 1727, and by Tschernoff in 1868 (see p. 118).

The mechanical properties of steel depend on both core and envelope of the cell. The core decreases in plasticity as the formation of beta iron is increased, which results from an increase of carbon content, or speed of quenching, or from cold work. Foreign materials, manganese for example, affect both the formation of beta and the dissociation of the carbide. Cold deformation ability depends also on the envelope, which is brittle. If the cores are only slightly plastic and only little deformation is possible before the adherence of the envelope to the core is broken, a brittle intergranular fracture results; whereas, if the whole elongates, the steel will be ductile. The tensile strength is constant per unit area of adhesion of the cement.

This paper is entirely a theory of *steel*. It is only slightly applicable to iron, and not at all to other metals. It does, however, clearly and definitely suggest the importance of the different states of carbon, the probability of allotropic transformations, and the variation in the scale and type of the distribution of the carbon in steel with heat treatment. To some degree all of these had been suggested before, but for the first time they begin to be properly integrated with each other to a consistent theory which has a spatial structural reality. All of Osmond's subsequent work can be traced to statements contained in this, his first, paper. To judge from his publications, he was remarkably narrow in his interests for a man of such great intellectual stature, though he was well read and even on occasion wrote poetry of a metallurgical nature!

Osmond's second paper was written by himself alone, after he had left the steelworks. It described work commenced in the laboratory of H. Le Chatelier at the École des Mines and completed in the laboratory of Troost at the Sorbonne. The paper bears the title *Transformations du fer et du carbone dans les fers, les aciers et les fontes blanches*, and was published both in a periodical in 1887 and separately the next year in a little book.[29]

Before Osmond, of course, the presence of a transformation in iron had been known principally as a result of Barrett's refinement of Gore's observations on recalescence, the anomalous specific heat measurements of Pionchon,[30] and the critical point observations of Tschernoff, confirmed by Brinell (see chaps. 10 and 14). However it was the development of the Le Chatelier thermocouple[31] which made thermal analysis practicable and Osmond was quick to seize upon it.* Prior to this, thermal analysis had been applied to the solidification of low-melting-point alloys by Rudberg,[34] and others, using glass thermometers, and to high-melting alloys by Roberts-Austen,[35] using the extremely tedious method of dropping a metal ball into a calorimeter of water, the increase of temperature of which was measured. Osmond used both the direct and, usually, the inverse-rate method of plotting thermal data. In iron he found three arrests in the heating or cooling rate and in steel three, two,

* Tait[32] was the first to use platinum iridium thermocouples pyrometrically, but Le Chatelier developed it as a really practical instrument. There is a useful early history of pyrometry by Carl Barus.[33] It is remarkable how few of the possible methods of temperature measurement have been fully exploited.

or one, depending on the carbon content. All these he associated with transformations, and he modified Tschernoff's point *a* by numerical subscripts to denote the three different ones, and with *c* and *r* to distinguish heating (*chauffage*) and cooling (*refroidissement*) respectively—a notation which except for typographical change has remained in use to this day, although the A_2 point is not regarded as an allotropic modification since our current narrow-mindedness excludes all but lattice change as allotropy. Osmond showed how the arrests varied with cooling rates and eventually disappeared if the cooling was fast enough. He made heating curves and found that quenched steel evolved heat around 350° C. as the iron returned to the alpha form and the carbon to the carbide. The portion of beta iron (which he believed to be quasi-vitreous) varied with the carbon content in the steel and with cooling rate. In resumé:

Steel annealed at a red heat is principally constituted as alpha iron, the elementary granules of which are enveloped with carbide of iron; quenched steel is beta iron in a *continuous* mass, holding carbon in solution; tempered steel is an intimate mixture of the preceding with a variable proportion of alpha iron and of iron carbide infinitely subdivided. All transitions and intermediate states are possible between these three types.

Osmond's establishment of the definite connection between the thermal critical points and hardening behavior was an important advance; nevertheless his theory that hardening was due to beta iron alone was extreme and rapidly drew attack from practical metallurgists and theorists alike.

The first attack was by the American Henry Marion Howe in his *Metallurgy of Steel*,[36] which was printed in book form in 1890 from the same plates used when much of it appeared in 1887–89 in serial form in the *Engineering and Mining Journal*. Howe did relatively little research himself, but he had an ability to seize upon the significance of other peoples' work and to synthesize both science and practice into real understanding, which he clearly expressed. He was strongly influenced by geological analogies.* He stated that

From the microscopic study of polished sections, iron appears to be constituted, like granite and similar compound crystalline rocks, of grains of several distinct crystalline materials, of which seven common ones have

---

* "The petrologist who attempted to deduce the mechanical properties of a granite from its ultimate composition would be laughed at. However, are metallurgical chemists in a much more reasonable position?"

already been recognized, through peculiarities of crystalline form and habits, color, lustre, hardness, and behavior toward solvents.

He suggested the adoption of the term "meterals" as a distinct class name suggesting their resemblance to minerals, but fortunately this did not take root. His proposal to assign names to the different constituents in the manner of mineralogists was, however, followed and it was from him that the descriptive names ferrite, cementite, pearlite (originally pearlyte, often spelled perlite), and hardenite derived.* The name "Sorbite" he suggested for an acicular compound reported by Sorby which was perhaps a nitride, but this term was not adopted by others because the phase in question was only rarely encountered

* The nomenclature of metallographic constituents, which seems to have confused the outsider as much as it has helped the initiated, was originally limited to steel structures, although some of the terms have been extended by analogy to any constituents of similar shape and/or origin. Following Howe, a second group of names was suggested by Osmond in his 1895 paper. These honored various individuals—martensite (after Adolf Martens), troostite (after Louis-Joseph Troost, in whose laboratory at the Sorbonne Osmond did most of his early thermal analysis. He did important work on the thermochemistry of compounds of metallurgical interest, but contributed only his name to the structural side of the science), and sorbite (after Henry C. Sorby). Osmond's nomenclature was made before the various phases were clearly established and they caused considerable confusion, though Osmond attempted to justify his naming of transitional structures in a paper[37] published immediately after Roozeboom's famous application of the "phase doctrine" to steel. There is an interesting discussion on terminology in the reports of the committee appointed by the Iron and Steel Institute (see its *Journal*, 1901 to 1903). Howe's descriptive names received none of the objections that were leveled at Osmond's. The confusion is heightened by the fact that ferrite, cementite, graphite, and the later austenite are phases that belong in the equilibrium diagram, while martensite is a transitional phase, and the others apply only to specific arrangements or origins of mixtures of two equilibrium constituents. Hardenite and martensite were identical, but they were distinguished for a time by limiting the former term to the lower carbon contents. Sorbite and troostite were merely finer and finer forms of pearlite and the terms have now practically disappeared from use. Austentite, now used for gamma iron and any solid solution based on it, was named by Osmond after its discovery in quenched steel of very high carbon content. The name first appears in print in the 1901 version of Osmond's *Méthode général* and is not in the 1895 paper. It is a well-deserved honor to the English metallurgist William Roberts-Austen. Howe's 1903 proposal of Osmondite for the same phase was unadopted because the earlier name had already taken hold. A transitional structure in the decomposition of martensite was also unavailingly named Osmondite by Heyn. Heyn himself was honored by an ephemeral constituent. Ferronite was once used by Benedicks for ferrite which contained some carbon in solution. Ledeburite (after A. Ledebur) has been used for temper carbon, but it usually denotes the iron-cementite eutectic in cast iron. Attempts have been made to honor many men by associating their names with micrographic constituents, but the only one to originate after 1903 which has been widely accepted is Bainite, first noted in 1934 and named in honor of E. C. Bain,[38] by members of the research staff of the United States Steel Corporation.

The word, "eutectic" had been used by F. Guthrie[39] in 1884 to represent the alloy of lowest melting point: it was a simple extension to use it adjectively for the characteristic structure. It was at first used by Osmond to refer to similar structures resulting entirely from a solid transformation. The word eutectoid was finally proposed by Howe[40] after a long correspondence on terminology in which Howe and Osmond independently coined and the term was later reintroduced by Osmond for another constituent. Howe's table summarizing the "minerals" which composed iron is reproduced in Table 2. This first appeared in *Engineering and Mining Journal* on August 18, 1888.[42] Howe's criticism of Osmond was based largely upon the discrepancies between his findings and those of Brinell. The change in hardness in medium-carbon steel in Brinell's experiments followed exactly the change in the condition of the carbon revealed by nitric acid test; and Howe confirmed this by a simple scratch-hardness test on a gradient-quenched bar of his own. This being so why invoke beta iron, any more than in cold working? He also rejected the production of beta on cold working unless one wanted to produce beta brass, bronze, etc. by working, for it was a general phenomenon. Though admitting the suddenness of the change, and not denying the possibility of allotropy, he (like Ledebur, Brinell, Tschernoff, Jullien and many others before him) attributed hardening to a change in condition of the carbon. "To sum up, Osmond's theory accords neither with our old nor his new facts: while the latter like the former harmonize well with the carbon-theory. The carbon-change being a fact, the $\alpha\beta$ allotropic change of iron as yet wholly unproved, the balance of present probability is readily seen."

There followed a series of arguments between the "allotropists" and the "carbonists," which aroused bitter personal and scientific antagonisms. The allotropists contended rightly that there was a transformation occurring in pure iron at a bright red heat, and wrongly that the high temperature form of iron could be retained at room temperature by quenching and was intensely hard; the carbonists that the hardness of steel directly depended upon the amount of carbon present

---

benmutic ("well-transforming"), though Howe for a time favored aeolic. The Widmannstätten structure was so named in meteorites long before the advent of microscopic metallography. Though brought to fame by Belaiew in 1909,[41] it had previously been applied to structures in artificial steel by both Sorby and Osmond.

Metallographers concerned with materials other than iron and steel have been less sensitive to nomenclature, and have not found it necessary to coin more mineralogical words. They refer to grains of the various phases known by Greek letters on the constitution diagram, to lamellae acicular or spheroidal particles, to intergranular constituents, or else they borrow that term from steel metallography which best describes the appropriate configuration or supposed origin of their constituents.

It is a pity that the greatest names in metallography—Sorby and Osmond—should not be fittingly preserved in its nomenclature. Osmondite has been proposed three times without being adopted, and the term sorbite was associated only with with an ill-defined aggregate which, being merely a fine or unresolvable pearlite, has ceased to need a distinguishing name, and is rarely used today. It would be a fitting tribute to a great man were the meaning of the term sorbite to be redefined to include all eutectoid structures and so replace pearlite in its broadest use, but custom is too well entrenched.

Table II. Minerals which compose iron. Howe, 1888.

Number	Name (Sorby's)	Name (Suggested here)	Probable composition	Occurrence (An important constituent of)	Occurrence (Little or none present in)	Occurrence (Occurs chiefly in)	Color by (Direct illumination)	Color by (Oblique illumination)	Lustre	Behavior on heating	Form, habit, etc.	Hardness	Relative solubility
1.	Free iron.	Ferrite.	Nearly pure iron.	Malleable iron, chief component. Open grey cast-iron, especially when annealed. Forms about 1/3 of pearlyte.		Malleable iron.				Crystallizes, segregates from thin plates to grains.	Crystals, probably interfering cubes or octahedra, homogeneous, malleable; malleable? Nearly or quite equiaxed after hot forging; elongated by cold-work; made equiaxed by reheating. Sometimes as shells surrounding and shooting into crystals of pearlyte: also as parallel plates within and dovelling the crystals together. In grey pig, probably as stout layer against graphite.	Comparatively soft.	More soluble than cementite.
2.	Iron combined with carbon = the intensely hard compound.	Cementite.	Iron with cement carbon.	About 1/3 of refined white cast-iron, and 1/2 of spiegeleisen. About 1/3 of pearlyte.	Open grey cast-iron, soft steels, weld and ingot-iron.		Intensely brilliant.	Perfectly black.		Changes little, serrating somewhat. Changes to ferrite on losing its carbon.	Usually structure-less, occasionally in flat plates, say, 90,001@·02 in. thick. In blister-steel as net-work surrounding and occasionally shooting into crystals of pearlyte.	Intensely hard.	Less soluble than ferrite.
3.	"The pearly constituent or compound" recrystallized.	Pearlyte.	A mixture of about 1/3 ferrite and 1/3 cementite.	Ingot- and weld-steel of all kinds unless hardened. Almost sole component of moderately hard steel ("70% carbon").	Very soft iron.		Dark on brilliant metallic ground in refined white pig.	Bright pearly on black ground in refined white pig.		Components combine at a high temperature to form hardenite.	Pearly, fine parallel plates, curved and straight, of ferrite, 1-40,000 in. alternated with cementite 1-80,000 in. thick. In soft ingots in irregular groups, often 1-30 in. diam., independent of ingot-structure. Also in ostrich-feather crystals in white pig iron.		
4.	"The pearly constituent or compound" uncrystallized.	Hardenite.	Iron and hardening carbon probably in all proportions up to 2 or possibly 3%.	In Bessemer and probably all other classes of steel when quenched. Arises from union of all minerals present.	Annealed or slowly cooled steel and cast-iron and in very soft iron under all conditions.					Separates (probably below 'W') into pearlyte and free ferrite or cementite.	Very minute grains, about 1-20,000 in.diam.	Intensely hard.	More soluble than cementite.
5.	Ruby and dark crystals.	Sorbite.	Perhaps silicon or nitride of titanium.		Weld iron and good cast-steel.	Cast-iron.	Ruby and deep blue.				Triangles, rhombs, hexagons, complex crosses, less than 1-1000 in. diameter.		
6.		Graphite.	Carbon.	Cast-iron.	Steel, ingot and weld iron.	Cast-iron.	Iron black.	Iron black.	Metallic	Changes little and slowly.	Comparatively large, somewhat irregular laminar, often bent, tapering edges. In grey Scotch pig, uniformly distributed, ·08@·05 in. broad, ·0005@·0010 in. thick. In No. 3 pig partly in irregular radiating groups.	1 to 2.	Insoluble.
7.		Slag.		Weld iron and steel.	Steel, ingot and weld iron.		Black.		Metallic		In hammered blooms is in irregular patches; in bar, plate, etc., iron in fine threads.		
8.		Undetermined residue.	Probably matrix or residue from formation of substances 1 to 6, and hence of widely varying composition.	Cast-iron.	Ingot iron and steel.						Very irregularly distributed.		
9.	More soluble metallic substance.	metallic	Components of meteoric iron.								A rhombic lattice- or net-work, orientation often uniform over a considerable area.		More soluble than 10.
10.	Less soluble metallic substance.	metallic									Often crystallized in relation to the orientation of the inclosing net-work of No. 10.		Less soluble than 9.
11.	Schreibersite.		Iron, 55·4@87·2%, Nickel, 4·42@28·1%, Phosphorus, 7·8@14·9%.	Components of meteoric iron.							Often a thin skin covering the net-work of No. 10.		

in it and the form in which carbon existed, which varied with heat treatment. Though they knew that carbon changed its nature rather suddenly at temperatures associated with heat evolution or absorption, they maintained (on the basis of the behavior of pure iron) that hardening had little to do with these effects.* The argument, with excellent men on both sides, inspired a great deal of experimental work which eventually led to the modern picture.

In 1890, partly as a result of Howe's criticism, Osmond[44] had slightly modified his stand—"the more affirmative theory I formally enunciated now appears to me after much investigation and reflection, to have been premature"—but he still insisted on associating hardness with an allotropic form of iron, with carbon necessary only through a secondary effect. He gives a new and more precise definition of recalescence as follows:

the liberation of heat which produces recalescence is the result of the chemical combination of the iron with the carbon; and conversely the absorption of heat during heating a_{c1} corresponds to the dissociation of the carbide Fe_3C, a compound which is stable at low temperatures.

Convinced of allotropic transformations in pure iron (although at that time a little uncertain as to the significance of Ac_2, which he thought marked merely the end of a range in which two phases coexisted), he concluded that "carbon, in the state of hardening carbon, maintains iron in the beta condition, during slow cooling, up [i.e., down] to a temperature which is in inverse proportion to the amount of carbon contained in the metal." Tracing recalescence to lower temperatures for more rapid cooling, and proving that hardened steel had a greater heat of solution in acid and therefore contained more "latent heat hardening" led him directly to the statement that

the influence of carbon is of the same character as that of the rate of cooling and both combine to produce the final result. The rate of cooling alone is not sufficient under the ordinary conditions in which hardening is effected to maintain an appreciable fraction of the iron in the beta condition. But, as, under the same conditions, it is easy to maintain the carbon in the state of hardening carbon, and as the hardening carbon imparts stability to beta

---

* Most modern metallurgists tend to consider themselves more nearly allotropists, yet much can be said for the carbonists since allotropy is only a means to the end—it is martensite (an extremely fine-grained and strained form of alpha iron with a large amount of carbon in inforced solid solution) which is hard, not gamma iron. Had there been no transformation, at some temperature alpha iron would have dissolved enough carbon to give a form of precipitation hardening on quenching and reheating, a viewpoint which Benedicks[43] suggested long before precipitation hardening in general was known.

iron, it is evident in what manner beta iron may be successfully preserved up to the ordinary temperature by the aid of carbon.

He states unequivocally, however, that "in opposition to the opinion generally accepted, I believe that it is not carbon, but the allotropic beta iron which is the principal cause of these new properties." The theory, he says could be proved beyond doubt if a steel with much hardening carbon could be soft, and one with very little hard, and he cites dubious experiments on both points. During tempering, he remarked, "The beta iron returns to the alpha condition and the hardening carbon to the condition of iron carbide," but whether simultaneously or successively was hard to decide. In this paper he also greatly extended his observations on the effect of alloying elements on the critical points, which was suggested by Roberts-Austen's (1888) hypothesis that the atomic volume, following the position in the periodic classification, determines whether the additive favors beta or alpha iron —tending to make iron assume or preserve the particular form in which it has itself its lowest atomic volume, or the converse for elements of higher atomic volume.

Osmond combined both the microscopic and the thermal approaches in his *Méthode générale pour l'analyse micrographique des aciers au carbone*, published in 1895,[45] which was less a report of research results than a masterly short textbook of experimental method and principles for students and workers of steel. In this he defines and describes the various metallographic constituents in steel, naming sorbite, troostite, and martensite for the first time. After showing the sequence of structure changes on each of a series of steels quenched from progressively higher temperatures (in which he was clearly influenced by Sauveur's orderly presentation[46] of similar data), he proceeds to a theoretical discussion of the observations. He gives a diagram (Fig. 101) showing how cementite decomposes above Ac_1 to yield an atmosphere of carbon which diffuses into adjacent ferrite to form martensite, which is iron with carbon in solution in the form of hardening carbon (*carbone de trempe*). The area occupied by martensite increases with temperature or by increasing the carbon content. The carbon-free iron can be in either the alpha, beta, or gamma form, depending on temperature, but when it contains carbon it is always gamma above A_1. Rapid cooling preserves the carbon as hardening carbon and, Osmond believed, this carbon served to retain some of the iron in the form of beta or gamma. Martensite he thought to be cubic, with the carbon unequally distributed between the body and the envelope of the needles. (It is a pity

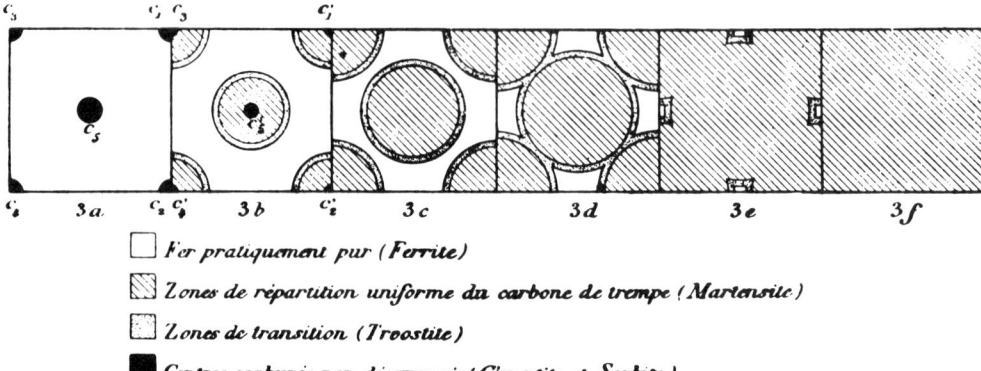

Fig. 101.—*Diagram showing solution and diffusion of carbon away from a carbide particle above A_{c1}. (Osmond, 1895.)*

that he did not live to see Bainite recognized.) Carbon, Osmond believed, played the role of a mineralizer: on cooling, the molecules would, after a momentary confusion, rapidly organize themselves and the new forms would grow progressively with temperature until mutual interference of the individuals produced a network of joints which provided surfaces of weakness. These became more marked the higher the temperature, eventually forming cracks. Alpha iron, he still thought of as a mass of globulites. The mass of alpha iron is divided into fairly regular polyhedra in contact with each other which "can be considered without too much fancy as pentagonal dodecahedra. Are these polyhedra crystals or grains? Are their surfaces cleavages or boundaries?* The answer does not appear in doubt to me. The mass [*pâte*] is perhaps crystalline, but the envelopes are not."

That Osmond still thought of allotropy as a purely molecular property rather than crystalline arrangement is shown by his elaborating Fe_3C into $Fe\alpha_3C$ and other forms because of its magnetic change. This is reminiscent of Matthiessen's allotropic alloys and is in keeping with the whole nineteenth-century molecular emphasis. In 1891 he was in fact arguing about molecules of beta iron being hard while the grains were soft.[47]

The *Méthode générale* was reprinted in 1901 as a part of the important volume, *Contribution a l'étude des alliages* published by the Commission des Alliages of the Société d'Encouragement pour l'Industrie Nationale.[48] The intervening six years has seen the determination

---

* *Joints*, the word still used in French for grain boundaries, and which Osmond employs in a sense analogous to that of the geologists.

of a good iron-carbon constitution diagram by Roberts-Austen, and the application by Roozeboom of the phase rule and general thermodynamic principles to Roberts-Austen's results to give a nearly correct diagram. Osmond uses the new diagram to explain the successive changes in steel structure that he had previously reported, and omits Figure 101 with its accompanying discussion. He describes new experiments with a steel with 1.57 per cent carbon in which austenite was found, and its transformation to martensite at liquid air temperature studied. (Austenite had been reported though not named in a brief paper of 1895.)[49] Osmond's metallographic technique, already good in 1895, had improved considerably by 1901. The classic photographs of pearlite and martensite (Figs. 102, 103) appear only in the second form of the *Méthode générale*.

Another change from the earlier version of the paper is the addition of a photograph showing geometric growth steps on iron crystals. Osmond's concern with crystallography followed the important paper by the English metallurgist John Stead[50] which marked the beginning of a new emphasis on crystallinity as distinct from mere chemistry or granularity. In 1895 Osmond did not think that grains were necessarily crystalline, although he knew that varying degrees of molecular order were involved. In the year 1900 he published two papers[51] on the crystallography of iron. He preferred to limit the term crystal to matter having a geometric shape that is a function of its internal symmetry, a fragment of crystallized matter not so shaped being merely a grain, piece, or chunk.* Amorphous matter could be geometric or non-geometric in shape, and crystalline matter might be shaped geometrically without regard to internal symmetry. Confusion had arisen, Osmond believed, from the tendency to use the word crystal as a synonym for crystalline matter. The papers are profusely illustrated with pictures of historical interest from many sources. Osmond analyzed old and new data on iron, concluding that Grignon's grains in burnt iron were pseudo-crystals, and that gamma iron is cubic with a preference to forming octahedra. He related the structures of manganese and nickel steels (austenite with annealing twins) to that of annealed brass, and extended Saniter's pioneering experiments on hot etching.[53] "Beta" iron he studied also in alloys with silicon or phosphorous which have no A_3

---

* A note by Howe, "What Is the Essence of Crystalhood?" published in 1902[52] took an opposing point of view and suggested the term crystal should refer only to internal order, which is approximately the current use, although the term grain always means crystalline grain.

Fig. 102.—*Pearlite in cemented steel containing*
1.60 per cent carbon. Etched by polish-attack, ×1000.
(Osmond, 1901.)

point, and which were supposed to be isomorphic with alpha at room temperature as there was no recrystallization on heating. Alpha was studied by means of etch pits,* which had previously been observed in iron by both Stead[55] and T. H. Andrews.[56] Osmond was puzzled by the lack of sharpness of A_2 (for the phase rule had now caught up with metallographers), and he suggested either isomorphism or cohesion between the phases to explain the observed temperature range of the transformation.

The second part of Osmond's paper, which was written in collab-

---

* It is interesting to note that etch pits, which were used effectively by the early metallographers, later came to be generally regarded as undesirable artifacts. The writer was trained to strive for smooth, clear grain surfaces only to have it announced in 1953 that etch pits can mark individual dislocations in the crystal lattice, and thus conventional metallography is brought once more in touch with the research frontier. Etch pits have long been regarded as important in mineralogy, for they give information on symmetry not discernible by external shape. A good discussion of them is to be found in the book by Baumhauer published in 1894.[54] This is illustrated with some very fine patterns in nepheline which are undoubtedly lines of dislocations forming low-angle boundaries and which would be at home in current research papers.

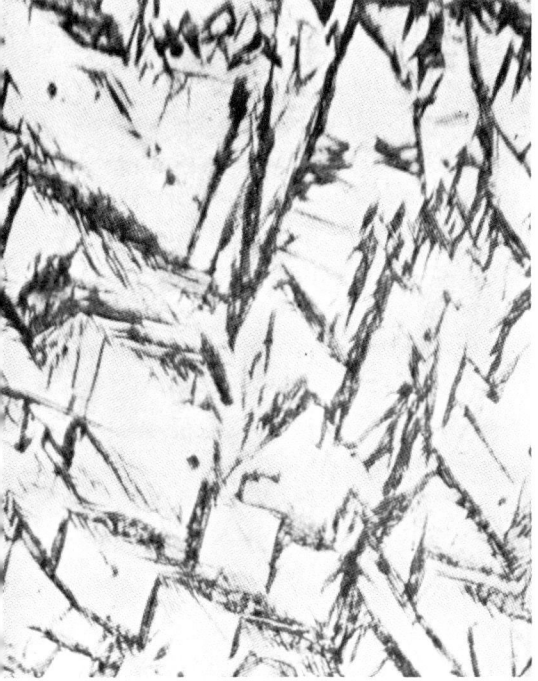

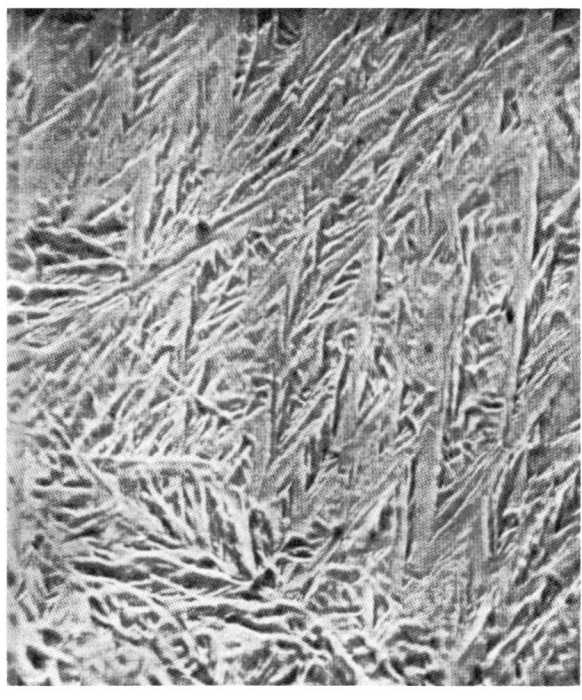

Fig. 103.—*Hardenite (i.e., martensite) needles in austenite matrix. Etched with dilute hydrochloric acid, ×250. (Osmond, 1901.)*

Fig. 104.—*Same sample as in Figure 103 after cooling in liquid air, ×250.*

oration with G. Cartaud, describes attempts to grow crystals of iron by reduction of ferric chloride vapor with hydrogen in the temperature of stability of each of the allotropic modifications. Alpha and beta gave identical crystal forms, and gamma showed some slight difference in detail of dendrite formation and face shape. Improving on Saniter's hot-etching method, the authors showed conclusively that alpha and beta in polycrystalline iron at 540° and 820° C., respectively, were structurally identical with ferrite at room temperature. They conclude that the joints between the polyhedra revealed by the action of acids on polished sections indicate the limits of crystalline development, and are not necessarily sections through tension surfaces (a concept tentatively borrowed from the geologists' mud cracks and basalt cleavages). Iron in the gamma stage, whether pure at high temperatures, or in high nickel or manganese steels at room temperatures showed grains with twins, quite distinct from alpha. In a steel of unspecified composition (but which must have been unusual), they found four different networks, visible either after hot-etching or after simply heating a polished surface in hydrogen. These were identified as the original alpha boundaries, martensite outlines, and both twins and grain boundaries in gamma iron. They recognized that the change from gamma to beta iron involved a change of symmetry, but that there was an orientation and grain size relationship between the two: most importantly, however, they saw that martensite represented a transformation of austenite during cooling and that it could be suppressed by appropriate alloying additions. Martensite occurred by transformation which began in definite crystallographic directions: they suggest erroneously (100) and (110) planes because these correspond to maximum molecular separation. They suggest that the transformation proceeds until the mechanical pressure due to change of volume added to the osmotic pressure of the dissolved carbon becomes enough to prevent the transformation.* With Osmond's papers of 1898 and 1899 on nickel steels[57] this work laid the basis for the theory of alloy steels which was developed later, particularly by Osmond's countrymen, Louis Grenet and Leon Guillet.

The collaboration with Cartaud continued until the younger man's death in 1907 and gave rise to a series of papers relating crystallography and mechanical behavior. Charles Frémont joined them in one paper[58] —the first to appear in the new *Revue de Métallurgie* founded in 1904 —and in later papers with Osmond alone[59] reported on the mechanical behavior of single crystals of iron. This and other twentieth-century developments are beyond the scope of the present book. After Cartaud's death Osmond, who was increasingly isolated by his deafness, virtually retired from metallurgy and contributed little more than a few remarks by way of discussion of other people's work until his death in 1912 at the age of sixty-three.

Osmond was a great scientist. By combining microscopy and thermochemistry he made metallography into a science. To him structure and energy contained the explanation of most metallurgical phenomena. He was a solitary man, working with collaborators only at the beginning and end of his career. If he took too seriously his allotropic theory of hardening of steel, this was partly an inevitable reaction to the too-violent attack made on it by others. Though he taught no mass of students, his ideas had influence wherever steel was thought about. It seems strange today how exclusively Osmond worked with steel. His only significant work on non-ferrous metals was that on gold alloys

* Osmotic pressure was then used, following Van't Hoff, almost as chemical potential would be today.

which was initiated by Roberts-Austen[60] to show the role of inter-granular constituents, and the related study of copper-silver alloys which provided classic support for the "solution theory" of alloys. Steel was his principal love. He was a metallurgist: he brilliantly perceived the relationship of general laws to the material he was interested in—an attitude in contrast with that of his colleague Le Chatelier, who used metallurgical experience to deduce general principles.

16. Outline of the Development of Metallography after 1890

In the preceding chapter the work of Osmond has been traced throughout his life. His work precipitated a metallurgical revolution and in large measure gave direction to it. His first two papers could be compared only with the work of Tschernoff and Brinell. Within ten years, however, metallography was almost popular. By 1890 a new science was discernible; by 1900 it was beginning to be formulated as an integrated body of knowledge to be absorbed by students and by industrial metallurgists. The outlines of this interesting but complex story are given by R. F. Mehl in his excellent *Brief History of the Science of Metals:*[1] here we will glance at only a few events, leaving a detailed account for another occasion.

The decade 1890–1900 saw immense activity in France, the United States, England, and (to a lesser extent) in Germany and Holland. Men observed the structures of metals under a wide variety of conditions and tried with varying success to build up something like a science of metals. The greatest activity continued to be centered on steel, that fascinating material whose complicated behavior encouraged a diversity of explanations. The service to metallography performed in 1894 by Behrens[2] of Delft in sketching the microstructures of a wide range of useful and interesting alloys passed almost unnoticed by his contemporaries, perhaps because the significance of the observations was obscured by their diversity. Though he has some remarkably good drawings of dendrites and of etch pits, many of his drawings have a curious air of unreality and his explanations of the structure are on the whole superficial. Nevertheless, he did observe many structures for the first time, and his observations on changes of structure with mechanical and heat treatment such as the homogenization and change of grain shape of bronze on working and annealing (Figs. 105, 106) were of basic importance.

Despite the fine start with Sorby and the prompt appreciation of both Tschernoff and Osmond, metallography was quite slow to develop in England. Its leading figures were William Roberts-Austen, J. O. Arnold, and John Stead, the most influential being the first named, whose work has been collected and discussed by one of his students and associates, S. W. Smith.[3] Roberts-Austen was associated with Thomas Graham at the Royal Mint for a few years before that great chemist's death, and first became interested in metals in connection with practical problems of liquation as involved in the preparation of a new set of assay trial plates for the mint. As soon as Osmond's method of thermal analysis was published, he rapidly seized upon it and greatly improved the technique, though he did not take up microscopic metallography until 1896, and his first photomicrographs do not appear until 1899. He was a close personal friend of Osmond and a champion of allotropy against the attacks of J. O. Arnold and Robert Hadfield in England. Most of his scientific work was supported by the Alloys Research Committee of the Institution of Mechanical Engineers and is reported in their *Proceedings*.[4] His first aim was to study alloys of gold and copper to support his theory that alloying behavior depended on position in the periodic table, but circumstances eventually made him abandon this idealized approach. After producing several copper-alloy constitution diagrams he embraced both steel and microscopy with enthusiasm to produce (in 1899) the first iron-carbon constitution diagram that resembles the one we accept today. A classic paper, "The Constitution of Metallic Alloys,"[5] related alloys to the general framework of chemical knowledge of solutions and provided a new basis for metallurgical thinking and teaching. Roberts-Austen was a fine lecturer and writer, and his lectures before the Royal Institution and the Royal Society of Arts did much to show the importance of metallography to scientist and industrialist alike. His book, *Introduction to the Study of Metallurgy*,[6] first published in 1891, was continually enlarged in later editions as the subject grew and is the first textbook to combine scientific metallography with the old metallurgy. Through his position

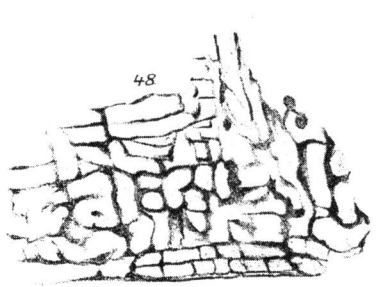

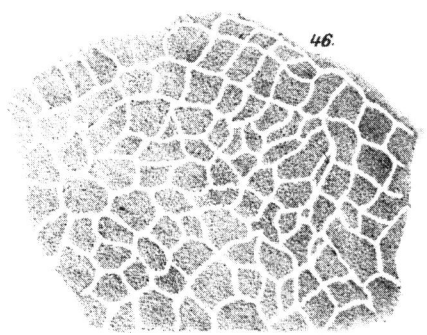

Fig. 105.—Bronze with 7 per cent tin, cast. Etched with sulphuric acid, ×60. (Behrens, 1894.)

Fig. 106.—Same as Figure 105, forged and annealed, ×60. (Behrens, 1894.)

as professor in the Royal School of Mines he influenced many students and set the pattern for British metallurgical education for more than a generation.

The soon-to-be-standard method of determining constitution diagrams by combined thermal analysis and microscopic studies developed rapidly in England. After work on copper-silver and gold-aluminum alloys, C. T. Heycock and F. H. Neville[7] produced a most elegant study of the copper-tin diagram which showed complete mastery of the techniques. Heycock had previously distinguished himself by producing the first microradiograph (Fig. 107)[8] by methods which he demonstrated at a meeting of the Royal Institution on April 2, 1897—less than two years after the discovery of X-rays.

Another English metallurgist, J. O. Arnold, was an important figure on the metallurgical scene. He was a professor of metallurgy at Sheffield and had access to Sorby's samples and some advice from that master. He is principally to be noted for his violent opposition to Roberts-Austen and Osmond and their allotropic theory of steel hardening. He was a good experimenter but weaker theoretically. He was right in pointing out[9] that pure iron quenched from the beta range was soft and that it did not have much greater solubility for carbon than did alpha, yet his supposed hard carbide $Fe_{24}C$ was hardly an advance over Osmond's allotropic theory. Arnold anticipated Roberts-Austen in showing the segregation of impurities to grain boundaries in gold[10]

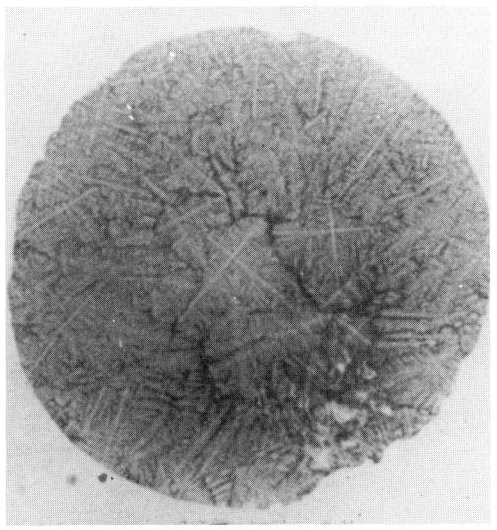

Fig. 107.—*Microradiograph of an aluminum-copper alloy, ×3. (Heycock, 1898.)*

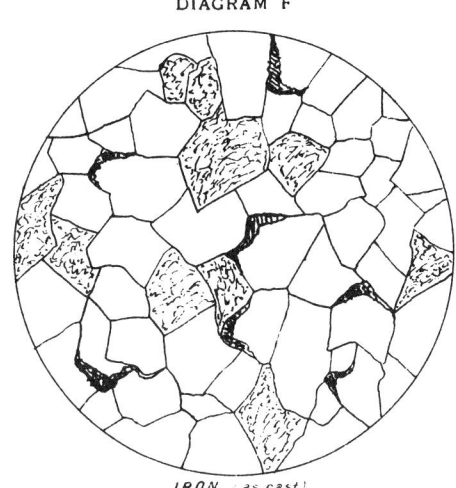

IRON (as cast)

Fig. 108.—*Drawing of the microstructure of iron (i.e., low-carbon steel) in cast state, ×600. (Arnold, 1894.) Slightly reduced.*

and describes most clearly the manner in which orientation-sensitive etching enables optical distinction between grains of different orientation. His early papers are illustrated with drawings rather than photomicrographs, and he had remarkable skill in depicting the significant parts of the structures (Fig. 108). The best presentation of his ideas on the nature of steel is in his "Influence of Carbon on Iron" published in 1896.[11]

John E. Stead of Middlesburgh was an important figure on the English scene. Temperamentally the opposite from Arnold, he avoided controversy. He began as a chemist and did work on color-carbon in 1883. He developed eclectically the preferred metallographic techniques, describing them in detail in an 1894 paper,[12] and continued thereafter to make significant contributions. His most important work is that on crystallization, recrystallization, and grain growth of low-carbon steel,[13] from which many later studies were to stem. An important paper was that of Saniter on the hot-etching of steel in calcium chloride[14] (Fig. 109), which disclosed the structure of steels at high temperatures and was later developed by Osmond.

Important roles in the unraveling of the mechanism of hardening of steel were played by the American metallurgists Howe and Sauveur. Howe's early criticism of Osmond was mentioned in chapter 15. He provided a compromise theory which is not far from the modern truth.[15] Sauveur provided some of the experimental facts for Howe

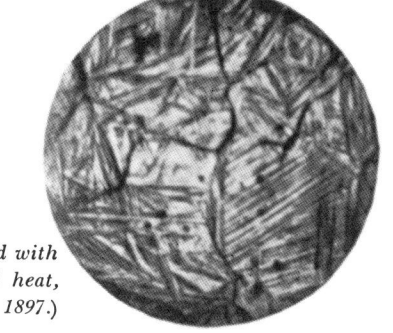

Fig. 109.—Surface of steel etched with calcium chloride at a bright red heat, ×200. (Saniter, 1897.)

and was the first, in 1896,[16] to carry out quantitative metallography. After measuring the volume percentages of various constituents, he showed how they varied linearly with composition (Fig. 110). His paper was a masterly survey of theories on the hardening of steel and was followed by another thirty years later in which he again studied the shades of prevailing opinion,[17] the results showing the increasing sophistication of metallurgical science. His *Metallurgical Reminiscences*[18] reflects the spirit of the early days of metallography with vividness.

In 1898 Sauveur personally founded a journal, *The Metallographist,* to further the new science. It was limited to papers on structural metallography (most of which were reprinted or translated from other journals), and it contained biographical information and lively discussions on many topics. Sauveur has said that the number of subscribers never exceeded five hundred and after a period of different format and title as the *Iron and Steel Magazine,* designed to attract industrial interest and advertising revenue, the journal ceased publication in 1906.

There was a particularly important meeting of the American Institute of Mining and Metallurgical Engineers as a part of the International Engineering Congress held in Chicago in August, 1893, as part of the World Columbian Exposition ceremonies. The record includes papers by Martens, Osmond, Sauveur, Howe, and Hadfield, followed by extensive discussion.

Despite the high level of industrial and scientific activity in Germany, the early work of Martens expanded but slowly. Ledebur, Baron von Jüptner, and others participated effectively in the discussions on the mechanism of steel hardening, but a fertile combination of experiment and theory failed to appear until the turn of the century, with E. Heyn. Gustav Tamman began his comprehensive series of determinations of constitution diagrams in 1903 and there were few metallurgical advances thereafter in which he and his collaborators were not in some way involved.

The success of the work of the English Alloys Research Committee stimulated in France the formation of the Commission des Alliages by the Société d'Encouragement pour l'Industrie Nationale. This started in 1893 and was active in offering prizes—the first of which appropriately went jointly to Osmond and Roberts-Austen—and in directly financing research at various universities and other institutions. The first paper to be published was by George Charpy in 1896. The commission's work terminated with the publication in 1901 of reports on all of the research it had financed, together with related papers by Roozeboom, Osmond, Le Chatelier, Guillaume, Carnot, Mme. Curie, and others. The handsome volume[19] is profusely illustrated with photomicrographs and is as valuable to the metallurgical historian of today as it was to the practitioners at the time it was published.

The work of George Charpy played an important role in the consolidation of metallography. In 1897 he systematically studied the relationship between structure and the type of cooling curve,[20] and he also studied the relationship between the structure of brasses and their behavior in service. Although Sorby had observed the elongation of iron grains on cold working and their recrystallization on heat treatment, and Sauveur[21] had related measured grain size to properties in 1893 and to annealing temperature in 1895, it was not until Charpy's 1897 paper that the recrystallization and grain growth of brass or copper —perhaps the simplest of all metallographic studies—was clearly related to property changes caused by cold working and annealing! Charpy is best known for his work on mechanical properties of alloys.[22] He observed slip bands on the surface of deformed brass, noting that a "polished bar of brass submitted to a very slight extension presents the same appearance as if it had been attacked chemically." Charpy had observed recrystallization in brass somewhat earlier,[23] and there was, of course, a background of related work such as that of Kalisher and Bolley on zinc and mechanical tests on variously annealed non-ferrous metals. The beginning of systematic metallographic studies of the deformation process is, however, in the work of Ewing and Rosenhain.[24] Rosenhain was then a student at Cambridge. He went on to become the leading scientific metallurgist of the time, and to give metallurgy the strong slant toward physics that still characterizes it.

As has so often been the case, metallurgists were both in advance of

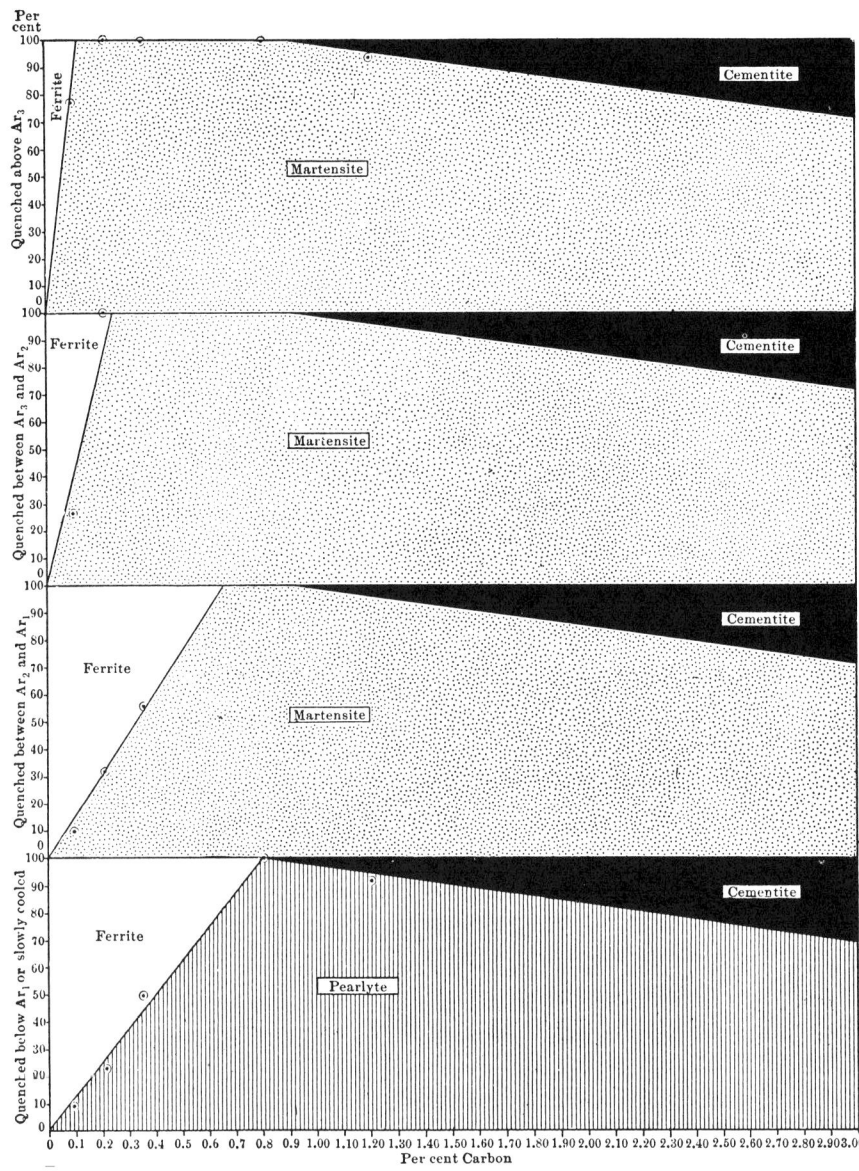

Fig. 110.—*Diagram showing fractions of various constituents as a function of composition and treatment. (Sauveur, 1896.) In 1897 Osmond published a similar diagram incorporating his measurements of retained austenite in quenched steels with over 1 percent carbon.*

and behind the pure science of their day. Curiously, the phase rule (that quintessence of thermodynamics of heterogeneous systems so perfectly adapted to the needs of the metallographer) had found its way into an elementary text for mineralogists[25] in 1888—long before it was discussed by metallurgists even in research papers. Le Chatelier[26] made useful thermodynamic generalizations as applied to chemical reactions in general, but they remained unrelated to structural metallography until 1897.[27] It was left for Roozeboom[28] to discuss constitution diagrams in general and, by slightly adjusting Roberts-Austen's iron-carbon constitution diagram, to show the great power of the phase rule. Nevertheless Le Chatelier was outstanding. His avowed purpose was to apply scientific methods to industrial problems, but in doing so he advanced the sciences themselves, particularly chemical thermodynamics, into which he was led by a study of the calcination of lime. He developed the first really practical thermocouple, essential to high-temperature chemistry and metallurgy, and devised a complete system of metallographic practice,[29] including the preparation of levigated alumina for use as a polishing material, and the design of an inverted microscope.* Le Chatelier classified alloys in terms of the freezing-point curves, and related these to structure and electrical conductivity. He made careful measurement of the dilation of steel and other alloys: in none of these was he the first, but in all, his combination of theoretical and experimental skills produced great advances.

At the turn of the century there was still much confusion as to the nature of allotropic transformations, as to the nature of solid solutions and on the mechanism of transformation and diffusion. Although crystallography and space-lattice theory was well developed, the relation between atom and crystal was not explicit and there seems to have been little thought on the relationship of atoms to each other in a continuous lattice. Just as it had once proved difficult to think of many small crystals in contact with each other differing only in orientation, so it seems to have been difficult to conceive of an arrangement of atoms

---

* The inverted microscope was, the writer believes, a disservice to metallographers. Solely to avoid the trivial step of mounting or holding the specimen, mechanical and optical complications were introduced that have continued to elaborate into that monstrosity, the modern "metallograph," a fit companion to the 1958 automobile. Eventually, we may hope, simplicity will return and manufacturers will market a microscope-camera combination to be used comfortably by an individual who does not have hands growing from the top of his head and on which photography is possible without clambering about. Few metallographic discoveries have been made without direct study of actual samples, yet no man can sit for thirty minutes at a modern inverted metallographic microscope without acquiring an aching neck, elbow, back, and head.

within the crystal lattice wherein the molecule essentially disappears because all atoms are equivalent. The molecule is meaningful in solution or in vapor; it is usually meaningless in a crystal of a metal or alloy, but this was not commonly reflected in metallurgical writings until the 1920's when the significance of X-ray structure analysis had become well accepted. Even at the present time the unit cell—a mere mathematical fiction—is given undue prominence, and we are only just beginning to realize that translational symmetry has been overemphasized and that an individual atom is (except for long-range strain effects) actually only concerned with a few shells of neighbors, and that the vacancy is often as important as the atom.

The first decade of the twentieth century saw an immense growth in knowledge of metallic structures and a widespread appreciation of the importance of the new techniques. The whole character of the science changed as the number of metallographic practitioners increased. Methods were refined and for a time standards of quality were raised. It became somewhat less exciting but a great deal more useful in industrial operations, both in the control of old products and the development of new ones. Metallography had not helped directly in the discovery of the first important alloy steels—manganese, silicon, and nickel steels, or even the first high-speed tool steels, or stainless steel—in fact, conversely, a study of these helped to develop metallography; but it was essential to the improvement of these materials and to the accumulation of the present knowledge of the role of alloy elements in controlling transformation rates.

Though there remained still much to be learned by classical methods, by 1912 metallography had become more a useful tool for tilling known fields than a means of exploring new terrain. In that year, however, the diffraction of X-rays from crystals was experimentally demonstrated by Friedrich and Knipping, following the prediction of Von Laue, and a real science of solids became possible. Though not the first, metals were among the earliest crystals to have their lattices measured by the Braggs in 1913,[30] yet metallurgists were soon involved in war production and were slow to realize the potentialities of the new method. Interest increased following Westgren and Phragmen's use of X-ray diffraction in 1922 as a precise tool for studying alloy constitution,[31] and when the structures of alpha, "beta," gamma, and delta iron were all found.[32] By 1925 the diffraction camera was a common tool in most research laboratories, and was used for the study of orientations, the identification of constituents, and changes in solubilities and it went some way toward replacing the microscope. The mysteries of order-disorder changes were unraveled, and found to be particularly interesting as an example of co-operative phenomena generally. Deformation was studied as never before, though with a temporary blindness until subgrains were observed microscopically. Hume-Rothery's great generalizations on the criteria for the formation of terminal and intermediate solid solutions was followed by the realization that many alloys are not strictly "metallic." Age-hardening (discovered empirically in 1906 in aluminum alloys, though previously unknowingly utilized in sterling silver and palladium-base alloys which were found to harden on slow cooling) was related to changes of solubility on the basis of metallographic studies, but X-ray diffraction in its most sophisticated form was needed before the changes in aggregation responsible for the hardening could be seen or understood.

However, X-ray diffraction, so useful for regular lattices and relatively large crystals, needs elaborate interpretation when it comes to structures with locally variable strain and composition. It is interesting to observe the present (1958) return to direct structural observation of metals in a most elegant form, using the electron microscope on extremely thin sections. Of recent years even optical metallography has partially returned to favor as a research tool; partly because it can give information on individual dislocations via the lowly etch pit, partly because the dihedral angle of microcrystals gives a key to interfacial energies, but principally because the scale of aggregation revealed to the microscope is still an important one and can no more be ignored than can structure on any other scale. In the interaction of forces and configurations on varying and interpenetrating scales lies, perhaps, the essence of all scientific explanation.

245

APPENDICES

A. "Experiments of Etching upon Iron and Steel"

*By Sven Rinman**

Kongliga Vetenskaps Akademiens Handlingar
Stockholm, 1774, **35,** pp. 1–14

*Iron
solution*

It is commonly known, and may be read in chemical books, how iron and steel are etched, bitten, eaten, and dissolved, by almost all liquids except expressed oils: but as this happens with remarkable differences, both from the properties of the dissolving matter & the different kinds of iron and steel, these diversities ought to be examined and known, on account of the uses resulting therefrom in particular arts and businesses. Thus solution of iron in vitriolic acid, or common green vitriol, is the best for writing ink, the solutions in other acids either corroding the paper or giving an unserviceable black.

different
menstrua
for difft.
purposes.

The curriers blacken their leather with a solution in animal acid, or sour milk, as the least corrosive. The tinned plate makers employ as a mordent for the iron vegetable acids, though with the mineral, as alum, it is quicker cleaned; but these last are accompanied with an inconvenience, that the tin will not so well fasten thereon. The etchers on iron and steel use a prepared etching water, as being less costly, and making an evener & less violent corrosion than aqua fortis. In a word,

* This translation is printed, with permission, from the manuscript preserved in the Wedgwood Papers, British Museum add. MS 28312, pp. 155–61 (fols. 82–85) . The manuscript is in the very clear handwriting of Alexander Chisholm, who for several years served as secretary and laboratory assistant to the chemist William Lewis and after the latter's death joined Josiah Wedgwood, who also acquired many of Lewis' chemical notebooks. For information on Chisholm's career see F. W. Gibbs, "William Lewis, M.B., F.R.S. (1708–1781)," *Annals of Science,* 1952, **8,** 122–51; "A Notebook of William Lewis and Alexander Chisholm," *Annals of Science,* 1952, **8,** 202–20.

in every kind of business wherein iron solution or etching occurs, it is necessary to know both the differences of its dissolvents and the differences in its own habitus. Were we sure of knowing fully both the principles of metals and their dissolvents different habitus in all possible varieties, we might be able to tell beforehand what effect any one ought to produce: but experience convinces us that we cannot thus judge with certainty of the event. It is proved indeed that aqua fortis or nitrous acid has a strong attraction to inflammable matter: but the rule, that the more inflammable matter a metal contains, the more powerfully it is acted upon by that acid, admits of exception: The grey or so called *nodsatt* cast iron has demonstrably more of the inflammable in it, than the white *hardsatt:* yet the grey dissolves much more difficultly in this acid, and leaves in solution a black sediment, sometimes of equal bulk and shape with the bit of cast iron itself, consisting of a matter like blacklead, which is found to be iron earth overcharged with phlogiston: Take away a part of this grey iron's redundant phlogiston, either by melting it into white iron, or by cementation with absorbents in Réaumur's way into malleable iron or steel, and it becomes soluble in aqua fortis completely without residuum. Other exceptions to this rule might be mentioned, which shew, that our judgment ought to be founded on actual experiment when we come to apply it to arts and businesses. While I was employed about the distinctions of iron and steel, I had opportunity of examining the following particulars. 1) What kinds of iron & steel are strongest attacked and dissolved by acids. 2) What alteration of colour and texture are observable in the different kinds thereof. 3) What kinds of corrosive is most serviceable for each kind of etching on iron and steel. These expts. are applicable to the damasking business described above & on other occasions.

(1) Filings of soft iron and hard steel, being put in two separate glasses, and aqua fortis poured on them, were so hastily and violently attacked, with strong ebullition & fumes, that little difference could be perceived in the time. The filings of burnsteel seemed indeed to be somewhat hastier acted on than those of iron, but cast steel on the contrary somewhat later.

(2) A piece of iron of the softest Osmund, weighing by an accurate set of proportional weights 64 lb. 30 loths, and a piece of unwrought & unhardened cast steel of the so called *kåm* steel, weighing 68⅓ lb., both filed clean, and of equally large surface, were put in a large glass, and a sufficient quantity of common parting water poured upon them, which seemed first

grey & white pig metal

iron and steel differently acted on

with bubbles to begin to work on the iron, but immediately after did the same on the steel. The strongest fuming having ceased, the glass was set in gentle warmth, on which the solution went on more violently, and a quantity of black & red brown iron calx fell to the bottom as usual. Both pieces were taken out, washed, and dried very clean. The iron was found to weigh 38 lb. 1 loth, so that it had lost nearly 40 per cent and was now much eaten, with deep furrows, and some elevated black fine striae, but between them white and silver bright. The steel weighed 51 lb. 16 loths, and so had lost only 24 per cent. It was now evenly coated over with a black colour which could by no means be washed off, and uniformly corroded with shallow furrows longitudinally. Both pieces were laid a second time in fresh aqua fortis, as also a third time in strong parting water diluted with half pure water, after which the iron was found as before of a shining bright white silver colour, destroyed all to a thin *skolla*, which was thickest in the middle, and weighed now only 8 lb., so that it had lost in all 87 per cent. The steel on the other hand retained a dark ash grey colour evenly all over, and weighed still 47 lb., so that in the same time, and in the same sharp waters, it had lost only 30 per cent, which shews plainly, that iron is acted on more than doubly stronger and quicker than steel, and that the surface of the iron proves always white & bright, but that of steel dark. Herein it may be observed, that after the steel has taken the dark colour or skin over it, it is difficultly attacked by the parting water. If the solution be boiled down till all becomes dry, the iron acquires also a black rusty surface, but is made white & bright again in fresh aqua fortis, which does not happen with steel.

(3) To examine the habitus of colour in less violent solution, I put 5 polished pieces of equal bulks in strong aqua fortis diluted with 2 parts of water: after strong ebullition in gentle warmth, during which the black sediment that settled on the surfaces was frequently scraped off, the pieces were taken out, and washed & dried clean. It was then found that (a) The soft Osmund Iron above mentioned was all over the polished side much eaten, silver white & bright, with some few elevated grey striae. (b) Dannemora iron from Österby was very little eaten, with a white broad line in the middle along the bar, and two light ash grey lines at the sides. (c) Burn-steel from the same iron was less acted on, had a dark grey colour, with many small somewhat elevated black points sprinkled evenly all over, and it could be seen how the parting water had dug a little pit round every point. A piece of English cast steel behaved in the same manner, becoming

only somewhat darker in colour. (d) Steel of cast iron softened by cementation, quite close and fine, without the least corrosion, took an uniform light ash grey colour. (e) Cold-short iron became quite dull & white, with fine black strokes (*ritsor*) which shewed incompactness.

(4) A damasked bar, compounded & very well welded of five sorts of iron & steel, equal quantities of each, in the following order, viz. good Norbergs iron: burn-steel from Dannemora iron; pure Dannemora iron from Österby; raw steel, or unwrought *smält* steel from Schisshyttan; soft and tough Osmund Iron—was polished on one side, and set with its end in strong parting water diluted with 2 parts of common water. After the acid had in an hour finished its principal action, the bar was taken out and washed clean. (a) The Norbergs iron shewed itself in a white dull slender line, with finer black rays, sharp & elevated. (b) The burn-steel was in a somewhat broader line of a dark grey colour, with wavy black specks here and there. And this line was remarkably higher, or less eaten down, than the iron line, nor had it any perceptible elevated rays in it. (c) The Österby iron in a somewhat slenderer line, of an ash grey colour, lighter than *b*, but darker than *a*, except [or without, *utan*] some few elevated veins. (d) The raw steel in a like broad line as *b*, but of a darker black grey colour, intersprinkled with more black shaded and wavy specks. (e) The Osmund Iron in a line of like breadth with *a*, silver white, bright & sparkling. Remarkably deeper and more craggy corrosions than in the steel line, with many fine rays, running lengthwise to the bar, partly elevated & partly eaten down, less black than in *a*. The same etching, on the same sorts, was repeated oftener than once, for greater certainty, with the same events.

Etching waters

(5) To try the differences in the etching of damask'd iron with different etching waters, were among others the following experiments made. (a) Strong parting water alone, stroked on one polished side of the above mentioned damask bar, begun immediately to boil, and after one or two minutes had produced the effect that when the bar was washed clean the damasking appeared very plain, with its dark grey, grey and lighter, and white veins, which had all pretty sharp edges, not the soft gradual shadings generally desired. When the acid begun to work, it was observed that the bubbles rose first from the steel lines, and that these were consequently first acted on, particularly the lines of burn-steel. (b) 1 part of the foregoing aqua fortis, diluted with 2 parts of water, gave a more agreeable etching, and exhibited the darker & lighter in

a better shading. A part of this etching water dried upon the bar, whereby the damask gained a mixture of brown veins which had no bad effect. (c) A [usual] water among those who etch figures on blades, consisting of pure water 1½ lb., copper vitriol ½ loth, alum 2 loths, commonsalt 1 loth, stroked on the same bar, did not discover the damasking sufficiently, nor can it have sufficient working unless the damasked piece be kept wholly immersed in the etching water for 6 or 8 hours, with gentle digestion heat; but then the damasking seems to become so much the better. (d) Still better does it appear, when to the foregoing composition is added ¼ part of A. fortis. Herewith the veins are discovered in an agreeable and elegant shading; so that this etching water seems justly preferable to the foregoing; nor is it so dear but that it can be had in sufficient quantity for immersing any work in it, in a convenient copper vessel. (e) An etching water of the same salts, but dissolved in vinegar, performed its office too slowly,

Copper pre-cipitated uncommonly hard on iron. and shewed only this remarkable, that the precipitated copper fastened itself here from more than ordinarily hard, particularly upon the steel lines, which in a different view may among certain artists be an enriching experiment. Several other etching materials, as Spt. of vitriol, Spt. salt, and some of the most rational compositions found in books, were likewise tried, but all were found inferiour in their effects to the foregoing.

(6) From these few experiments, but made with all possible accuracy, I think we may draw the following conclusions:

1. That though aqua fortis is found mostly to have a stronger action on steel than iron, in virtue of its great attraction to the inflammable matter, of which steel contains something more than iron does; yet at the same time it sooner loses its power on the steel, and deposites thereon a sediment, consisting partly of inflammable matter, & partly of iron calx, which covers the steel and hinders the action of the etching water, and likewise gives the steel a more or less black surface, according as it contains more or less of the inflammable matter, or as it is more or less hard, insomuch that we are in condition to judge in some degree of the steels hardness from the degrees of lighter or darker grey colour which it takes in the etching. On the contrary this sediment never fastens at all on iron, whereby the etching water has more freedom to act evenly thereon, and this seems to be the reason of its being so remarkably stronger eaten and quicker dissolved than steel, not only by Aqua fortis but by all acids, especially when they work upon it for some length of time. Hence it seems to come, that iron rusts also more than steel, and that we often find

what remains of rusted iron to be mostly of the steel kind, later subject to change. In like manner we may judge of iron; that the less it is eaten, and the more it inclines to grey colour, so much the harder it is; but contrariwise, the faster it dissolves, and the whiter & brighter it becomes, so much the softer. Also, the more uniform whiteness the surface acquires, and the less hollowed furrows or raised darker striae are seen thereon, so much the more close & uniformly hard is the same iron. This all the experiments evince: and therefore I think I cannot be deceived, when, on a damask'd & newly etched work, all the waves & veins which seem a little elevated, and of an ash grey or dark colour, I pronounce of the steel kind, and the whiter somewhat depressed ones to be pure iron, as the over-sent little damasked proof bits sufficiently shew. Some differences are also herein observable according as the kind of steel or iron is good, red short, or cold short, and according as the etching is done warm or cold.

2. That remarkable differences in the lighter and darker colours, and in the deeper or shallower erosions, are observable also even in irons of different kinds, as we may see on comparing the Österby iron with the softest Osmund (3.a.b) besides other proofs; whence may be concluded, that from well welded bar irons, without any addition of steel, a kind of so called damask work may likewise be made, which for my part I should prefer for gun barrels to that in which the hard and brittle steel is alternately interlaid with the iron.

3. That varieties in the colour & appearance of the veins upon damask'd iron & steel are producible also in some degree by differences in the etching water & in the manner of applying it. In ¶**5.**b. we find, that if weakened Aqua fortis is used, & suffered to work out and dry upon the piece, there are formed, along with the darker and lighter, brown veins also.

4. That where the etchings are required somewhat deep, as the Turkish commonly are, and which seems indeed to be expedient especially on firearms where the damask would otherwise wear off and become indistinct, it is necessary that the work be immersed in the etching water and set in the warmth for some hours: it is not difficult, by smearing with varnish, or with a mixture of chalk & oil olive, to hinder the acid from touching any other parts than those where it is intended to act. Where figures are to be etched on blades, heat the blade to a yellow color, and with a little cotton dipt in linseed oil smear it over quite thin: the oil immediately drying by the heat makes a good etching ground, in which figures may be

varnish

drawn, and the blade afterwards set as far as the drawing reaches for half a day in the etching water **5**.c. till the graving is sufficiently deep.

5. That etching affords an easy way of distinguishing the different iron and steel kinds, in regard to their hardness, compactness, and uniform or dissimilar internal properties; and that thereby we may learn to know the matters which are to be sought after for damasking, and thence judge in what order they ought to be laid together.

6. and lastly, That the darker grey surface which steel receives from etching water, is either rubbed off by scowering, or wears away in time, and then the steel exhibits its true colour, which is always whiter than that of iron when both are unpolished. So that it is not wonderful if the higher lines on an old damasked work, as being the most elevated, become brighter by wear than the depressed veins of iron. But in the foregoing we speak only of the colour which the etching water produces.

B. "On the Microscopical Structure of Iron and Steel"

By H. C. Sorby, LL.D., F.R.S., &c.*

The microscopical study of fractured surfaces is unsatisfactory, not only on account of the optical difficulties, but because a fracture shows the line of weakness between the crystals, and not their internal structure. All my results are based on the examination of flat sections. These should be finished by grinding with Water of Ayr stone, and polished so as not to alter the true structure of the extreme surface. Anything approaching to a burnished surface or polished scratches is fatal to good results. In general, after having been polished with the finest rouge and water, so as to show few or no scratches, the surface was acted on by very dilute nitric acid, and repeatedly examined in a small trough of water, until it was found that the acid had properly developed the structure. In some cases it is, however, best to polish with dry rouge on parchment, and not to use acid. Thin glass covers were afterwards mounted over the surface with Canada balsam. Some of my preparations have kept perfectly well for above twenty years, but others have deteriorated considerably. Objects thus prepared must be examined by means of two special kinds of surface illumination, viz., first, the side parabolic reflector now common, but I believe originally made for this purpose, which gives oblique light, and, secondly, a small silver reflector, covering half the object glass, which throws the light directly down on the object, and from this it is reflected back through the other half of the lens. With the oblique illumination, a polished surface looks black, but with the direct illumination it looks bright and metallic. A truly black substance appears black in both cases. A magnifying power of about sixty linear is most generally suitable, but the sections will bear a higher perfectly well.

* This is the preprint issued for the Spring, 1885, Meeting of the Iron and Steel Institute (see p. 180), which differs considerably from the final version of the paper as published in the *Journal* for 1887. It is reprinted with permission from a copy preserved as No. 79 in Vol. II of Sorby's *Collected Works* in the University of Sheffield Library.

The following is a summary of the chief results:—

Iron containing little or no carbon and of uniform character shows little if any change when acted on by the dilute acid, and no well marked structure is developed.

Hammered bloom shows an intimate mixture of varying crystals of iron with minute or larger portions of slag.

Iron bars rolled *hot* show that the slag is drawn out into long thin rods, which, in some cases, are so numerous as to form a very considerable part of the whole bulk. Possibly there is a drawing out of the particles of iron, but, when examined cold, longitudinal sections show little or no elongation of the ultimate crystals, which are more or less equiaxed in all directions, as through the metal recrystallised in cooling. When acted on by dilute nitric acid, the crystals change in a varying manner, and by direct light show varying tints of brown, and by oblique light varying tints of blue, which colours are probably to a great extent due to interference, as in thin plates.

When such a bar is hammered *cold* the crystals are compressed, broken up, distorted, and elongated in the line of the bar. When kept some time at a red heat, these distorted crystals recover their ultimate regular structure by recrystallising, the tendency being to often grow somewhat longer in a line perpendicular to the direction in which they were previously elongated.

This recrystallisation of distorted crystals probably plays an important part in worked irons.

When reheated in large masses it seems as though large crystals were formed, each crystal having a more or less uniform composition, and when rolled out hot these are distorted and compressed. On cooling they recrystallise as a patch of smaller, more or less equiaxed, crystals, all of one general character.

Planes of welding are often well shown by variations in the structure of the iron.

Many specimens of malleable iron show clearly that two constituents are present, viz., iron, and the subsequently described compound of iron and carbon, which has a pearly structure.

Rolled bars often show a much more irregular mixture of layers containing varying amounts of slag and steel than would be suspected from the appearance of surfaces dressed in the usual manner.

The alternation of layers of iron and steely iron is well shown by many Swedish irons. The difference between the two constituents is better seen after the bar has been kept some time at a red heat, when there is a more complete separation into larger crystals of two perfectly well-marked substances, having a totally different structure. One of these is just like the main constituent of such bar iron as contains little or no carbon, having no trace of linear marking, after being acted on by dilute acid; whereas the other constituent shows linear markings, varying in distance, but often about 1/20000 inch apart, which, when the acid has acted to a proper extent, give rise to all the splendid colours of mother-of-pearl, the tints being raised when the section is seen in water, and still more so when mounted in balsam. By oblique and direct illumination the colours are nearly complimentary.

The existence of three totally different metallic substances is well seen in the case of a thick square bar of Swedish iron partially converted into blister steel by cementation. In the centre the structure is just as in the case of a bar kept red hot, being a mixture of well-grown and distinct crystals of free iron and of the pearly compound. Passing towards the outside, the introduction of a certain amount of carbon has altered the whole of the free iron, so as to give rise to a ring consisting entirely of the pearly compound, showing colours of great variety and beauty. Still nearer the surface we come to a part in which occurs a network of veins of an extremely hard substance, giving an intensely brilliant reflection and no trace of colour, and as distinct as possible from either of the other two substances. Taking all the facts into consideration, it seems almost certain that this hard substance must contain more combined carbon than the pearly constituent. It occurs not only in thick veins, but also penetrates between the layers of the pearly compound, which thus becomes more and more dense towards the outside of the bar, until it appears as if at length it broke up into crystals of the hard substance, and a structureless residue.

The existence of these three well-marked constituents is also well borne out by what is seen when white cast iron is decarbonised by heating with oxide of iron. The exterior is changed into what has just the same structure as bar-iron, and the centre into the pearly constituent of steel, whilst in the intermediate portion there is a mixture of the two. The third and very hard constituent of blister steel corresponds in all essential particulars with the hard constituent of the white iron before being decarbonised.

The three constituents just described are totally distinct from one another. There is no more passage from one to the other than there is between the mica, felspar, and quartz of granite.

The varying character of the ingots of soft and hard steel to a great extent depends on the varying proportion of the three principal constituents. Soft Bessemer steel is seen to be a mixture of free iron and the pearly compound. In medium steel this latter occurs almost alone, whereas in hard steel there is little, if any, free iron, but numerous thin plates of the very hard compound.

An interesting feature in certain samples of both cast steel and cast iron is, that the pearly compound crystallises out from fusion in fairly large feathery crystals, easily seen with the naked eye when an ingot is broken; but when suitable sections are examined with a higher power, it is seen that ultimate structure of the original crystals is more or less completely independent of their form. Radiate groups or small irregular crystals seem to show that after solidification at a high temperature the pearly constituent recrystallises on further cooling. The large crystals thus appear to be analogous to the paramorphs met with in some rocks. If we were to suppose that during the cooling the crystals acquire strains which are relieved by recrystallisation, all the facts would be fairly well explained. At the same time this may possibly occur independent of any strain. That such a recrystallisation does occur in the case of hammered or rolled steel admits of no doubt. The hard plates in hard steel, and the small crystals of free iron in soft, are arranged or drawn out in the length of the bar, like the slag in bar iron; but, if the pearly compound be the chief constituent, it occurs as small crystals in no way drawn out, as if they crystallised after the rolling.

The grain of hardened steel is so fine and uniform that so far I have not been able to unravel its ultimate structure, but it is sometimes easy to see that the damascening is due to variation in quality, depending on the original crystallisation.

The absence of slag from bar steel and Bessemer metal distinguishes them in a most striking manner from bar iron, which usually contains such a large quantity.

The three metallic constituents of steel also occur in different proportions in different qualities of cast iron, but this latter contains in addition two other substances. On the whole, the most characteristic constituent is the graphite, and to a large extent the structure and properties depend on the manner in which it has crystallised. In very grey pig it crystallises out the first in scattered plates. Then over it sometimes occur small crystals, which may be silicon, and a moderately thick and fairly uniform layer of what seems to be free iron. Next over this comes the pearly compou. forming the chief bulk of the whole, and finally a distinct metallic substance is left as a sort of residue. In the case of some medium pig the graphite has often crystallised in radiate groups, over which there is little or no free iron, but the amount of the final residue is comparatively large. This residual substance does not occur as a visible constituent of cast steel unless the very thin layer between the large crystals be looked upon as analogous. Its true nature is somewhat doubtful, but the most probable explanation of the various facts is that it is not a special compound, but a variable residual mixture, possibly containing silicon and divers impurities. In other qualities of pig the chief constituent is the pearly compound, which crystallised first in large feathery crystals, throwing off the graphite and recrystallising on cooling.

In white refined chilled iron the pearly compound also crystallises first, leaving a large amount of the intensely hard substance and some of the residue just described. Graphite occurs only in the small concretions of grey iron, seen here and there.

As far as mere structure is concerned, some specimens of Styrian steel correspond closely with white refined iron.

Finally, we come to *Spiegeleisen*. The specimen which I have specially studied consists mainly of the intensely hard compound, crystallised in large plates; and the interspaces are filled up with a mixture of very much smaller crystals with a little of the pearly substance, so as to have a most beautiful and fine-grained structure, not well seen with a power under fifty linear.

Taken, then, as a whole, the various kinds of iron and steel are seen to be varying mixtures of six or seven substances, having very different properties, viz.:—

1. Free iron.
2. The pearly compound with carbon.
3. The intensely hard compound, probably with more carbon.
4. The residual, probably variable, substance.
5. Graphite.
6. Possibly crystallised silicon.
7. Slag, including fused iron oxide.

It is quite probable that the individual character of some of these constituents may be modified by the presence of small quantities of sulphur, phosphorus, or other impurities, or the manner in which they separate out from one another changed thereby. This question would require special investigation. My chief aim now is to show that the various kinds of iron and steel are usually varying complex mixtures of two, three, four, and sometimes of even five different substances, which can be distinctly recognised with the microscope. Except in a few very special cases, iron and steel are therefore not analogous to *simple minerals,* but to *complex rocks.* Such being the case, the microscopical examination of properly prepared sections throws much light on the causes of their varying properties. Very much still remains to be learned about the details. What I have done so far is little more than to show how such an inquiry could be carried out, and to describe some of the most conspicuous structures met with in ordinary varieties of the metals.

The sections of iron and steel examined were prepared above twenty years ago. The general conclusions then derived from their study were described, and the microscopical photographs taken directly from the preparations were exhibited, at the meeting of the British Association at Bath in 1864.

C. Additional Illustrations
and Notes (1988)

p. 13: Many other uses of superficial chemical attack on metals are described in my articles "Some Constructive Corrodings," included in *A Search for Structure* (Cambridge, Mass., 1981), pp. 332–342.

p. 24: An excellent discussion of Damascus swords, unfortunately with only a brief summary in English, is J. Piaskowski, *O stali Damascenskiej* (Wroclaw, 1974). Two papers appeared in 1987, one on their metallurgy by J. D. Verhoeven and L. L. Jones in *Metallography* **20**, 145–180, and a full discussion of the history and ethnology of wootz by Bennet Bronson in *Archaeomaterials* **1**, 13–51. Modern steels of wootzlike composition but extensively reworked to achieve a much finer scale of heterogeneity and improved properties are discussed in historical perspective by O. D. Sherby and J. Wadsworth in *Scientific American*, February 1985.

p. 150: There is an excellent treatment of the structure of meteorites in John Burke, *Cosmic Debris: Meteorites in History* (Berkeley, 1986). This includes a reproduction of Thompson's 1804 engraving of the "Widmanstätten" structure in Krasnojarsk. The earliest account of the structure to be published in a widely accessible journal was that of Gillet de Laumont in the *Journal des Mines*, 1915, **38**, 232–37.

p. 169: The centenary of Sorby's discovery of the true microstructure of steel was marked by conferences in both England and the United States. See Harold Moore's biographical article in *Jour. Iron and Steel Inst.*, 1963, **201**, 305–9; the book *Metallography 1963, Iron and Steel Institute Special Report No. 80* (London, 1964); and C. S. Smith, editor, *The Sorby Centennial Conference on the History of Metallurgy* (New York, 1965). The last named carries important articles on Sorby's work in metallography and petrography, which preceded and to some extent inspired his metallurgical discoveries, together with a list of Sorby's publications and papers by and about some of the leading contributors to the developing science of metals. Norman Higham's biography, *A Very Scientific Gentleman: The Major Achievements of Henry Clifton Sorby* (Oxford, 1963), gives a fine account of his manner of life and work and of his contributions to many different fields of science.

p. 183: A. Clifford, *Cut Steel and Berlin Iron Jewelry* (London, 1971), gives an account of the steel jewelry with highly polished facets that was popular in the early nineteenth century. The techniques used for polishing these "gems" (described in detail in *Jour. Franklin Inst.*, 1830, **6**, 265–67) were probably more directly antecedent to metallographic polishing than were those used in the graphic arts. The polishing of gemstones, of course, is behind them all.

p. 211: See also F. Wever, "On the background and beginning of metallography in Germany," in Smith, editor, *The Sorby Centennial Symposium,* pp. 163–69, and R. Pusch, "Die Anfang der Metallographie," *Stahl und Eisen,* 1964, **84,** 808–16.

p. 215: In 1886, at a meeting of the U.S. Association of Charcoal Iron Workers, Wedding contributed a paper comparing charcoal- and coke-smelted irons with some better sketches of microstructure. These were reproduced in the Association's *Journal,* 1886, **7,** 120–34, and elicited discussion by Henry Bessemer and others which well reflects the views of industrial metallurgists on the new microscopy. In 1890, Wedding published a book, *Das Kleingefüge des Eisens,* containing many photomicrographs, but his work attracted little attention.

p. 228: The arguments pro and con the existence and nature of beta iron are discussed by M. Cohen and J. M. Harris in chapter 15 of *The Sorby Centennial Symposium,* pp. 209–34. An important debate on allotrophy between H. M. Howe and K. Honda is reported in *Jour. Iron and Steel Inst.,* 1919 (i), 460ff.

p. 244: The paper by E. S. Davenport and E. C. Bain, "Transformation of austenite at constant subcritical temperatures," *Trans. AIME,* 1930, **90,** 117ff., marks the apex of metallographic investigations made using the optical microscope. The frontier of research had moved to another level of structure. A paper by G. H. Quincke, *Proc. Royal Soc.,* 1907, **78A,** 60–67, pointing out that polycrystalline materials must contain nonconformities, followed by Beilby's observation of amorphous surfaces produced by polishing and his suggestion in a paper entitled "The hard and soft states in metals," *Jour. Inst. Metals,* 1911, that deformation produced internal layers of amorphous metal provoked much discussion among metallurgists. (One is reminded of Robert Hooke's suggestion that vitreous matter was present in hard steel.) Walter Rosenhain suggested in 1912 that grain boundaries were necessarily amorphous. The arguments that followed led to intensive reexamination of the nature of structure in general and the role of imperfection in crystalline materials. In 1914, Cecil Desch published the first analysis of grain shape on the basis of interface tension equilibrium. From these metallurgical discussions early in the present century eventually arose concepts that would change the nature of science quite as profoundly as did the new physics of structures within the atom.

p. 245: The complete revolution in understanding of structure that followed the application of x-ray diffraction to the study of materials has caused the center of scientific interest to move to structure on a scale far below that discussed herein. Nevertheless optical metallography has become of increasing importance in the development of both new and old materials and in the control of their production. A new field called "stereology" has emerged, the principles of which were independently developed by metallurgists in the United States and the Soviet Union studying the metric relationships between the microscopist's two-dimensional section of a polycrystalline aggregate and the true structure in three dimensions. (See S. A. Saltikov, "The method of sections in metallography" (in Russian), *Foundry Laboratory,* 1946, **12,** 816–35, and C. S. Smith and L. Guttman, "Measurement of internal boundaries in three dimensional structures by random sectioning," *Trans. AIME,* 1953, **197,** 81–87.) Computerized equipment is now available for optical and electronic scanning of structures that automatically yields numbers reflecting the shape and size of microconstituents in three dimensions. The present state of the field is well summarized in John C. Ross, *Practical Stereology* (New York, 1986).

In "An interpretation of microstructure," *Trans. AIME*, 1948, **175**, 15–51, I introduced the concept of dihedral angle and showed the importance of interface energy equilibrium in determining the shapes of microconstituents. In "Some elementary principles of polycrystalline microstructure," *Metallurgical Reviews*, 1964, **9**, 1–48, a classification of shapes was attempted.

An excellent overview of the field of physical metallurgy is given in R. W. Cahn and P. Haaser, eds., *Physical Metallurgy* (3rd ed.; Amsterdam, 1983). The *Encyclopedia of Materials Science and Engineering* edited by Michael Bever (Oxford and Cambridge, Mass., 1986) covers the science and engineering of all materials, metallic, organic and inorganic.

Fig. B.—Spotted metal organ pipes in the church at Witney, Oxfordshire. The insert shows, natural size, the untouched surface of a modern cast sheet of organ-pipe metal. (See page 62.)

Fig. C.—"*Instantaneous light box*" *encased in green* moiré métallique, *made in London about 1820. Photo courtesy of Bryant and May Ltd. and the Science Museum, London. (See page 65.)*

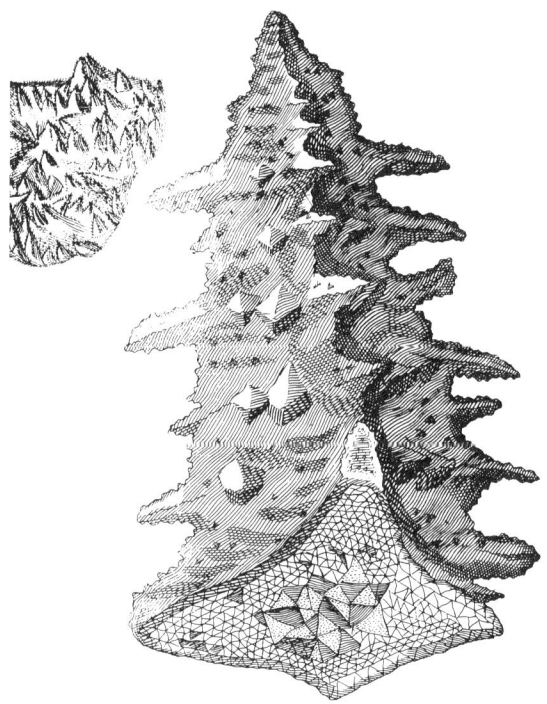

Fig. D.—*The earliest published illustration of a metal dendrite. From J. H. Zannichelli,* **De ferro ejusque Nivis preparatione** . . . (*Venice, 1713*), *a study of many iron-bearing crystals inspired by "iron snow," a form of ferrous sulphate for medicinal use.*

INTRODUCTION TO SECTION 1

1. Charles Singer, E. J. Holmyard, and A. R. Hall (eds.), *A History of Technology*, Vols. I and II (Oxford, 1954 and 1956).

2. J. R. Partington, *Origins and Development of Applied Chemistry* (London, 1935).

CHAPTER 1: *The Pattern-Welded Blade*

1. Herbert Maryon, Personal communication.

2. E. Salin and A. France-Lanord, *Le Fer à l'époque Mérovingienne* (Paris, 1943). See also E. Salin, *La Civilisation Mérovingienne, Troisième Partie, Les Techniques* (Paris), 1957.

3. A. France-Lanord, "La Fabrication des épées damassées aux époques Mérovingienne et Carolingienne," *Le Pays Gaumois*, 1946, Nos. 1, 2, 3; "Les Techniques métallurgiques appliquées à l'archeologie," *Revue de Métallurgie*, 1952, **49**, 411–22; "Les Épées damassées de vᵉ au xᵉ siècle étude technique et archéologique," *Bulletin Archéologique*, 1950, pp. 193–202.

4. A. Liestøl, "Blodrefill og mål," *Viking*, 1951, **15**, 71–96.

5. France-Lanord, "La Fabrication des épées damassées aux époques mérovingienne et carolingienne."

6. Liestøl, *loc. cit.*

7. Magnus Aurelius Cassiodorus, *Varia* v, No. 1. In T. Mommsen (ed.), *Monumenta Germaniae historica, I* (Berlin, 1894), p. 143.

8. H. H. Coghlan, *Notes on Prehistoric and Early Iron in the Old World* (Oxford, 1956). See also the same author's paper on *La Têne* swords in *Sibrium*, 1956/57, **3**, 129–42.

9. C. Panseri, "Richerche metallografiehe sopra una spada da guerra del XII secolo," *Documenti e contributi per la storia della metallurgia*, No. 1, Milan, 1954.

10. J. W. Anstee and L. Biek, "A Study in Pattern-welding," *Medieval Archaeology*, 1961, **5**, 71–93.

11. A. Leroi-Gourhan, "Notes pour une histoire des aciers," *Techniques et Civilisations*, 1951, **2**, 4–11.

12. T. A. Wertime, *The Coming of the Age of Steel* (forthcoming book to be published by the University of Chicago Press, 1961).

Additional References, 1965:

L. Beck, *Die Geschichte des Eisens*, I (2d ed.; Brunswick, 1890–91), 711–20 and 1026–30.

C. Böhne, "Die Technik der damaszierten Schwerter," *Arch. Eisenhüttenwesen*, 1963, **34**, 227–34.

A. E. Collins, "On Pattern-welding," *Man*, 1950, **50**, No. 159.

H. R. E. Davidson, *The Sword in Anglo-Saxon England* (Oxford, 1962).

A. France-Lanord, "La fabrication des épées de fer gauloises," *Revue d'Histoire de la Sidérurgie,* 1964, **5,** 315–27.

H. Maryon, "Pattern-welding and Damascening of Sword-Blades," *Studies in Conservation,* 1960, **5,** 25–37 and 52–60.

R. E. Oakeshott, *The Archaeology of Weapons* (London, 1960).

C. Panseri, "L'acciaio di Damasco nella leggenda e nella realtà," *Armi Antiche* (Boll. dell'Accad. S. Marciano-Torino), 1962. English translation to appear in *Gladius,* Vol. V (in preparation).

J. Piaskowski, "The Manufacture of Mediaeval Damascened Knives," *Jour. Iron and Steel Inst.,* 1964, **202,** 561–68.

E. Schürmann, "Untersuchungen añ Nydam-Schwerten," *Arch. Eisenhüttenwesen,* 1959, **30,** 121–26.

CHAPTER 2: *The Etching of Armor*

1. James G. Mann, "The Etched Decoration of Armour," *Proc. British Academy,* 1942, **28,** 17–44.

2. *Experimenta de coloribus.* Manuscript transcribed in 1409 and included in compilation by Jehan de Begue, 1431. Eng. trans. by M. P. Merrifield in *Original Treatises on the Art of Painting* (London, 1849).

3. H. Williams, "The Beginning of Etching," *Technical Studies in the Field of the Fine Arts,* 1934, **3,** 16–18.

4. *Stahel und Eysen künstlich weych unnd hart zu machen* (Mainz, 1532). This was included in the *Drey schöner künstreicher Büchlein* (Leipzig, 1532). An annotated Eng. trans. from a 1539 edition is given by H. Williams, *Technical Studies in the Field of the Fine Arts,* 1935, **4,** 64–92. Etching recipes are also to be found in *T Bouck va Wondre,* 1513, ed. H. G. T. Frencken (Roermond, 1934), and in the early sixteenth-century manuscript edited by Otto Johannsen, *Peder Månssons Schriften über technische Chemie und Hüttenwesen,* 1941.

5. Vannoccio Biringuccio, *De la pirotechnia* (Venice, 1540 [Eng. trans., New York, 1942]).

6. Alessio Piemontese (pseud.), *Secreti del Alessio Piemontese . . .* (Venice, 1555).

7. *Loc. cit.*

8. Abraham Bosse, *Traicté de manières de graver en taille douce sur l'airain* (Paris, 1645).

CHAPTER 3: *The Damascus Blade*

1. K. A. C. Creswell, *Bibliography of Arms and Armour in Islam* (Royal Asiatic Society: London, 1956).

2. E. von Lenz, "Uber Damast," *Zeit. für historische Waffenkunde,* 1906/8, **4,** 132–42.

3. N. T. Belaiew, "Damascene Steel," *Jour. Iron and Steel Inst.,* 1918, **97,** (i), 417–37; "Sur le 'Damas' Oriental et les lames damassées," *Métaux et Civilisation,* 1950, **1,** 10–16.

4. *Idem.,* "Les Précurseurs de la métallographie," *Revue de Métallurgie,* 1914, **11,** 221–27.

5. B. Zschokke, "Du Damassé et des lames de damas," *Revue de Métallurgie,* 1924, **21,** 635–69.

6. J. von Hammer, "Beitrag zur Geschichte der Luftsteine, aus turkischen und arabischen Werken," *Fundgruben des Orients*, 1814, **4**, 277–87.

7. S. V. Granscay, "The New Galleries of Oriental Arms and Armor," *Metropolitan Museum of Art Bull.*, New York, 1958, **16**, 241–56.

8. P. Oberhoffer, "Über das Gefüge des Damaszenerstahls," *Stahl und Eisen*, 1915, **35**, 140.

9. K. Harnecker, "Beitrag zur Frage des Damaszenerstahls," *Stahl und Eisen*, 1924, **44**, 1409–11.

10. *Loc. cit.*

11. *Loc. cit.*

12. —. Massalski, "Préparation de l'acier damassé en Perse," *Ann. du Journal des Mines de Russie*, 1841 (published 1843), pp. 297–308.

13. H. T. P. J. duc de Luynes, *Mémoire sur la fabrication de l'acier fondu et damassée* (Paris, 1844). German extract in *Polytechnische Centralblat*, 1845, **1**, 315–21.

14. Henry Wilkinson, "On the Cause of the External Pattern or Watering of the Damascus Sword-Blades," *Jour. Royal Asiatic Soc.*, 1837, **4**, 187–93; "On Iron," *ibid.*, 1839, **5**, 383–89. See also the same author's useful *Engines of War* (London, 1841).

15. J. M. Heath, "On Indian Iron and Steel," *Jour. Royal Asiatic Soc.*, 1839, **5**, 390–97.

16. Massalski, *loc. cit.*

17. Francis Buchanan, *A Journey from Madras through the Countries of Mysore . . .*, Vol. II (London, 1807).

18. J. W. Needham, *The Development of Iron and Steel Technology in China* (Newcomen Society: London, 1958). This will form a chap. in Vol. VI of *Science and Civilisation in China* (Cambridge University Press: 1954—). The author is indebted to Dr. Needham for an opportunity to see the manuscript of this important work before publication and for stimulating discussions of Oriental metallurgy generally.

19. Vannoccio Biringuccio, *De la pirotechnia* (Venice, 1540 [Eng. trans.; New York, 1942]).

20. G. Agricola, *De re metallica* (Basle, 1556). (Eng. trans., London, 1912 [reprinted; New York, 1950]).

21. David Mushet, "Experiments on Wootz or Indian Steel," *Phil. Trans. Royal Soc.*, 1804, **95**, 175; also *Phil. Mag.*, 1805, **22**, 40–48. Reprinted in Mushet's *Papers on Iron and Steel* (London, 1840), pp. 650–78.

22. *Loc. cit.*

23. *Loc. cit.*

24. G. B. della Porta, *Magiae naturalis libri XX* (Naples, 1588). (Eng. trans., London, 1658 [reprinted, New York, 1958]).

25. Balthazar de Monconys, *Journal des voyages* (Paris, 1666), III, 65.

26. J. Jacquin, "Chemische Versuche mit dem Sagh," *Fundgraben des Orients*, 1816, **5**, 44. Eng. trans. in *Asiatic Jour. and Monthly Register*, 1818, **5**, 571–72.

27. *Loc. cit.*

28. *Loc. cit.*

29. J. Barker, "Method of Renewing the *Giohare* or Flowery Grain of Persian Swords Commonly Called Damascus Blades," *Fundgraben des Orients*, 1816, **5**, 40–43. Reprinted (under name S. Baker) in the *Annual Register*, 1818, pp. 599–602.

30. *Loc. cit.*

31. R. A. F. de Réaumur, *L'Art de convertir le fer forgé en acier . . .* (Paris, 1722 [Eng. trans., Chicago, 1956]).

32. Joseph Moxon, *Mechanick Exercises, or the Doctrine of Handiworks,* Vol. I, No. 3 (London, 1677).

C. Panseri, "L'acciaio di Damasco nella leggenda e nella realtà," *Armi Antiche* (Boll. dell'Accad. S. Marciano-Torino), 1962. English translation to appear in *Gladius,* Vol. V (in preparation).

CHAPTER 4: *Attempts To Duplicate Damascus Steel*

1. J. Stodart, "Account of the Method of Making Steel in Mysore Country." Letter quoted by Benjamin Heyne in *Tracts Historical and Statistical of India* (London, 1814), pp. 356–64.

2. J. J. Perret, *Mémoire sur l'acier* (Paris, 1779).

3. J. R. Bréant, "Description d'un procédé à l'aide duquel on obtient une espèce d'acier fondu, sembable à celui des lames damassées Orientales," *Bull. Société d'Encouragement pour l'Industrie Nationale,* 1823, **22,** 222–27. Reprinted in *Annales de Mines,* 1824, **9,** 319–28. Eng. trans. in *Repertory of Arts,* 1824, **45,** 306–14; *Technical Repository,* 1824, **6,** 49–55; and *Annals of Philosophy,* 1824, **8,** 267–71.

4. H. Damemme, *Essai pratique sur l'emploi de l'acier* (Paris, 1835).

5. C. S. Smith, *Four Outstanding Researches in Metallurgical History* (Philadelphia: American Society for Testing and Materials, 1963).

6. H. de Thury, "Recherches sur le dessin ou le Moiré des anciens Damassés," *Bull. Société d'Encouragement pour l'Industrie Nationale,* 1822, **21,** 84–92.

7. Reported by Hericart de Thury, *Bull. Société d'Encouragement pour l'Industrie Nationale,* 1821, **20,** 37–47, 351–85.

8. M. Faraday, "Analysis of Wootz or Indian Steel," *Quarterly Jour. Science,* 1819, **7,** 288–90.

9. J. Stodart and M. Faraday, "Experiments on the Alloys of Steel Made with a View to Its Improvement," *Quarterly Jour. Science,* 1820, **9,** 319–30.

10. P. Anossoff, "Mémoire sur l'acier damassé," *Annuaire du Journal des Mines de Russie,* 1841 (published 1843), pp. 192–236.

11. E. von Lenz, "Über Damast," *Zeit. für historische Waffenkunde,* 1906/8, **4,** 132–42.

12. N. T. Belaiew, "Les Précurseurs de la métallographie," *Revue de Métallurgie,* 1914, **11,** 221–27.

13. H. T. P. J. duc de Luynes, *Mémoire sur la fabrication de l'acier fondu et damassée* (Paris, 1844). German extract in *Polytechnische Centralblat,* 1845, **1,** 315–21.

14. [K. Schib and R. Gnade], *The Metallurgist, J. C. Fischer, and His Relations with Britain* (Schaffhausen, 1947); See also the same authors' *Johann Conrad Fischer, 1773–1854* (Schaffhausen, 1954).

CHAPTER 5: *Welded Damascus Steel*

1. W. Moorcroft and G. Trebeck, *Travels in Himalayan Provinces 1819–1825,* Vol. II (London, 1841).

2. —. Massalski, "Préparation de l'acier damassé en Perse," *Annuaire du Journal des Mines de Russie,* 1841 (published 1843), pp. 297–308.

3. P. Wäsström, "Beskrifning på damascherade skjut-gevär af järn och stål," *Kogl. Vetenskaps Academiens Handlingar,* 1773, **34,** 311–18.

4. Sven Rinman, "Rön om etsning på järn och stål," *Kogl. Vetenskaps Akademiens Handlingar* (Stockholm), 1774, **35,** 3–14.

5. Wäsström, *op. cit.;* Henry Wilkinson, *Engines of War* (London, 1841).

6. W. Greener, *The Gun or a Treatise on . . . Small Arms* (London, 1835).

7. Moorcroft and Trebeck, *loc. cit.*

8. *Loc. cit.*

9. J. J. Perret, *L'Art du coutelier,* Part I (Paris, 1771).

10. *Idem., Mémoire sur l'acier* (Paris, 1779).

11. B. F. J. Hermann, "Expériences sur l'acier damassé," *Nova Acta Acad. Sci. Imp. Petropolitanae,* 1798 (published 1802), **12,** 352–63.

12. —. Clouet, "Instruction sur la fabrication des lames figurées, ou des lames dites Damas," *Journal des Mines,* Ann. 12, 1803/4, **15,** 421–35.

13. Reported by Hericart de Thury, *Bull. Société d'Encouragement pour l'Industrie Nationale,* 1821, **20,** 37–47, 351–85.

14. A. Crivelli, *Sull'arte de fabricare le sciabole di damasco* (Milan, 1821).

15. B. Zschokke, "Du Damassé et des lames de damas," *Revue de Métallurgie,* 1924, **21,** 635–69.

16. K. Harnecker, "Beitrag zur Frage des Damaszenerstahls," *Stahl und Eisen,* 1924, **44,** 1409–11.

17. P. Holstein, *Contribution a l'étude des armes Orientales* (2 vols; Paris, 1930); Vaclav Šolc, *Swords and Daggers of Indonesia,* with photographs by W. Forman (London, n.d.); A. H. Hill, "The Malay Kĕris and Other Weapons," *Malay Museum Popular Pamphlet,* No. 5 (Singapore, 1956). See also *Jour. Malayan Branch Royal Asiatic Soc.,* 1956, Vol. **29.**

18. W. Rosenhain, "Notes on Malay Metal Work," *Jour. Anthropological Soc.,* 1901, **31,** 161–66.

19. H. T. P. J. duc de Luynes, *Mémoire sur la fabrication de l'acier fondu et damassée* (Paris, 1844). German extract in *Polytechnische Centralblat,* 1845, **1,** 315–21.

Additional References, 1965:

J. P. Frankel, "The Origin of Indonesian 'Pamor,'" *Technology and Culture,* 1963, **4,** 14–21.

B. F. J. Hermann, "Von der Bereitung des Damascenerstahls," *Chemische Annalen* (Crell), 1792 (ii), pp. 99–108.

See also references to chapter 1.

CHAPTER 6: *The Japanese Sword*

1. A. Dobrée, "Japanese Sword Blades," *Archaeological Journal,* 1905, **62,** 1–19, 218–55.

2. H. L. Joly and Hogitaro Inada, *The Sword Book in Honcho Gunkiko . . .* (Reading, 1913). Commonly known as "Sword and Samé." Reprinted London, 1962.

3. Inami Hakusui, *Nippon-To, the Japanese Sword* (Tokyo, 1948).

4. J. M. Yumoto, *The Samurai Sword: A Handbook* (Tokyo, 1958). See also Junji Homma, *Japanese Sword,* Japanese Art Series (Tokyo: The National Museum, 1948).

5. K. Tawara ("Investigation of Japanese Swords"), *Kikkai Gakkai Shi (Jour. Soc. Mech. Eng.),* 1918, **22,** 1–39. In Japanese.

6. *Idem., Nippon-to no kagateki kenkyu* ("Scientific Study of Japanese Swords") (Tokyo, 1953). In Japanese.

7. M. Chikashige, *Alchemy and Other Chemical Achievements of the Orient* (Tokyo, 1936).

8. C. S. Smith, "A Metallographic Examination of Some Japanese Sword Blades," *Doc. e contributi per la storia della metallurgia* No. 2 (Milan: Assoc. Ital. di Metallurgia, 1957), pp. 42–68.

9. J. W. Needham, *The Development of Iron and Steel Technology in China* (Newcomen Society: London, 1958). See chap. 3, n. 18.

10. Smith, *loc. cit.*

11. Quoted by Joly and Inada, *loc. cit.*

12. See nn. 1–8 above.

13. *Loc. cit.*

14. *Loc. cit.*

15. *Loc. cit.*

16. See nn. 2, 3 above.

17. Junji Homma, *Meito zufu* ("Atlas of Famous Swords") (Tokyo, 1935). In Japanese.

18. Le Marquis de Tressan, "L'Évolution de la garde de sabre japonaise," *Bull. Soc. Franco-Japonais*, Nos. 18, 19, 20, 22, 25 (Paris, 1901–12).

19. H. L. Joly, "Introduction à l'étude des montures de sabres au Japon," *Bull. Soc. Franco-Japonais*, No. 14, Paris, 1909, pp. 31–84; *Japanese Sword Mounts: Descriptive Catalogue of the Collection of J. C. Hawkshaw* (London, 1910); *Catalogue of the G. H. Naunton Collection of Japanese Sword Furniture* (London, 1912).

20. H. Gunsaulus, *Japanese Sword Mounts in the Collections of the Field Museum* ("Field Museum Publication No. 216" [Chicago, 1923]).

21. J. Homma (ed.), *Masterpieces of Japanese Swordguards*, Folio of 159 plates (Tokyo: The Society for the Preservation of Japanese Art Swords, 1952). Text in Japanese and English.

22. Raphael Pumpelly, "Notes on Japanese Alloys," *American Jour. Science,* 1866, **42,** 43–45.

23. J. J. Rein, *The Industries of Japan* (London, 1889).

24. W. Gowland, "Metals and Metal Working in Old Japan," *Trans. Japan Society*, 1915, **13,** 20–100.

25. W. C. Roberts-Austen, "Cantor Lectures on Alloys," *Jour. Soc. Arts*, 1888, **36,** 1137–46; 1893, **41,** 1022–43.

26. F. Binkley, *Japan, Its History, Arts and Literature* (8 vols; London, 1903–4). Metallurgy is mostly in Vol. VIII.

27. H. Wilson, *Silverwork and Jewellry* (2d ed. with chap. on Japanese methods by Unno Bisei; London, 1951).

28. *Loc. cit.*

29. *Loc. cit.*

30. *Loc. cit.*

31. See n. 20 above.

32. *Loc. cit.*

CHAPTER 7: *Moiré Métallique*

1. —. Baget, "Sur le moiré métallique," *Journal de Pharmacie,* 1818, **4,** 25.

2. J. F. Daniell, "On Some Phenomena Attending the Process of Solution and . . . the Laws of Crystallization," *Jour. Science and Arts*, 1816, **1,** 24–49.

3. C. F. Peschel, "Verfertigung des Atlas-Blech der lakirten Waaren," *Annalen der Physik*, 1818, **58,** 438–39.

4. L. W. Gilbert, "Moirirtes Metall und Malerei darin, oder das Moiré Métallique," *Annalen der Physik*, 1820, **64,** 279–302.

5. —. Berry, "Mémoire sur le moiré métallique applicable aux feuilles d'étain," ("par M. Berry, peintre à la Rochelle"), *Bull. Société d'Encouragement pour l'Industrie Nationale*, 1821, **20,** 11–12.

6. [—. Alard], *Repertory of Arts, Manufactures and Agriculture*, 1818, **33,** 189–90.

7. R. Blakemore and J. James, "Specification . . . for a Method . . . of Crystallizing or Rendering Crystallizable the Surface of Tin Plates . . . ," *Repertory of Arts*, 1819, **34,** 198–201.

8. C. S. Smith, "Architectural or Decorative shapes of Cu-Al Alloys," U.S. Patent No. 2,050,069, 1936.

CHAPTER 8: *The Corpuscular Philosophers*

1. Marie Boas, "The Establishment of the Mechanical Philosophy," *Osiris* (1952), **10,** 412–541. See also G. B. Stones, "The Atomic View of Matter . . . ," *Isis*, 1928, **10,** 445–65.

2. D. Sennert, *Hypomnemata physica* (Frankfurt, 1636).

3. J. C. Magnen, *Democritus reviviscenes sive de atomis* (Pavia, 1646).

4. P. Gassendi, *Syntagmata philosophicum* (1649). Quotations are from Eng. trans. by W. Charleton, *Physiologia* (London, 1654), p. 337.

5. Nicolas Hill, *Philosophia epicurea . . .* (Paris, 1601).

6. Sebastian Basso, *Philosophia naturalis adversus Aristotelem* (Geneva, 1621).

7. René Descartes, *Oeuvres,* ed. Tannery (1898), II, 486.

8. *Idem. Principia philosophia* (Paris, 1644).

9. *Idem. Oeuvres,* ed. Tannery (1904), IX, 276–78.

10. A. Le Grand, *An Entire Body of Philosophy According to Descartes.* Eng. trans. by R. Blome (London, 1694).

11. Claude Perrault, *Essais de physique* (Paris, 1680).

12. N. Hartsoeker, *Principes de physique* (Paris, 1696). The sections quoted are from chap. v, pp. 86–112.

13. Jacques Rohault, *Traité de physique* (Paris, 1671).

14. *Idem., Traité de physique.* Eng. trans. by J. Clarke (London, 1723), II, 156–57.

15. Johannes Kepler, cited by M. von Laue, *History of Physics* (New York, 1950), p. 116.

16. Robert Hooke, *Micrographia* (London, 1665).

17. Christian Huygens, *Traité de la lumière* (Leiden, 1690).

18. Isaac Newton, *Opticks* (2d ed.; London, 1718), pp. 363–64.

19. *Ibid.,* pp. 243–44.

20. *Ibid.,* p. 370.

21. A. R. Hall, *The Scientific Revolution, 1500–1800* (London, 1954).

22. Isaac Todhunter and Karl Pearson, *A History of the Theory of Elasticity and Strength of Materials* (2 vols. in 3; London, 1888–93).

23. Emanuel Swedenborg, *Principia rerum naturalium* (Dresden and Leipzig, 1734).

24. Herbert Dingle, "The Scientific Work of Emanuel Swedenborg," *Endeavour,* 1958, **17,** 127–32.

CHAPTER 9: *The Early Microscopists*

1. R. S. Clay and T. H. Court, *The History of the Microscope* (London, 1932).

2. Henry Baker, *Employment for the Microscope* (London, 1753).

3. Henry Power, *Experimental Philosophy in Three Books, Containing New Experiments* (London, 1664).

4. Robert Hooke, *Micrographia* (London, 1665).

5. Thomas Birch, *History of the Royal Society* (London, 1756), III (1757), 5.

6. Jacques Rohault, *Traité de physique* (Paris, 1671). Eng. trans. by J. Clarke (2 vols.; London, 1723).

7. Isaac Newton, *Notebook* (Cambridge University Library, Ad. MS. No. 3975). Private communication from Marie Boas.

8. J. F. Grindel van Ach, *Micrographia nova* (Nuremburg, 1687). Latin and German eds. issued simultaneously.

9. Antoni van Leeuwenhoek, "A Letter from Mr. Antony van Leeuwenhoek, F.R.S. to Mr. Chamberlaine Concerning the Causes of the Different Tastes of Waters and Edge of Razors," *Phil. Trans. Royal Soc.*, 1702, **22**, 899.

10. Leeuwenhoek, "Observations on Animalculae in Water, the Dissolution of Silver etc., *ibid.*, 1703, **23**, 1430.

11. G. Homberg, "Réflections sur différentes vegetations métallique," *Mém. Acad. Sci.*, 1692, pp. 145–52.

12. Leeuwenhoek, "Observations upon the Edge of Razors, etc.," *Phil. Trans. Royal Soc.*, 1709, **26**, 493–98.

13. W. Lewis, *Commercium philosophico-technicum, or the Philosophical Commerce of Arts* (London, 1763).

CHAPTER 10: *The Study of Fracture*

1. G. F. Hill, "Ancient Methods of Coining," *Numismatics Chronicle*, 1922, **2**, 1–42.

2. Sidney P. Noe, "Two Hoards of Persian Sigloi," *Numismatic Notes and Monographs*, No. 136 (American Numismatic Society, 1956).

3. Vannoccio Biringuccio, *De la pirotechnia* (Venice, 1540 [Eng. trans., New York, 1942]).

4. Lazarus Ercker, *Beschreibung allerfürnemsten mineralischen Ertzt unnd Berckwercksarten* (Prague, 1574). Eng. trans. from Ger. ed. of 1580 by Anneliese G. Sisco and C. S. Smith (Chicago, 1951).

5. Louis Savot, *Discours sur les médailles antiques* (Paris, 1627).

6. Mathurin Jousse, *Fidelle ouverture de l'art de serrurier* (Paris, 1627). Partial translation by C. S. Smith and A. G. Sisco in *Technology and Culture*, 1961, **2**, 131–45.

7. André Félibien, *Principes de l'architecture* (Paris, 1676).

8. Joseph Moxon, *Mechanick Exercises, or the Doctrine of Handy-Works* (London, 1677).

9. Dud Dudley, *Metallum Martis, or Iron Made with Pit-Coale, Sea-Coale &c . . .* (London, 1665). Reprint, ed. J. N. Bagnall (West Bromwich, 1854).

10. Robert Boyle, *Essay about tne Origine and Virtue of Gems* (London, 1672). The quotation is from Vol. III of the collected works, ed. Thos. Birch (London, 1744).

11. R. A. F. de Réaumur, "De l'Arrangement que prennent les parties des

matières métalliques et minerales, lorsqu'après avoir été mises en fusion, elles viennent à se figer," *Mém. Acad. Sci*, 1724, pp. 307–16.

12. *Idem.*, *L'Art de convertir le fer forgé en acier* (Paris, 1722 [Eng. trans., Chicago, 1956]).

13. *Idem.*, *Nouvel art d'adoucir le fer fondu . . .* (Paris, 1762) . Part III, memoir 4, pp. 80–85. See n. 12, Eng. trans. pp. 371–75.

14. Johann Rudolph Glauber, *Furni novi philosophici* (Amsterdam, 1648).

15. C. F. Geoffroy, "Observations sur un métal que résulte de l'alliage du cuivre & du zinc," *Mém. Acad. Sci.*, 1725, pp. 57–66.

16. Simon Pierre Fournier, *Manuel typographique* (2 vols.; Paris, 1764–66). Eng. trans. by H. Carter (London, 1930).

17. Karl Franz Achard, *Recherches sur les propriétés des alliages métalliques* (Berlin, 1788). This book is of great rarity and is unknown to metallurgical historians. A detailed analysis of it will be published by the present author.

18. Robert Mallet, *Physical Conditions Involved in the Construction of Artillery* (London, 1856).

19. C. Schafhaeutl, "On the Combinations of Carbon with Silicon and Iron and Other Metals . . . ," *Phil. Mag.*, 1839, **15**, 417–28; 1840, **16**, 44–50, 297–304, 426–34, 514–23, 570.

20. E. F. Dürre, *Über die Constitution des Roheisens und der Werth seiner physikalischen Eigenschaften* (Leipzig, 1868). Preliminary version in *Berg- und Hüttenmannischen Zeitung* (1865 and 1868) and in *Zeitschrift für das Berg-Hütten- und Salienen-wesen*, 1868, **16**, 70–131, 271–301.

21. Dimitri K. Tschernoff, " 'Kriticheskii Obzor Statei gg. Lavrova y Kalakutzkago o Stali v Stalnikh Orudiakh' i Sobstvennie ego Izsledovanie po Etomuje Predmetu" ("Critical Review of Articles by Messrs. Lavrov and Kalakutzkii on Steel and Steel Ordnance, with Original Investigations on the Same Subject") , *Zapiski Russkago Tekhnicheskago Obshestva*, 1868, pp. 399–440. Eng. trans. by William Anderson of Tschernoff's original contribution (pp. 423–40 only), "On the Manufacture of Steel and the Mode of Working It," *The Engineer*, 1876, **2**, 1–4. *Proc. Instn. Mech. Engrs.*, 1880, pp. 286–307. French trans., *Revue Universelle des Mines*, 1880, **7**, 129. See also the note on Tschernoff's work by A. F. Golovine in *Actes du Huitième Congrès International d'Histoire des Sciences*, 1956, **3**, 1961–64, and the important re-evaluation by V. D. Sadovsky, "Tschernoff and the Development of the Theory of Heat Treatment of Steel," in *The Sorby Centennial Conference on the History of Metallurgy*, ed. C. S. Smith (New York, 1965), pp. 189–208. A Tschernoff bibliography by J. A. Melnikova was published in Leningrad in 1951.

22. *Idem.*, "Izsledovanie, Otnosiashchiasia po Struktury Litikh Stalnykh Bolvanok" ("Investigations on the Structure of Cast Steel Ingots"), *Zapiski Imperatorskago Russkago Tekhnicheskago Obshestva*, 1879, pp. 1–24. Eng. trans. by William Anderson, *Proc. Instn. Mech. Engrs.*, 1880, pp. 152–83.

23. J. A. Brinell, *Jernkontorets annaler*, 1885, **40**, 9–38. Ger. trans., "Über die Texturveränderungen des Stahls bei Erhitzung und bei Abkühlung," *Stahl und Eisen*, 1885, **11**, 611–20. English abstracts in *Jour. Iron and Steel Inst.*, 1886 (i), pp. 365–67, and 1898, **53**, 145–89 (by J. E. Stead); *The Metallographist*, 1899, **2**, 129–36.

24. H. M. Howe, *The Metallurgy of Steel* (New York, 1890), p. 172.

25. W. A. Lampadius, *Grundriss einer allgemeinen Hüttenkunde* (Göttingen, 1827).

26. *Report of the Commissioners Appointed to Inquire into the Application of Iron to Railway Structures* (2 vols., one containing plates; London, 1849).

27. Charles Hood, "On Some Peculiar Changes in the Internal Structure of Iron . . . ," *Instn. Civil Engrs., Minutes of Proceedings,* 1842, pp. 180–84.

28. J. O. York, "Account of a Series of Experiments on the Comparative Strength of Solid and Hollow Axles," *Instn. Civil Engrs., Minutes of Proceedings,* 1843, p. 89.

29. W. J. M. Rankine, "On the Causes of Unexpected Breakage of Journals on Railway Axles . . . ," *Instn. Civil Engrs., Minutes of Proceedings,* 1843, pp. 105–8.

30. J. E. McConnell, "On Railway Axles," *Proc. Instn. Mech. Engrs.,* 1849, pp. 13–27.

31. R. Stephenson, "Discussion on Deterioration of Railway Axles," *Proc. Instn. Mech. Engrs.,* 1850 (April), pp. 3–14.

32. J. E. McConnell, "On Railway Axles," *Proc. Instn. Mech. Engrs.,* 1849 (October), pp. 13–27; also *ibid.,* 1850 (January), pp. 5–19.

33. J. A. Roebling, *Report of John A. Roebling . . . to the . . . International Bridge Companies, on the Condition of the Niagara Railway Suspension Bridge. August 1, 1860* (Trenton, N.J., 1860).

34. A. Schrötter, "Ist die krystallinische Textur des Eisens von Einfluss auf sein Vermögen magnetisch zu werden?" *Sitzungsberichte Akad. Wissenschaften,* Vienna, 1857, **23,** 472–81.

35. A. Wöhler, "Bericht über die Versuche, welche . . . mit Apparaten zum Messen der Biegung und Verdrehung von Eisenbahnwagen-Achsen während der Fahrt, angestellt werden," *Zeit. für Bauwesen,* 1858, **8,** 642–52.

36. David Kirkaldy, *Results of an Experimental Inquiry into the Tensile Strength and Other Properties of . . . Wrought Iron and Steel* (Glasgow, 1862).

37. John Percy, *Metallurgy: Iron and Steel* (London, 1864), pp. 8–13.

38. ". . . Bessemer Steel Manufacture and Some Points in the Metallurgy of Steel," *Practical Mechanics Jour.,* 1865, **1,** 225–28.

39. H. C. Sorby, "On the Microscopical Structure of Iron and Steel," *Jour. Iron and Steel Inst.,* 1887 (i), p. 265.

40. A. Martens, "Über das Kleingefüge des schmiedbaren Eisens, besonders des Stahles," *Stahl und Eisen,* 1887, **7,** 235–42.

41. H. M. Howe, *Metallurgy of Steel* (New York, 1890), p. 196.

42. R. W. Raymond *et al.,* "Discussion on Paper of T. A. Rickard," *Trans. AIME,* 1893, **23,** 560, 573–77; 1895, **24,** 809; 1896, **26,** 1026.

CHAPTER 11: *Idiomorphic Crystals, etc.*

1. Thomas Fuller, *The Worthies of England* (London, 1662).

2. J. E. Stead, in discussion on a paper by Roberts-Austen, *Proc. Instn. Mech. Engrs.,* 1897, pp. 77–78.

3. Hermann Boerhaave, *Institutiones et experimenta chemiae* (Paris, 1724). Published from lecture notes without the author's authorization. Eng. trans. Timothy B. Dallowe (London, 1735); Authorized version, *Elementa chemiae* (Leyden, 1732). Eng. trans. Peter Shaw (London, 1727 [enlarged, 1741]).

4. Basilius Valentinus, *Currus triumphalis antimonii* (Leipzig, 1604).

5. Nicolas Lémery, *Traité de l'antimoine* (Paris, 1707).

6. Samuel Zimmerman, *Probierbuch* (Augsburg, 1573).

7. Pierre Joseph Macquer, *Elemens de chimie-pratique* (Paris, 1751 [Eng. trans., Edinburgh, 1758]).

8. Antoine Baumé, *Chymie experimentale et raisonée* (Paris, 1773).

9. B. G. Sage, "Observations sur les cristallisations des substances métalliques, par l'interméde du mercure," in his *Mémoires de chimie* (Paris, 1777). Sage's work was known to Grignon before 1775.

10. John Hill, *A History of Fossils* (London, 1748).

11. Pierre Clement Grignon, *Mémoires de physique, sur le fer . . .* (Paris, 1775).

12. *Idem.*, "Mémoire sur les métamorphoses du fer, ou réflexions chymiques et physiques, sur les différentes situations du fer dans le terre dans son traitement jusqu'a sa perfection et sa destruction . . . ," *Mémoires de physique . . .* , pp. 56–79.

13. Grignon, "Mémoire sur des cristallisations métalliques, pyriteuses et vitreuses artificielles, formée par le moyen du feu," *ibid.*, pp. 476–81.

14. J. B. Romé de l'Isle, *Essaie de cristallographie* (Paris, 1772), p. 321.

15. L. B. Guyton de Morveau, "Observation de la cristallisation du fer," *Journal de Physique*, 1776, **8,** 348–53.

16. *Idem.*, *Journal de Physique*, 1777, **9,** 303–5.

17. *Idem.*, "Sur les cristallisations métallique," *Journal de Physique*, 1779, **13,** 90–92.

18. Jean Andrez Mongez, "Response à M. de Cissay (Sur les cristallisations métalliques)," *Journal de Physique*, 1781, **18,** 74–76.

19. David Mushet, "On the Crystallisation of Iron," *Phil. Mag.*, 1823, **61,** 22–24, 83–87.

20. Frank H. Storer, "On the Alloys of Copper and Zinc," *Memoirs American Acad. Arts and Sciences*, 1861, **8,** 27–56.

21. C. W. C. Fuchs, *Die künstlichen dargestellten Mineralen* (Haarlem: Hollandsche Maatschappij des Wetenschappen, 1872).

22. H. Vogelsang, "Sur les cristallites," *Arch. Neerlandisches*, 1872, **7,** 42; also *Die Krystalliten* (Bonn, 1875).

23. G. A. Daubrée, *Études synthétiques de géologie expérimentale* (Paris, 1879).

24. Léon Bourgeois, "Reproduction, par voie ignée, d'un certain nombre d'espèces minérales appartenant aux familles des silicates, etc." (Thesis, Paris, 1883), p. 65.

25. F. Fouqué and A. M. Lévy, *Synthèse des minéraux et des roches* (Paris, 1882).

26. "Izsledovanie, Otnosiashchiesia po Struktury Litikh Stalnykh Bolvanok" ("Investigations on the Structure of Cast Steel Ingots"), *Zapiski Imperatorskago Russkago Tekhnicheskago Obshestva*, 1879, pp. 1–24. Eng. trans. by William Anderson, *Proc. Instn. Mech. Engrs.*, 1880, pp. 152–83.

27. J. S. Jeans, *Steel, Its History, Manufacture, Properties and Uses* (London, 1880), pp. 644–54.

A good summary of the knowledge in 1820 is given by Haussmann, "On Metallurgic Crystallography," *Edinburgh Philosophical Journal*, 1821, **5,** 155–64, and 344–51. He cites Zanichelli as the first to observe the octahedral crystallization of iron and refers to the octahedral form of meteoric iron revealed by fracture and by etching.

CHAPTER 12: *Etching for Macroscopic Study*

1. P. C. Grignon, "Mémoire sur l'unité du fer," in his *Mémoires de physique* (Paris, 1775), pp. 38–55.

2. J. A. Cramer, *Elementa artis docimasticae* (Leyden, 1739 [Eng. trans., London, 1741]).

3. See n. 7 below.

4. Sven Rinman, "Rön, om etsning på järn och stål," *Kongl. Vetenskaps Akademiens Handlingar* (Stockholm), 1774, **35,** 3–14. For an English translation see Ap-

pendix A. This and the following paper by Wäsström are included in the volumes of the German translation of the *Handlingar* published under the editorship of A. G. Kastner in 1780 and 1781.

5. Peter Wäsström, "Beskrifning på damascherade skjut-gevär af järn och stål," *Kongl. Vetenskaps Akademiens Handlingar,* 1773, **34,** 311–18. Discussion by Sven Rinman, pp. 318–21.

6. Sven Rinman, *Försöck till järnets historia* (2 vols.; Stockholm, 1782). German trans. by C. J. B. Karsten (Liegnitz, 1814–15).

7. Tobern Bergman, *De analysi ferri* (Upsala, 1781). Reprinted with a few changes in Bergman's *Opuscula physica et chemica,* Vol. III (Upsala, 1783). French trans. with critical notes by P. C. Grignon (Paris, 1783).

8. *Ibid.*

9. J. J. Perret, *L'Art du coutelier* (Paris, 1771).

10. *Idem., Mémoire sur l'acier* (Paris, 1779).

11. Guyton de Morveau, "Acier," *Encyclopédie Méthodique* (Paris, 1786) I, 420–65.

12. C. A. Vandermonde, C. L. Berthollet, and G. Monge, "Mémoire sur le fer considere dans ses differens états métalliques," *Mém. Acad. Sci.,* 1786 (pub. 1788), pp. 132–200.

13. C. A. Vandermonde, G. Monge, and C. L. Berthollet, *Avis aux ouvriers en fer sur la fabrication de l'acier* (Paris, An. II [i.e., 1793]). Reprinted in *Journal de Physique,* 1793, **43,** 373–86, and *Annales de Chimie,* 1797, **19,** 13–46.

14. J. Stodart and M. Faraday, "Experiments on the Alloys of Steel Made with a View to Its Improvement," *Quarterly Jour. Science,* 1820, **9,** 319–30; "On the Alloys of Steel," *Phil. Trans. Royal Soc.,* 1822, **112,** 253–70.

15. W. Wade, in *Reports on the Strength and Other Properties of Metals for Cannon . . . by Officers of the Ordnance Department, U.S. Army* (Philadelphia, 1856), p. 299. Written in 1854.

16. W. Nicholson, "Report on the Art of Making Fine Cutlery," *Nicholson's Journal,* 1804, **4,** 127–32.

17. G. [i.e., William] Thomson, ". . . Sul ferro malleabile trovato da Pallas in Siberia . . . ," *Atti dell'Accademia della Scienze di Siena,* 1808, **9,** 37–51. See R. T. Gunther, "Dr. William Thomson F.R.S. . . . ," *Nature,* 1939, **143,** 667–68, and C. S. Smith, "Note on the History of the Widmannstätten Structure," *Geochimica et Cosmochimica Acta,* 1962, **26,** 971–72.

18. Carl von Schreibers, *Beyträge zur Geschichte und Kenntniss meteorischer Stein- und Metall-massen* (Vienna, 1820). This was intended as a supplement to E. F. F. Chladni, *Über Feuer-Meteore . . .* (Vienna, 1819).

19. Alessio Piemontese (pseud.), *Secreti del . . . Alessio Piemontese* (Venice, 1555). Notes from French trans. (Antwerp, 1561).

20. P. M. Partsch, *Die Meteoriten* (Vienna, 1843).

21. M. W. Haidinger, *Handbuch der Mineralogie* (Vienna, 1845).

22. *Idem.,* "Bemerkungen über die zuweilen in geschmeidigen Eisen entstandene Krystallinische Struktur . . . ," *Sitzungsberichte, K. Akad. Wiss. Wien,* 1855, **15,** 354–60.

23. F. Leydolt, *Jahrbuch, K. K. Geologisch Reichanstalt,* Vienna, 1851, **2,** 124–32. Reprinted in Auer, see n. 25 below.

24. Alois von Auer, *Der polygraphische Apparat* (Vienna, 1853).

25. *Idem., Die Entdeckung des Naturselbstdrucken* (Vienna, 1854). Polyglot ed. in German, English, Italian, and French.

26. Martin Hardie, *English Coloured Books* (London, 1906).

27. Ernst Fischer, "Zweihundert Jahre Naturselbstdruck," *Gutenberg Jahrbuch,* 1933, pp. 186–213.

28. A. Smith, "Notice of a Mass of Meteoric Iron . . . ," *Edinburgh New Phil. Jour.,* 1862, **16,** 108–24.

29. W. H. Wollaston, "Observations and Experiments on the Mass of Native Iron Found in Brazil," *Phil. Trans. Royal Soc.,* 1816, **106,** 281–85.

30. J. F. Daniell, "On Some Phenomena Attending the Process of Solution and . . . the Laws of Crystallization," *Jour. Science and Arts,* 1816, **1,** 24–49.

31. *Idem.,* "On the Mechanical Structure of Iron Developed by Solution . . . ," *Jour. Science and Arts,* 1817, **2,** 278–93.

32. *Idem., Introduction to the Study of Chemical Philosophy* (London, 1839).

33. W. Lüders, "Über die Äusserung der Elasticität an Stahlartigen . . . und Stahlstäben . . . ," *Dinglers Polytechnische Journal,* 1860, **155,** 18–22.

34. [David Kirkaldy], "Description of Experiments on the Comparative Strength, etc., of Steel and Wrought Iron, by Messrs. Robert Napier and Sons," *Trans. Instn. of Engineers in Scotland,* 1859, **2,** 134–63.

35. David Kirkaldy, *Results of Experimental Enquiry into . . . Wrought Iron and Steel* (Edinburgh, 1862).

36. *Idem., Experimental Inquiry into the Relative Properties of Wrought Iron Plates Manufactured at Essen . . . and Yorkshire* (London, 1876).

37. Peter Barlow, *Treatise on the Strength of . . . Materials* (3d ed.; London, 1867).

38. H. Tresca, "On the Flow of Solids, with the Practical Application in Forgings etc.," *Proc. Instn. Mech. Engrs.,* 1867, pp. 114–43. Tresca had submitted a number of papers to the French Academy of Sciences but this paper and that cited in n. 41 below provide the best summary of his experiments and viewpoint.

39. R. Mallet (Discussion on work cited in n. 38 above), *Proc. Instn. Mech. Engrs.,* 1867, p. 143.

40. *Idem., The Physical Conditions Involved in the Construction of Artillery* (London, 1856), p. 214.

41. H. Tresca, "On Further Applications of the Flow of Solids," *Proc. Instn. Mech. Engrs.,* 1878, pp. 301–44.

CHAPTER 13: *Metallography in Sheffield*

1. D. W. Humphries, "Biographical Notes on H. C. Sorby." Unpublished manuscript prepared for *Conversazione* at time of British Association Meeting, August, 1956, Sheffield.

2. Cecil H. Desch, "The Services of Henry Clifton Sorby to Metallurgy," Second Sorby Lecture, Sheffield, 1921.

3. D. T. Ansted, J. Tennant, and W. Mitchell, *Geology, Mineralogy, and Crystallography . . . ,* in Orr's *Circle of the Sciences,* Vol. V (London, 1854).

4. John Holland, *38th Annual Report of Sheffield Literary and Phil. Soc.,* 1860, pp. 10–12.

5. William Baker, *41st Annual Report of Sheffield Literary and Phil. Soc.,* 1863.

6. Henry Clifton Sorby, *Diary.* Manuscript in custody of the Registrar, University of Sheffield. The *Diary* consists of eleven small notebooks filled with neat but not very legible handwriting in pencil. They cover the years 1859 to 1908. Some notebooks are missing, covering most of the periods 1870–83, 1894–96, and 1902–6.

7. John Percy, *Metallurgy: Iron and Steel* (London, 1864), p. 736.

8. *Sheffield Daily Telegraph,* September 4, 1863. A clipping is preserved in *Col-*

lected Papers of H. C. Sorby, II, 98, University of Sheffield Library, Shelf No. B508(S). I am indebted to Dr. A. R. Entwisle for the identification of the source of this clipping and also those referred to in Notes 9, 16, and 17, below.

9. *Sheffield and Rotherham Independent,* September 4, 1863. *Collected Papers . . . ,* II, 98.

10. John Percy, Letter to Sorby dated January 20, 1864. *Sorby Letters,* Sheffield Public Library, No. 51-249. Sorby's diary records that he wrote to Percy the previous day.

11. Walter White, Letter to Sorby dated January 23, 1864. *Sorby Letters,* Sheffield Public Library, No. 51–250.

12. H. C. Sorby, "On Microscopical Photographs of Various Kinds of Iron and Steel," *Report of 34th Meeting of the British Assn. Adv. Science* (Bath, 1864), Part II, p. 189.

13. George Gore, Letter to Sorby, dated January 17, 1865, *Sorby Letters,* No. 51–269.

14. W. Vivian, Letter to Sorby, dated July 27, 1866, *Sorby Letters,* No. 51–297.

15. Lionel S. Beale, *How To Work with a Microscope* (4th ed.; London, 1868), pp. 181–83.

16. *The Reader,* 1864, **2,** 301 (March 5). Clipping in *Collected Papers,* II, 99, verso.

17. *The Reader,* 1864, **2,** 331 (March 12); correspondence, pp. 399, 460. Clippings in *Collected Papers,* II, 97.

18. *Loc. cit.*

19. J. C. Bayles, "Microscopic Analysis of the Structure of Steel and Iron," *Trans. American Inst. of Mining and Metallurgical Engrs.,* 1882–83, **11,** 261–74.

20. H. C. Sorby, "The Structure of Iron and Steel" (abstract), *Jour. Iron and Steel Inst.,* 1882 (ii), pp. 702–3.

21. *Idem., Diary,* Entry for October 14, 1884.

22. *Idem.,* "The Microscopical Structure of Iron and Steel," *Iron and Steel Inst.* preprint, 8 pp. Undated but undoubtedly issued as a preprint for the May, 1885, meeting of the Institute in London. Copy in *Collected Papers,* Vol. II, No. 79. Reproduced in Appendix B.

23. *Idem.,* "On the Microscopical Structure of Iron and Steel," *Jour. Iron and Steel Inst.,* 1887 (ii), pp. 254–88.

24. *Idem.,* "On the Application of Very High Powers to the Study of the Microscopical Structure of Steel," *Jour. Iron and Steel Inst.,* 1886 (i), pp. 140–47.

25. John Percy, "President's Address," *Jour. Iron and Steel Inst.,* 1886 (i), pp. 8–32.

26. W. P. Beale, Letter to Sorby dated November 10, 1863, *Sorby Letters,* No. 51–248. Filed with Beale's letter is an undated, but probably contemporary, "List of Irons etc." in Sorby's handwriting (No. 51–248A), which mentions many other samples, including converted blister, annealed cast steel, decarburized cast iron, etc.

27. See n. 20 above.

28. H. C. Sorby, "On Some Remarkable Properties of the Characteristic Constituent of Steel," *Proc. Yorkshire Geological and Polytechnical Soc.,* 1886, **9,** 145–46.

CHAPTER 14: *19th Century Chemistry and Physics*

1. H. M. Leicester, *The Historical Background of Chemistry* (New York, 1956). Chap. xvi contains a good brief history of the rise of modern atomic theory.

2. M. von Laue, *History of Physics.* Eng. trans. R. Oesper (New York, 1950).

3. O. Lehmann, *Molekularphysik, mit besonderer Berücksichtigung mikroscopischer Untersuchungen und Anleitung zu solchen* (2 vols.; Leipzig, 1888–89).

4. P. Groth, *Molekularbeschaffenheit der Krystalle* (Munich, 1888).

5. W. H. Wollaston, "On the Elementary Particles of Certain Crystals," *Phil. Trans. Royal Soc.,* 1813, pp. 114–18.

6. P. Barlow, "Probable Nature of the Internal Symmetry of Crystals," *Nature,* 1883, **29,** 186–88, 205–7, 404.

7. L. Sohncke, *"Entwickelung einer Theorie der Krystallstruktur . . ."* (Leipzig, 1879); P. Groth, *Einleitung in die chemische Krystallographie* (Leipzig, 1904).

8. W. Thomson, "Molecular Constitution of Matter," *Proc. Royal Soc. Edinburgh,* 1890, **16,** 693–724.

9. A. Matthiessen, "Report on Chemical Nature of Alloys," *Report of 33rd Meeting of the British Assn. Adv. Science,* 1863, pp. 37–47.

10. *Idem.,* "On Alloys," *Journ. Chem. Soc.,* 1867, **20,** 201–20.

11. A. Levol, "Mémoire sur les alliages, considérés sous le rapport de leur composition chimique," *Ann. Chim. Phys.,* 1852, **36,** 193–224.

12. C. J. B. Karsten, "Der Saigerhüttenprozess," *Archiv für Bergbau und Hüttenwesen,* 1825, **9,** p. 22.

13. J. Percy, *Metallurgy, Silver and Gold,* Part I (London, 1880), p. 350.

14. P. A. Bolley, "Zur Kenntniss der Molekulareigenschaften des Zinks," *Ann. Chim. und Pharmacie,* 1855, **95,** 294–306.

15. S. Kalisher, Über der Einfluss der Wärme auf die Molekularstruktur des Zinks," *Ber. deutsche chemische Gesellschaft,* 1881, **14,** 2747–53; "Über dei Molekularstruktur der Metalle," *ibid.,* 1882, **15,** 702–12.

16. J. N. von Fuchs, "Theoretische Bemerkungen über die Gestaltings-zustande des Eisens" (read in 1851), *Abhl. K. Bayerischen Akademie der Wissenschaften* (Munich), Class II, 1853–55, **7,** 3–15. Reprinted in *Dinglers Polytechnische Journal,* 1852, **154,** 346–55.

17. C. E. Jullien, "Théorie de la constitution des fontes, fers et aciers," *Mém. Société des Ingenieures Civil,* 1865, pp. 85–88, 148–51, 182–86. Reprinted in *Revue Universelle des Mines,* 1866, **20,** 159–88. See also the same authors, *Carbures de fer* (Paris, 1852), and *Recherches sur l'aciération* (Paris, 1868).

18. *Idem., Traité théorique et pratique de la métallurgie de fer* (Paris, 1861).

19. E. Nairn, "An Account of the Effect of Electricity in Shortening Wires," *Phil. Trans. Royal Soc.,* 1780, **70,** 334–37.

20. G. Gore, "On a Momentary Molecular Change in Iron Wire," *Proc. Royal Soc.,* 1869, **17,** 260–65.

21. W. F. Barrett, "On Certain Molecular Changes Occurring in Iron at a Low Red Heat," *Phil. Mag.,* 1873, **46,** 472–78.

22. R. Norris, "On Certain Molecular Changes Which Occur in Iron . . . ," *Proc. Royal Soc.,* 1878, **26,** 123–33; Preprinted in *Phil. Mag.,* 1877, **4,** 389–95.

23. P. G. Tait, "First Approximation to a Thermo-electric Diagram," *Phil. Trans. Royal Soc. Edinburgh,* 1872, **27,** 125–40.

24. J. B. Biot, *Traité de physique* (3 vols.; Paris, 1816). The discussion on constitution of matter is in Vol. I, esp. pp. 466–516.

25. F. Savart, "Récherches sur la structure des métaux," *Ann. Chim. Phys.,* 1829, **41,** 61–75. Abridged Eng. trans. in *Edinburgh Jour. Science,* 1830, **2,** 104–11.

26. M. L. Frankenheim, *Die Lehre von der Cohäsion* (Breslau, 1835).

27. I. Todhunter and K. Pearson, *A History of the Theory of Elasticity and the Strength of Materials* (2 vols. in 3; Cambridge, 1886, 1893).

28. A. E. H. Love, *A Treatise on the Mathematical Theory of Elasticity* (1st ed.; Cambridge, 1892–93).

29. R. J. Boscovich, *Theoria philosophiae naturalis* (Vienna, 1758 [Eng. trans. from enlarged 3d ed., Chicago and London, 1922]).

30. L. Boltzmann, "Zur Theorie der elastischen Nachwirkung," *Sitzungsberichte der K. Akademie der Wissenschaften,* Vienna, 1874, **70,** 275–366.

31. Clark Maxwell, "Constitution of Bodies," *Encyclopaedia Britannica* (9th ed., 1878), VI, 310–13.

32. C. Barus, "The Viscosity of Steel and Its Relation to Temperature," *American Jour. Science,* 1887, **34,** 1–9. See also *Bull. U.S. Geological Survey,* Nos. 73 (1891) and 94 (1892).

33. J. H. Poynting and J. J. Thompson, *Properties of Matter* (9th ed.; London, 1922).

34. G. F. C. Searle, *Experimental Elasticity, a Manual for the Laboratory* (Cambridge, 1908).

35. P. Ewald, T. Pöschl, L. Prandtl, and H. Senftleben, *Elastizität und Mechanik der Flüssigkeiten und Gase,* in Müller-Pouillets, *Lehrbuch der Physik,* Vol. I, Part II (Brunswick, 1929). Eng. trans. F. Dougall and W. M. Deans, *The Physics of Solids and Fluids, etc.* (London and Glasgow, 1930).

36. C. J. B. Karsten, "Über die Veränderungen des Zustandes der Mischungen durch die Temperature Verschiedenheit," *Archiv für Mineral. Geog. Bergbau und Hüttenkunde,* 1830, **2,** 179–86. See also same author's *Handbuch der Eisenhüttenkunde* I (Berlin, 1841), 577 ff.

37. Sven Rinman, *Geschichte des Eisens . . . .* Trans. C. J. B. Karsten (2 vols.; Liegnitz, 1814–15).

38. P. Berthier, "Récherche du carbone et du silicium dans différentes variétes de fonte et d'acier." *Annales des Mines,* 1833, **3,** 209–30.

39. A. Gurlt, "Über die Kohleneisenverbindungen und ihren Einfluss auf die Roheisenbildungen," *Polytechnical Centralblatt,* 1856, **10,** 366–79. Eng. trans. in *Chemical Gazette,* 1856, **14,** 230–36, 254–60.

40. F. A. Abel, "Memorandum on Results of Preliminary Experiments Made with Thin Discs of Steel," *Proc. Instn. Mech. Engrs.,* 1881, pp. 696–705. See also *ibid.,* 1883, pp. 56–71 and 1885, pp. 30–57.

41. F. C. G. Müller, "Grundzüge einer Theorie des Stahls," *Stahl und Eisen,* 1888, **8,** 281–97.

42. H. Caron, "Études sur l'acier," *Comptes Rendus, Académie des Sciences,* 1863, **56,** 43–46.

43. L. Rinman, "Undersökningar om qväfvehalten i stål och tackjern, samt om beskaffenheten af kolet i härdadt och ohärdadt stål" ("Investigations on the Nitrogen Content in Steel and Cast Iron, and on the Form of Carbon in Hardened and Unhardened Steel"), *Öfversigt af K. Vetenskaps-akademiens Förhandlingar* (Stockholm, 1865), **22,** pp. 443–50.

44. H. Caron, "Études sur l'acier," *Comptes Rendus,* 1863, **56,** 211–15.

45. R. Akerman, "On Hardening Iron and Steel," *Jour. Iron and Steel Inst.,* 1879 (ii), pp. 504–42.

46. J. W. Spencer, Discussion on above paper, *Jour. Iron and Steel Inst.,* 1880 (ii), pp. 443–46.

47. F. Osmond and J. Werth, "Théorie cellulaire des propriétés des l'acier," *Annales des Mines,* 1885, **8,** 5–84.

CHAPTER 15: *Martens and Osmond*

1. W. A. Lampadius, *Grundriss einer allgemeinen Hüttenkunde* (Göttingen, 1827), p. 415.

2. [F. Kohn], "Steel under the Microscope," *Engineering*, 1867, **4**, 417.

3. *Ibid.*, p. 553. Reprinted in F. Kohn, *Iron and Steel Manufacturing* (London, 1869), pp. 114–16.

4. [F. Kohn], "The Berlin Castings," *Engineering*, 1867, **4**, 14–15. Reprinted in F. Kohn *Iron and Steel Manufacturing* (London, 1869), pp. 56–59.

5. [Eduard Schott], "Art Castings at the Vienna Exhibition," *Engineering*, 1873, **15**, 426–27, 466–67; **16**, 73–74, 231–232, 332.

6. Eduard Schott, *Die Kunstgiesserei in Eisen* (32 pp.; Brunswick, 1873). Pp. 25–32, dealing with coloring and plating, are not in the English version.

7. Adolf Martens, "Über die mikroskopische Untersuchung des Eisens," *Zeit. des Vereines deutscher Ingenieure*, 1878, **22**, 11–18. Eng. trans. in *Engineering*, 1879, **28**, 88–90.

8. [A. Sauveur], "Adolf Martens," *The Metallographist*, 1900, **3**, 177–81.

9. A. Martens, "Zur Mikrostructur des Spiegeleisens," *Zeit. des Vereines deutscher Ingenieure*, 1879, **22**, 205–14, 481–87.

10. *Idem.*, "Über Festigkeitversuche mit Eisen und Stahl," *Glasers Ann. für Gewerbe und Bauwesen*, 1880, **6**, 119–28; **7**, 475–78.

11. *Idem.*, "Über das Mikroskopische Gefüge . . . des Roheisens . . . ," *Zeit. des Vereines deutscher Ingenieure*, 1880, **24**, 397–406.

12. *Idem.*, *Zentral-zeitung für Optik und Mechanik*, 1880, No. 4. Eng. trans., "Microscope for Examining Metals," *Engineering*, 1882, **34**, 338.

13. *Idem.*, "Über das Kleingefüge des schmiedbaren Eisens, besonders des Stahles," *Stahl und Eisen*, 1887, **7**, 235–42; Über die mikroskopische Untersuchung des Eisens," *Sitzungsberichte des Vereins für Beförderung des Gewerbfleiss*, 1882, pp. 233–42; "The Microstructure of Ingot Iron in Cast Ingots," *Trans. AIME*, 1893, **23**, 37–63.

14. E. Heyn, "Einiges über das Kleinfüge des Eisens," *Stahl und Eisen*, 1899, **19**, 709–14, 768–71.

15. H. Wedding, "Properties of Malleable Iron Deduced from Its Microscopic Structure," *Jour. Iron and Steel Inst.*, 1885 (i), pp. 187–204. See also "Mikroskopische Gefüge des Eisens," *Stahl und Eisen*, 1887, **7**, 82–93 and "The Difference in the Microscopical Structure of Charcoal and Coke Pig Irons," *Jour. U.S. Assn. Charcoal Iron Workers*, 1887, **7**, 120–34.

16. A. F. Hill, "On the Crystallization of Iron and Steel," *Mechanics*, December 30, 1882, pp. 433–36. Reprinted in *Iron Age*, January 4, 1883, pp. 1, 3; "Steel in Construction," *Proc. Engs. Soc. of Western Pennsylvania*, 1880, **1**, 106–20.

17. D. K. Tschernoff. See footnote, p. 140.

18. J. C. Bayles, "Microscopic Analysis of the Structure of Iron and Steel," *Trans. AIME*, 1882/83, **11**, 261–74.

19. F. L. Garrison, "The Microscopic Structure of Iron and Steel," *Trans. AIME*, 1885/86, **14**, 64–75.

20. F. Osmond and J. Werth, "Théorie cellulaire des propriétés de l'acier," *Annales des Mines*, 1885, **8**, 5–84 and in *Mémorial de l'artillerie de la marine*, 1885, **15**, 225–310.

21. A. Portevin, "Floris Osmond, l'homme et l'oeuvre," "Commemoration du centenaire de la naissance de Floris Osmond," *Revue de Métallurgie*, 1950, **47**, 1–64.

22. J. Barba, *Étude sur l'emploi de l'acier dans les constructions* (Paris, 1874).

23. F. Osmond, "Microscopic Metallography," *Trans. AIME*, 1893, **22**, 243–65.

24. *Idem.*, "Transformations du fer et du carbone dans les fers, les aciers, et les fontes blanches," *Mémorial de l'artillerie de la marine*, 1887, **17**, 573–714. Also published as a separate book of same title (Paris, 1888).

25. *Idem.*, "Méthode générale pour l'analyse micrographique des aciers au carbone," *Bull. Société d'Encouragement pour l'Industrie Nationale*, 1895, **10,** 480–518. Reprinted with some important changes in *Contribution à l'étude des alliages* (Société d'Encouragement pour l'Industrie Nationale: Paris, 1901), pp. 277–326. Eng. trans. of the 1895 version "by a member of the Carnegie Steel Corporation" reprinted as "General Method for the Micrographic Analysis of Steel," *Proc. Engs. Soc. Western Pennsylvania*, 1908, **18,** 503–51. Eng. trans. of the later edition with other important papers by Osmond in the book by F. Osmond and J. E. Stead, *Microscopic Analysis of Metals* (London, 1901).

26. F. Osmond, "Sur la cristallographie de fer," *Annales des Mines,* 1900, **17,** 110–66. Eng. trans. in *The Metallographist,* 1900, **3,** 181–219, 275–90; F. Osmond and G. Cartaud, "Sur la cristallographie du fer," *Annales des Mines,* 1900, **18,** 113–52. Eng. trans. in *The Metallographist,* 1901, **4,** 119–49, 236–52.

27. H. Tresca, "Note sur l'écrouissage et la variation de la limite d'élasticité," *Comptes Rendus,* 1884, **99,** 351–59. See also chap. 12, n. 38.

28. M. Lévy, "Mémoire sur les divers modes de structure des roches éruptives," *Annales des Mines,* 1875, **8,** 337–438.

29. See n. 24 above.

30. —. Pionchon, "Étude calorimetrique du fer . . . ," *Comptes Rendus,* 1886, **102,** 1454–57.

31. H. Le Chatelier, "De la Mesure des températures elévées par les couples thermoélectriques," *Journal de Physique,* 1887, **6,** 23–31.

32. P. G. Tait, "First Approximation to a Thermo-electric Diagram," *Trans. Royal Soc. Edinburgh,* 1872/73, **27,** 125–40.

33. C. Barus, "On Thermo-electric Measurements of High Temperatures," *Bull. U.S. Geological Survey,* No. 54, 1889.

34. F. Rudberg, "Über eine allgemeine Eigenschaft des Metalllegierungen," *Poggendorfs Annalen der Physik,* 1830, **18,** 240; *Ann. Chim. Phys.,* **48,** 353 (from a Swedish publication of the previous year).

35. W. C. Roberts, "On the Liquation, Fusibility, and Density of Certain Alloys of Silver and Copper," *Proc. Royal Soc.,* 1875, **23,** 481–95.

36. H. M. Howe, *Metallurgy of Steel* (New York and London, 1890). More than half of the book appeared in serial form in the *Engineering and Mining Jour.,* beginning on March 5, 1887, and continuing at the rate of two to three pages per week thereafter until May 11, 1889, whereafter it appeared only as a supplement and in the bound book, which was printed from the same plates.

37. F. Osmond, "Remarques sur le mémoire de H. W. Bakhuis Rooreboom," *Bull. Société d'Encouragement pour l'Industrie Nationale,* 1900, **6,** 652–60. Eng. trans. in *The Metallographist,* 1901, **4,** 150–61.

38. E. C. Bain, private communication to the author, November 4, 1958. Dr. Bain has in his possession a photograph of the new structure which was presented to him, bearing the inscription, ". . . first noted and so named in 1934 in honor of E. C. Bain, by the Research Staff of the United States Steel Corporation." This is signed by José R. Vilella, John G. Zimmerman, E. S. Davenport, E. L. Roff, and Robert H. Aborn.

39. F. Guthrie, "On Eutexia," *Phil. Mag.,* 1884, **17,** 462–82.

40. H. M. Howe, "Eutectic or Aeolic?" *The Metallographist,* 1903, **6,** 245–49.

41. N. Belaiew, "Swords and Meteors," *Mining and Metallurgy,* 1939, **20,** 69–70.

42. H. M. Howe, "The Metallurgy of Steel," *Engineering and Mining Jour.,* 1888, **46,** 131–32.

43. C. Benedicks, "On Allotropy in General and That of Iron in Particular," *Jour. Iron and Steel Inst.*, 1912, **86,** 242–74.

44. F. Osmond, "On the Critical Points of Steel," *Jour. Iron and Steel Inst.*, 1890 (i), 38–71.

45. See n. 19 above.

46. A. Sauveur, "Microstructure of Steel," *Trans. AIME*, 1893, **22,** 546–57.

47. F. Osmond, "Note on the Microstructure of Steel," *Jour. Iron and Steel Inst.*, 1891 (i), pp. 100–101.

48. Société pour l'Encouragement de l'Industrie Nationale, *Contribution à l'étude des alliages* (Paris, 1901), pp. 277–326.

49. F. Osmond, "Sur la trempe des aciers extra-durs," *Comptes Rendus*, 1895, **121,** 684–86.

50. J. E. Stead, "The Crystalline Structure of Iron and Steel," *Jour. Iron and Steel Inst.*, 1898 (i), pp. 145–89. Reprinted in *The Metallographist*, 1898, **1,** 289–341.

51. See n. 26. above.

52. H. M. Howe, "What Is the Essence of Crystalhood?" *The Metallographist*, 1902, **5,** 52–56.

53. E. H. Saniter, "Carbon and Iron," *Jour. Iron and Steel Inst.*, 1897 (ii), pp. 115–29; "Allotropic Iron and Carbon," *ibid.*, 1898 (i), pp. 206–19.

54. A. Baumhauer, *Die Resultate der Ätzmethode in die Krystallographischen Forschung* (1 vol. plus atlas of plates; Leipzig, 1894).

55. See n. 50 above.

56. T. H. Andrews, "Micro-metallography of Iron," *Proc. Royal Soc.*, 1895, **58,** 59–64.

57. F. Osmond, "Sur la microstructure des alliages de fer et de nickel," *Comptes Rendus*, 1898, **126,** 1352–54; "Sur les alliages de fer et de nickel," *ibid.*, 1899, **128,** 304–6.

58. F. Osmond, C. Frémont, and G. Cartaud, "Les Modes de déformation et de rupture dans le fer et les aciers doux," *Revue de Métallurgie*, 1904, **1,** 11–45.

59. F. Osmond and C. Frémont, "Les Modes de déformation et de rupture dans le fer et les aciers doux," *Revue de Métallurgie*, 1904, **1,** 198–206; "Les Propriétés méchanique du fer en cristaux isoles," *Comptes Rendus*, 1905, **141,** 361–63; "Les Propriétés méchanique du fer en cristaux isoles," *Revue de Métallurgie*, 1905, **2,** 801–10.

60. W. C. Roberts-Austen and F. Osmond, "Microscopic Structure of Gold and Gold Alloys," *Engineering*, 1899, **67,** 254; "The Structure of Metals, Its Origin and Changes," *Phil. Trans. Royal Soc.*, 1896, **187,** 417–32; also "Récherches sur la structure des métaux, sa genèse et ses transformations," *Bull. Société d'Encouragement pour l'Industrie Nationale*, 1896, **1,** 1136–51.

Note added 1965: There is a discussion by F. Wever, "On the Background and Beginning of Metallography in Germany," and by P. G. Bastien, "On the Beginnings of Microscopic Metallography in France . . . ," in *The Sorby Centennial Symposium on the History of Metallurgy*, ed. C. S. Smith (New York, 1965), pp. 163–69 and 170–88, respectively. This volume also contains a paper on "The Beta Iron Controversy," by M. Cohen and J. M. Harris, pp. 209–34.

CHAPTER 16: *Metallography after 1890*

1. R. F. Mehl, *A Brief History of the Science of Metals* (New York, 1948).

2. H. Behrens, *Das mikroskopische Gefüge der Metalle und Legierungen* (Hamburg and Leipzig, 1894).

3. S. W. Smith, *Roberts-Austen, a Record of His Work* (London, 1914). See also

the same author's "Centenary Lecture on the Life and Work of Sir William Roberts-Austen," *Metallurgia*, March, 1943.

4. W. C. Roberts-Austen, "Report to the Alloys Research Committee": *Proc. Instn. Mech. Engrs.*, I, 1891, 543–604; II, 1893, 102–98; III, 1895, 238–97; IV, 1897, 31–100; V, 1899, 35–102; VI, 1904, 7–214; Summary, 1904, 1319–52.

5. W. C. Roberts-Austen and A. Stansfield, "The Constitution of Metallic Alloys (Alloys Considered as Solutions)," Read in Paris, 1900. *Proc. Instn. Mech. Engrs.*, 1904, 175–202.

6. W. C. Roberts-Austen, *Introduction to the Study of Metallurgy* (London, 1891).

7. C. T. Heycock and F. H. Neville, "On the Constitution of the Copper Tin Series of Alloys," *Phil. Trans. Royal Soc.*, 1903, **202,** 1–69.

8. C. T. Heycock, "X-ray Photographs of Solid Alloys," *Proc. Chem. Soc.*, 1897, **13,** 105–7. Microradiography was actually demonstrated to the audience at a Friday Evening Discourse at the Royal Institution on April 2, 1897, entitled, "Metallic Alloys and the Theory of Solution." Other fine microradiographs by Heycock were published in *Trans. Chem. Soc.*, 1898, **73,** 714–23, and in *Phil. Trans.*, 1900, **194,** 201–82.

9. J. O. Arnold, Discussion on Roberts-Austen's "Report to the Alloy Research Committee," *Proc. Instn. Mech. Engrs.*, 1891, 583–87. See also Arnold's papers, "The Physical Influences of Elements in Iron," *Jour. Iron and Steel Inst.*, 1894 (i), pp. 107–48, and "The Chemical Relations of Carbon and Iron," *Jour. Chem. Soc.*, 1894, **65,** 788–801.

10. J. O. Arnold and J. Jefferson, "Influence of Small Quantities of Impurities on Gold and Copper," *Engineering*, 1896, **61,** 176–78.

11. J. O. Arnold, "The Influence of Carbon on Iron," *Proc. Instn. Civil Engrs.*, 1896, **123,** 127–62.

12. J. E. Stead, "Methods of Preparing Polished Surfaces of Iron and Steel for Microscopic Examination," *Jour. Iron and Steel Inst.*, 1894 (i), pp. 292–318.

13. *Idem.*, "The Crystalline Structure of Iron and Steel, *Jour. Iron and Steel Inst.*, 1898 (i), **53,** 145–89; "Brittleness Produced in Soft Steel by Annealing," *Jour. Iron and Steel Inst.*, 1898 (ii), **54,** 137–54.

14. E. H. Saniter, "Carbon and Iron," *Jour. Iron and Steel Inst.*, 1897 (ii), pp. 115–29.

15. H. M. Howe, "The Hardening of Steel," *Jour. Iron and Steel Inst.*, 1895 (ii), pp. 258–309; H. M. Howe and A. Sauveur, "Further Notes on the Hardening of Steel," *Jour. Iron and Steel Inst.*, 1896 (i), pp. 170–79.

16. A. Sauveur, "Microstructure of Steel and the Current Theories of Hardening," *Trans. AIME*, 1896, **26,** 863–906.

17. *Idem.*, "The Current Theories of the Hardening of Steel Thirty Years Later," *Trans. AIME*, 1926, **73,** 859–908.

18. *Idem.*, *Metallurgical Reminiscences* (New York, 1937).

19. Société d'Encouragement pour l'Industrie Nationale, *Contribution à l'étude des alliages par MM. H.-W. Bakhuis Roozeboom, Ad. Carnot, G. Charpy, H. Le Chatelier, H. Gautier, Ed. Goutal, Guillaume, F. Osmond, Roberts-Austen, Mme. Sklodowska Curie*, Sous la direction et le controle de la Commission des Alliages 1896–1900 (Paris, 1901).

20. G. Charpy, "Étude microscopique des alliages métalliques," *Bull. Société d'Encouragement pour l'Industrie Nationale*, 1897, **2,** 384–419. Eng. trans. in *The Metallographist*, 1898, **1,** 87–106, 192–210. A. Sauveur appends to the latter (p. 103)

a note showing how quantitative measurements of various microconstituents relate to the constitution diagram. Reprinted in *Contribution à l'étude des alliages* (Paris, 1901), pp. 119–56.

21. A. Sauveur, "Microstructure of Steel," *Trans. AIME,* 1893, **22,** 546–57.

22. G. Charpy, "Récherches sur les alliages de cuivre et de zinc," *Bull. Société d'Encouragement pour l'Industrie Nationale,* 1896, **1,** 180–234. Reprinted in *Contribution à l'étude des alliages* (Paris, 1901), pp. 1–62.

23. *Idem.,* "Influence de la température de recuit sur les propriétés méchaniques et la structure du laiton," *Comptes Rendus,* 1893, **116,** 1131–33.

24. J. A. Ewing and W. Rosenhain, "The Crystalline Structure of Metals," *Phil. Trans. of the Royal Soc.,* 1900, **193,** 353–77. *Ibid.,* (Second paper) 1900, **195,** 279–310. Reprinted in *The Metallographist,* 1900, **3,** 94–130, and 1902, **5,** 81–109.

25. A. Fock, *Einleitung in die chemische Krystallographie* (Leipzig, 1888). Eng. trans., enlarged by W. J. Pope (Oxford, 1895).

26. H. Le Chatelier and G. Mouret, "Les Équilibres chimiques," *Revue Générale des Sciences,* 1891, **2,** 97–102, 138–44.

27. H. Le Chatelier, "L'État actuel des théories de la trempe de l'acier," *Revue Générale des Sciences,* 1897, pp. 11–22.

28. H. W. Bakhuis Roozeboom, "Eisen und Stahl vom Standpunkte der Phasenlehre," *Zeit. physikalische Chemie,* 1900, **34,** 437–87. Eng. trans. "Iron and Steel from the Point of View of the 'Phase-Doctrine,'" *Jour. Iron and Steel Inst.,* 1900 (ii), **58,** 311–16.

29. H. Le Chatelier, "La Technique de la métallographie microscopique," *Contributions a l'études des alliages* (Paris, 1901), pp. 421–40. Eng. trans. in *The Metallographist,* 1901, **4,** 1–22.

30. W. L. Bragg, "On the Crystalline Structure of Copper," *Phil. Mag.,* 1914, **28,** 355–60.

31. A. Westgren and G. Phragmen, many papers, particularly, "X-ray Analysis of Copper-Zinc, Silver-Zinc and Gold-Zinc Alloys," *Phil. Mag.,* 1925, **50,** 311–41.

32. *Idem.,* "X-ray Studies on the Crystal Structure of Steel," *Jour. Iron and Steel Inst.,* 1922, **105,** 241–73.

Although the author has been interested in metallurgical history throughout most of his professional life—he began to collect source books in 1932 when his eye was caught by the title of John Webster's *Metallographia* in one of his historian-wife's book catalogues—the opportunity to begin the present book did not arise until 1955–56 when the author was able to spend an unencumbered year in London under a fellowship from the John Simon Guggenheim Foundation and a research grant (NSF-G1674) from the U.S. National Science Foundation. This support is gratefully acknowledged.

The writer's interest in Oriental metal work, so apparent in Section I, began largely as a result of a conversation with Mr. B. W. Robinson of the Victoria and Albert Museum and was developed during discussions with Mr. Herbert Maryon of the British Museum, whose combination of practical craftsmanship with a knowledge of art and history is unmatched. But most of the sources used in this study are printed books, and most of the time has been spent in solitary research following hints and hunches to find pertinent material of widely different kinds in many libraries. A large fraction of the sources quoted have not previously been used by metallurgical historians and are not in any bibliography. Quotations from works in foreign languages are from published translations whenever these are listed in the bibliographic notes, otherwise they are by the author except for some important ones by Mrs. Anneliese Sisco, whose contribution is most gratefully acknowledged.

The writer wishes to thank particularly the following libraries which have granted access to their collections and have provided reproductions.

The British Museum
The Royal Institution of Great Britain
The Patent Office (London)
The Science Library (South Kensington)

*Acknowledg-
ments*

The Joint Library of the Iron and Steel Institute and the Institute of
    Metals (London)
The Victoria and Albert Museum (London)
The Sheffield University Library
The Sheffield City Library
Cambridge University Library
The Eisenbibliothek (Schaffhausen)
Yale University Library
New York Public Library
The Engineering Societies Library (New York)
The Library of Congress
The Patent Office Library (Washington, D.C.)
John Crerar Library (Chicago)
Midwest Inter-Library Center (Chicago)
The University of Chicago Library

Particular mention should be made of Miss Helen M. Smith of the
University of Chicago Library, who has located and arranged for loans
of many books, and Miss Charlotte Shivvers who devoted many hours to
putting in order and typing the final draft of the manuscript, checking
references, and collecting illustrations. Mrs. Betty Nielsen and Mr. B.
W. Larsen of the Institute for the Study of Metals made many of the
photomicrographs and photographs not otherwise ascribed.

[Numbers refer to pages. In some cases the key word will not be found on the page referred to, but the topic is discussed in other terms.]

291

twins, 98, 233–34
type metal, 113, 153

United States, early metallography in, 215, 226, 227, 239–40
urine, crystallization, 93

Valentine, Basil, 129
Vandemonde, C. A., 148
Van't Hoff, J. H., 234
vibration-induced crystallization, 122–27, 173
Victoria, Queen of England, 122
Vienna, 150, 152–53, 155–56
Viking swords, 3, 4, 7
vitreous metal, 93
Vivian, W., 175
Vogelsang, H., 139, 221
Vulcan, 7

Wade, W., 149
Wäsström, Peter, 30, 145, 210
Wallace Collection (London), 3, 12, 18
Water-of-Ayr stone, 175, 183, 256
wax, 10, 104
Weber, Wilhelm E., 202
Webster, John, 281
Wedding, H., 183, 215
Wedgwood Papers, 145
Weiss, C. S., 192
welded "Damascus" steel, 30–39, 142–47, 160, 249–55
Werth, J., 216, 217, 220, 221, 222
Wertime, Theodore A., 9
Westgren, Arne, 244
Wever, F., 278
White, Walter, 174
white cast iron, 101
    see also cast iron

Widmannstätten, Alois von, 150, 152
Widmannstätten structure, 150–57, 172
Wilkinson, Henry, 22, 27
Williams, H., 10
Williamson, W. C., 170
Wilson, H., 57, 61
wire drawing, 69, 79
Wöhler, A., 125, 127
Wolff, Walter, 35
Wollaston, W. H., 82, 157, 158, 193
wood, fracture, 105, 106
woodburytype, 185
Woolwich Arsenal, 207
wootz, 20–26, 28, 149, 160
    see also Damascus steel
work hardening, 78, 93, 205, 226
wrought iron, 105, 106, 124–27, 134, 135, 161–65, 172, 183, 256–59
    see also fiber, iron

X-ray diffraction, 150, 192, 193, 244, 245

yakiba, 49–56
York, J. O., 124
Yorkshire Geological and Polytechnical Society, 181
Young, J., 150
Yumoto, John M., 40

zag (Persian etchant), 22, 23, 27, 29
Zapffe, Carl, 121
Zeiss, Carl, 214
Zimmerman, Samuel, 130
zinc, 103, 195, 200, 241
Zirkel, Ferdinand, xix, 170
Zschokke, B., 14, 20, 23, 34